Ravi Nath
Industrial Process Plants

Also of Interest

Product-Driven Process Design.
From Molecule to Enterprise
Zondervan, Almeida-Rivera, Camarda, 2023
ISBN 978-3-11-101490-6, e-ISBN 978-3-11-101495-1

Integrated Chemical Processes in Liquid Multiphase Systems.
From Chemical Reaction to Process Design and Operation
Kraume, Enders, Drews, Schomäcker, Engell, Sundmacher (Eds.), 2022
ISBN 978-3-11-070943-8, e-ISBN 978-3-11-070991-9

Process Systems Engineering.
For a Smooth Energy Transition
Zondervan (Ed.), 2022
ISBN 978-3-11-070498-3, e-ISBN 978-3-11-070520-1

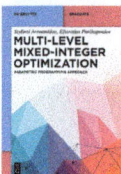

Multi-level Mixed-Integer Optimization.
Parametric Programming Approach
Avraamidou, Pistikopoulos, 2022
ISBN 978-3-11-076030-9, e-ISBN 978-3-11-076031-6

Ravi Nath

Industrial Process Plants

Global Optimization of Utility Systems

DE GRUYTER

Author
Dr. Ravi Nath, Ph D
Houston, Texas
United States of America
ravi.nath1@outlook.com

ISBN 978-3-11-101531-6
e-ISBN (PDF) 978-3-11-102067-9
e-ISBN (EPUB) 978-3-11-102269-7

Library of Congress Control Number: 2023947477

Bibliographic information published by the Deutsche Nationalbibliothek
The Deutsche Nationalbibliothek lists this publication in the Deutsche Nationalbibliografie;
detailed bibliographic data are available on the internet at http://dnb.dnb.de.

© 2024 Walter de Gruyter GmbH, Berlin/Boston
Cover image: loonger/iStock/Getty Images Plus
Typesetting: Integra Software Services Pvt. Ltd.
Printing and binding: CPI books GmbH, Leck

www.degruyter.com

To

Rosa

Audrey,
Ezra,
&
Aila,

and the future generation of movers and shakers . . .

Preface

There needs to be more clarity regarding optimization and related words. Ideally, in simple terms, optimization should mean finding the very best; however, this meaning has become diluted, and optimization among practicing engineers has become synonymous with finding some improvements, any improvement, not necessarily finding the best.

Mathematics, which is supposed to be an exact science, has unwittingly added to this confusion. Mathematically, optimization could be local or global. However, this distinction is often not made so that the same word could mean either a local or the global optimum. And the difference between the two could be huge, depending on the situation.

Another contributing factor to this confusion is using equipment efficiencies to describe equipment performance in utility systems, such as boilers. Such efficiency curves are almost always upside-down parabolic-shaped, which is a nonlinear function. So, it seems like we are dealing with a nonlinear problem. But are we? Not necessarily, why? Let us take a closer look at what efficiency is. It is a derived quantity; generally, it is a ratio of "energy usefully absorbed" to "total energy consumed," which for a boiler would be proportional to the ratio of steam produced to the fuel consumed. So, how about looking at the relationship in terms of the fundamental variables: steam production and fuel consumption? Whoa! Now, that relationship looks linear for the most part, except for a little upward slant at higher production rates, so a line with two or more segments would be a very close approximation. Consider an optimization technique called Mixed Integer Linear Programming (MILP), a variant of Linear Programming (LP). It can handle segmented linear functions using integer variables and still give guaranteed global optimum. In summary, if you rely on the efficiency curve to model a boiler, the problem is nonlinear. Still, if we revert to more fundamental quantities, the relationship suddenly becomes segmented linear, amenable to global optimization using MILP.

An implied criticism of using integers in modeling is that the problem computation time could grow exponentially with integers, i.e., if you introduce a single binary (0/1) variable, the computation time could double. That sounds bad, but what if we had 38 binary variables? The compute time now could be 2^{38} or over 200 billion times, and even if the base problem without integers was solved very fast, say in 1/100 of a second, that could take well over 87 years. It's a scary thought experiment; *let us think of an alternative method* that would be a reasonable reaction. But wait, that doubling of time with each additional binary integer is not the expected or the average but the absolute worst-case scenario . . . the reality is much different.

In this book, we will develop an MILP optimization model of a sample utility system; it has 38 integers, actually 37 binaries and one integer that could take an integer value between 0 and 4; it takes less than a second to solve it on a three-year-old laptop. The worst case is just a possibility. Yes, it does happen, but more for contrived

https://doi.org/10.1515/9783111020679-202

problems. However, operations optimization problems of utility systems are not those kinds; I can confidently say so, having done many such optimization projects.

And so on, there are many myths, and we will address some of those in the book.

Yes, granted, most problems, especially those in the process operations optimization arena, are nonlinear, and local optimization is all that can be achieved. Still, this idea that only local optimization is possible for any process plant operations optimization has become the norm. This attitude has become so pervasive that it applies even to problems where global optimization is a definite possibility.

One such problem is the operations optimization of utility systems embedded within a process plant, which is the subject matter of this book. Global optimization of utility systems is possible but seldom sought because of the prevailing attitude. This is sad as this general attitude is losing many opportunities to conserve significant energy and help lessen global warming.

This book aims to make this knowledge of global optimization of process plant utility systems widely available and help you model plant utility systems by giving a complete example from start to finish, modeling through implementation. The book does not assume prior knowledge of modeling or optimization; it starts with concepts and slowly builds the first model of the boiler, gradually refines the model, and then tackles other unit operations models, subsystems, and, ultimately, the entire utility system. The book develops deployment strategies for the optimization model for offline and online closed-loop usage. The pace increases steadily so as not to repeat things ad nauseam and to keep the book manageable in size.

The book is suitable for self-study, as an optimization project resource, or an upper-level one-semester course. The software used to develop these models is affordable and readily available. All model files mentioned in the book are available from the publisher's website.

I would be remiss not to take this opportunity to thank so many who have led me in my professional life journey. First and foremost, my gratitude to many professors in general and in particular to the late Prof. Rudy Motard for kindling the spirit of research; late Prof. Angelo Miele, the best teacher I ever had for demystifying optimization; the person from the University of Chicago at Union Carbide in Y. P. Tang's group, whose name I cannot recall, who introduced me to LINDO, a modeling system in which you could write model equations in algebraic form, a far cry from the dreaded MPSX format popular at that time. John Holiday for introducing me to the problem of utility system optimization; utility system engineers and utility system managers at various Union Carbide plants for supporting utility optimization projects at their plant sites; my colleagues and friends who encouraged me in all professional endeavors, in particular, Sanjay Sharma; my supervisors at Union Carbide (now Dow), Linnhoff March (now KBC Yokogawa), Setpoint (now Aspen Tech), and Honeywell, who gave me ample opportunities to work in the area of operations optimization of processes and utility systems. Prof. Linus Schrage and Kevin Cunnigham, the creators of LINDO, for taking an interest in my early modeling efforts; folks at LINDO systems, particularly Mark Wiley, for supporting any LINDO/LINGO software issues. Dane Takemoto, at Frontline Systems, for providing a complimentary license for the Premium Solver for the duration of this book project. Bob Esposito for introducing me to the world of publishing; Karin Sora and Stella Muller at De Gruyter for guiding me in preparing this manuscript. And last but not least, my family for bearing with me and supporting me all these years through thick and thin. Thank you all from the bottom of my heart.

Although I have been careful in developing the manuscript, I wouldn't be surprised if you find errors in the book. I am solely responsible for those; please do not hesitate to bring them to my attention via the publisher.

Ravi Nath
Houston, Texas
September 2023

Contents

Part II: **Model deployment**

Units of Measure (UOM)

Property	UOM	Comment
Currency	$	
Energy	MWh	In particular electrical energy
Energy, thermal	MWh.th or MWh(th)	
Energy, mechanical	MWh.sp	
Enthalpy	kJ/kg	Specific enthalpy
Length	m	Meter
Mass	ton or t	Metric ton, 1000 kg
Mass flow	t/h or tph	Metric ton per hour
Power	MW	In particular electrical power
Power, peak	MW.pk	Electrical power peak demand
Power, thermal	MW.th or MW(th)	
Power, mechanical	MW.sp	
Pressure	bar(A)	Bar (Absolute)
Temperature	C	Centigrade
Time	h or hr	Hour
Volume	m^3 or m3	Cubic meter
prefix		
	k	Kilo (multiplier of 10^3)
	M	Mega (multiplier of 10^6)
	G	Giga (multiplier of 10^9)

https://doi.org/10.1515/9783111020679-204

1 Introduction

This book is about the global optimization of utility systems; more specifically, it is about the global operations optimization of utility systems operations. There are several keywords here; let us clarify them to avoid confusion later. Note that clarifications in this section will focus on conceptual understanding; later sections will provide details. This section will also briefly discuss the justification and barriers to global optimization of utility systems operations.

Firstly, what is **optimization**? Webster's dictionary definition is: "Optimization: noun: is an act, process or methodology of making something as fully perfect, functional, or effective as possible." In the context of global optimization of utility systems operations, we can modify this definition to, *"Optimization: noun: is a methodology of making the operations of a utility system as cost-effective as possible."* In practice, optimization proceeds in two steps: the first is to determine the least cost operations policy (the optimum policy) and the second is to implement the optimum policy to achieve the benefits. This book will discuss both aspects of optimization.

Secondly, what is **global optimization**? In straightforward terms, global optimization refers to a methodology for determining the absolute best (per a prespecified criterion) within specified constraints; anything short of it is **local optimization**.

Thirdly, what is **operations optimization**? Operations optimization refers to the optimization of the operations of an existing facility. In an existing facility, all equipment is installed, and the design characteristics of each piece of equipment are known; equipment layout and interconnections are also fixed. In addition, key operating conditions such as temperature, pressure for material headers, and voltage for electrical headers have also been determined and, therefore, are known values. Note that dynamically changing quantities, such as the demand for utilities by the process units and the cost structure of purchased energy (fuel and electrical power), must also be specified to define the optimization problem fully. Optimization then determines the selection of the optimal equipment slate among the available equipment and the loadings of various equipment. Optimization results are also referred to as the optimal operations policy.

The operating cost of a typical plant utility system of a typical industrial process plant is enormous, often in tens if not hundreds of millions of dollars per annum [1, 2]. With so much money at stake, one would expect that heroic optimization efforts would be made to reduce the operating cost; however, such is usually not the case.

One reason for this complacency is that utility systems are almost always "cost centers" in process plants, i.e., their operating cost is prorated among the various process units, so it suffers from what is sometimes referred to as "the tragedy of the commons" [3]. Another reason for this complacency is that the structure of the utility system significantly differs from process units. For flexibility and safety reasons, utility systems have a large surplus capacity to meet increased utility demands encountered during startups, shutdowns, and emergencies. This surplus capacity is manifested in the form of spare

https://doi.org/10.1515/9783111020679-001

equipment. For example, a utility system may have four steam boilers, whereas only two may be required during normal operation. As another example, some of the boiler feed water (BFW) pumps may have the option to be driven either by steam turbines or by electric motors. Determining which boilers and which BFW pump drives to use necessitates optimizing discrete decisions where traditional optimization techniques do not readily apply. Part of the problem is that the conventional engineering curriculum primarily emphasizes an optimization method called Nonlinear Programming (NLP) [4]. Although NLP can theoretically address large classes of optimization problems, unfortunately, it has stringent requirements that all describing relationships be continuous with continuous derivatives. The continuity requirements limit the applicability of NLP, especially to utility systems with surplus equipment and equipment that can operate in one of the several operating modes.

Additionally, NLP generally guarantees only a local and not the global optimum, which is a significant shortcoming. However, other optimization methods exist: one particularly well-suited for optimizing utility systems is Mixed Integer Linear Programming (MILP) [5]. MILP does not impose function and derivative continuity requirements and thus allows relationships that incorporate discrete variables (in particular binary On or Off variables or general integer variables) and relationships spanning multiple operating modes and time periods.

One possible reason for the unpopularity of MILP methods could be that MILP problems are classified as "NP-complete" [6]. For such problems, in the worst case, the compute time increases exponentially, i.e., such problems suffer from the so-called "curse of dimensionality" [7], which discourages its use for utility systems that typically have many discrete decision variables. However, this is less of an issue with advances in computing capabilities, especially for well-formulated, real-life utility system operations optimization problems. Unlike NLP, MILP methods guarantee global optimum, which is reassuring. MILP, however, imposes a linearity requirement that all describing relationships be linear; this, is not a handicap, as we will discuss later in this book and detail techniques to overcome this limitation.

This book aims to remove the barriers to the global optimization of utility systems operations by providing practical tools and techniques to deal with the unique challenges in the global optimization of utility systems operations.

1.1 Systems view of industrial process plant and plant utility system

An industrial process plant is a chemical or petrochemical manufacturing facility at a particular physical location (site); it typically comprises one or more process units and a central utility system. There are too many process plants to list here; the interested reader is referred to an excellent book that discusses the most important industrial process plants [8].

The process units convert feedstock materials to value-added products and by-products and, in doing so, consume utilities (or energy) in the form of steam, fuel, electrical power, cooling water, and compressed air. The utility system, also called the energy system, is responsible for reliably providing utilities to the process units at all times, even as the process utility needs change significantly with changes in the operations of the process units.

From a systems perspective, we can view an industrial process plant comprising two distinct systems: the process system and the utility system. The process system is responsible for converting feedstocks to value-added products and byproducts, and the utility system is responsible for reliably providing the utilities demanded by the process system. Interconnections between the process and the utility systems are via multiple physical headers for material streams and electrical headers (or networks) for electrical power. Fig. 1.1 is a systems view of an industrial process plant.

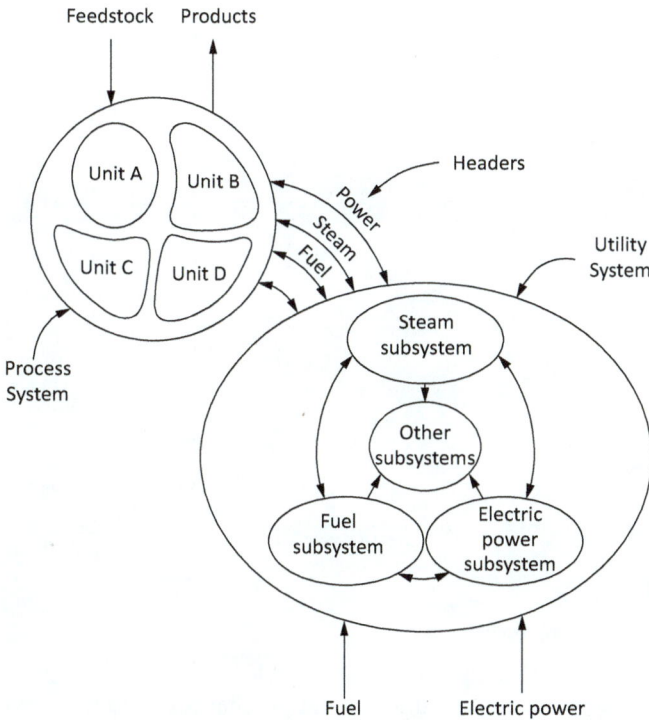

Fig. 1.1: A systems view of an industrial process plant.

This book focuses on the utility system only, so we need not dwell on the details of the process system as it will suffice to view the process system as a "black box" and focus only on the aggregate demands it makes on the utility system. In the next section, we will detail a typical utility system. For ease of understanding, we will think of a utility system in functional terms as comprising several subsystems, viz., the steam subsystem for generation and distribution of steam at various pressure levels, the fuel subsystem for purchase and distribution of fuel, the electric power subsystem for purchase and distribution of electrical energy, the cooling water subsystem for generation and distribution of cooling water, and the air compression subsystem for generation and distribution of compressed air. These subsystems do interact among themselves: for example, the steam subsystem needs fuel that comes from the fuel subsystem; it may also need electric power, or it may generate electrical power and thus interact with the electrical power subsystem. Figure 1.2 conceptualizes a utility system showing the utility subsystems, interactions among the subsystems, and interactions with the external energy suppliers and the process units.

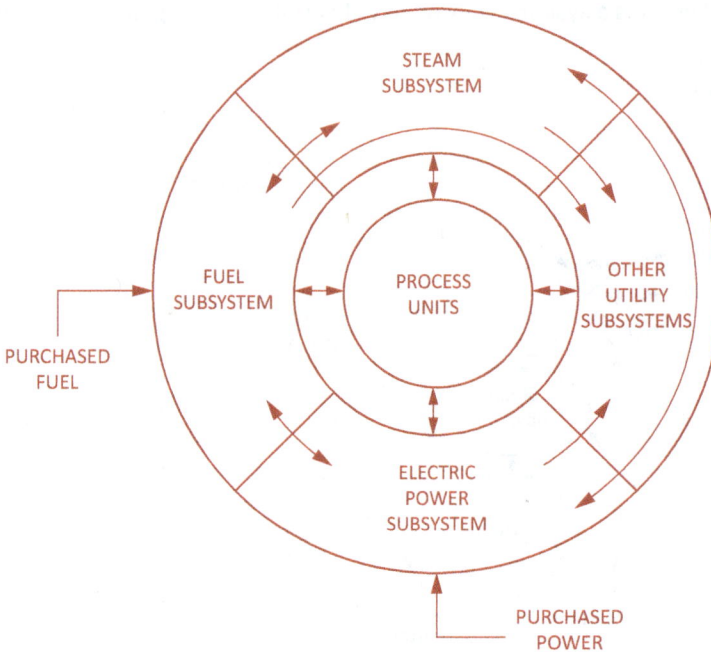

Fig. 1.2: Conceptual representation of a utility system.

Typically, a utility system self-generates some utilities and purchases others from external sources. Usually, the cooling water, compressed air, and much of the steam are self-generated. Some electrical energy may be self-generated, but almost always, at least some, is purchased from the local power grid for reliability reasons. One or more local fuel suppliers supply most, if not all, of the fuel. Steam, if purchased, is usually from a neighboring process or cogeneration facility. Binding legal contracts govern external energy purchases, which tend to be complex with provisions for a minimum "take or pay," "tier pricing," and "time of use rates."

1.2 Overview of utility subsystems

For ease of understanding, we will consider a utility system in terms of its functional subsystems. Each subsystem is described below, first in general terms and then in detail for a sample utility system that will be used in subsequent chapters to illustrate the optimization methodology and to create a functioning global optimizer for the operations of the sample utility system.

1.2.1 Steam subsystem

The steam subsystem comprises boilers or steam generators, steam turbogenerators, letdown stations, boiler feed water (BFW) pumps, boiler combustion air blowers and fans, deaerator(s), blowdown flash units, and a network of steam headers that transport steam to and from the process units.

Figure 1.3 shows a simple yet representative steam subsystem with three steam headers – high-pressure steam (HPS), medium-pressure steam (MPS), and low-pressure steam (LPS); a demineralized water (DMW) header and a process condensate (PC) header. Boiler BL1 generates MPS, and boilers BL2 and BL3 generate HPS. A backpressure turbogenerator STG and a condensing turbogenerator CTG produce electrical power. Boiler #1 auxiliaries are powered by an MPS/LPS backpressure steam turbine BL1AT; an electric motor BL2AM powers boiler #2 auxiliaries, and boiler #3 auxiliaries are powered either by an electric motor BL3AM or an HPS/LPS backpressure steam turbine BL3AT. A pressure-reducing station PRV1 lets down HP steam to the MPS header; note that boiler feed water de-superheats the letdown steam to the MPS header conditions. Another pressure-reducing station, PRV2, lets down MP steam to the LPS header, and there is a pressure relief valve VENT to prevent over-pressurization of the LP header. A blowdown flash unit, BD FLASH, recovers LP steam from the blowdown streams from the three boilers, and the discharge water from BD FLASH is sent to the wastewater treatment facility. A low-pressure deaerator unit, DEA, generates BFW for the boilers and PRV1. DEA consumes LP steam to bring BFW to saturation temperature to remove dissolved gases. BFW from the deàerator is pumped to high pressure using one or more of the four BFW pumps: BP1, BP2, BP3, and BP4. BP1 and BP3 are driven by electrical motors BP1M and BP3M, respectively. BP2 is driven either by HPS/LPS backpressure stream turbine BP2T or electric motor BP2M, and BP4 is driven by HPS/MPS backpressure turbine BP4T or electric motor BP4M. Fuel to BL2 and BL3 is supplied from both the low-pressure fuel gas header LPFG and the high-pressure fuel gas header HPFG. Energy to BL1 comes from the fuel oil header FO. All electrical connections are to the electric power header ELEC. In addition to the process plant steam consumers, several steam consumers within the utility system, such as pump and compressor drivers in the cooling water and compressed air subsystems, are not shown here.

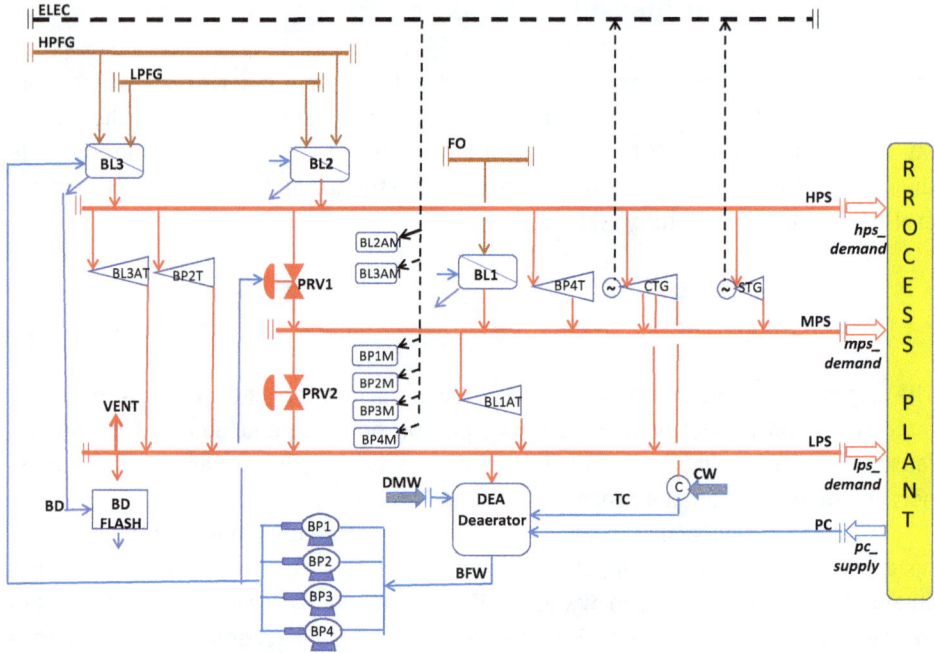

Fig. 1.3: Schematic of the sample steam subsystem.

Note that throughout this book, we will use solid line connections to show material flow streams and dotted line connections for electrical energy flow.

1.2.2 Fuel subsystem

The fuel subsystem comprises a network of fuel headers transporting fuel to and from process units and other utility subsystems. Usually there are multiple fuel headers. For each fuel header, there may be multiple suppliers. With each supplier, there is usually a different contractual agreement; for example, one may be a long-term supplier with a minimum "take or pay" agreement, another may be a short-term supplier with a tiered pricing agreement, and a third may be a spot market supplier with price tied to the current fuel price index. Among the fuel consumers, there may be certain restrictions, such as some could consume only a particular fuel type while others could consume a mix of fuels.

Figure 1.4 shows a simple yet representative fuel subsystem with an HPFG, an LPFG, and an FO fuel headers. In addition to the process plant fuel consumers, there are also several fuel consumers within the utility system, such as steam boilers (BL1, BL2, and BL3), a proposed gas turbogenerator (GTG), and internal combustion engines AC1E, 2E, 3E, and 4E. A pressure-reducing station PRVFG lets down HP fuel gas to the LPFG header, and a fuel compressor unit FC can uplift the LP fuel gas to the HPFG header. Boiler BL1 consumes fuel oil from the FO header; boiler BL2 can consume a mix of fuel gas from LPFG and HPFG headers, while boiler BL3 can consume either LPFG or HPFG, but not both simultaneously. GTG can produce electrical power by consuming HP fuel gas, and the drivers of Air Compressors AC1, AC2, AC3, and AC4 consume LP fuel gas to power internal combustion engines AC1E, AC2E, AC3E, and AC4E, respectively. Safety relief valves connect the LPFG and FO headers to the site flare to prevent over- pressurization of respective headers.

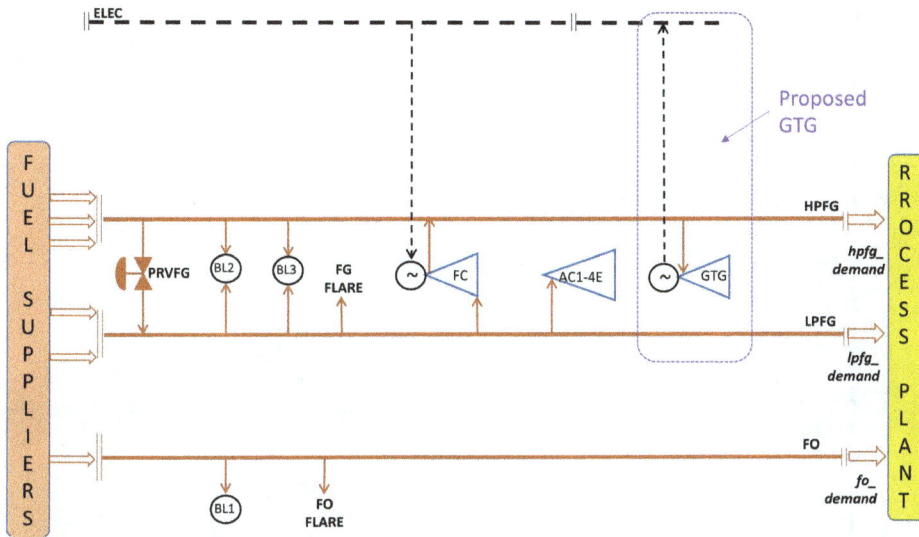

Fig. 1.4: Schematic of the sample fuel subsystem.

1.2.3 Electrical power subsystem

The electrical power subsystem is almost always connected to the local power grid for reliability. The supply contract with the local power supplier is usually complicated, as discussed in Chapter 8. An electrical power subsystem is the simplest to conceptualize, and usually, a simple balance model (IN = OUT) is sufficient unless electrical power networks are constraining.

Figure 1.5 shows a simplified schematic of an electric power subsystem showing high-voltage (HV) and low-voltage (LV) networks. High-voltage electrical energy (HVE) is distributed from HVE circuit breaker box(es) to various HV consumers and producers via multiple electrical circuits. Similarly, LV electrical energy is distributed from the LVE circuit breaker box(es) to the various LV electric power consumers via multiple electrical circuits. The figure shows the different electrical power producers and consumers in the utility subsystems; note that process plant electrical demand is shown as

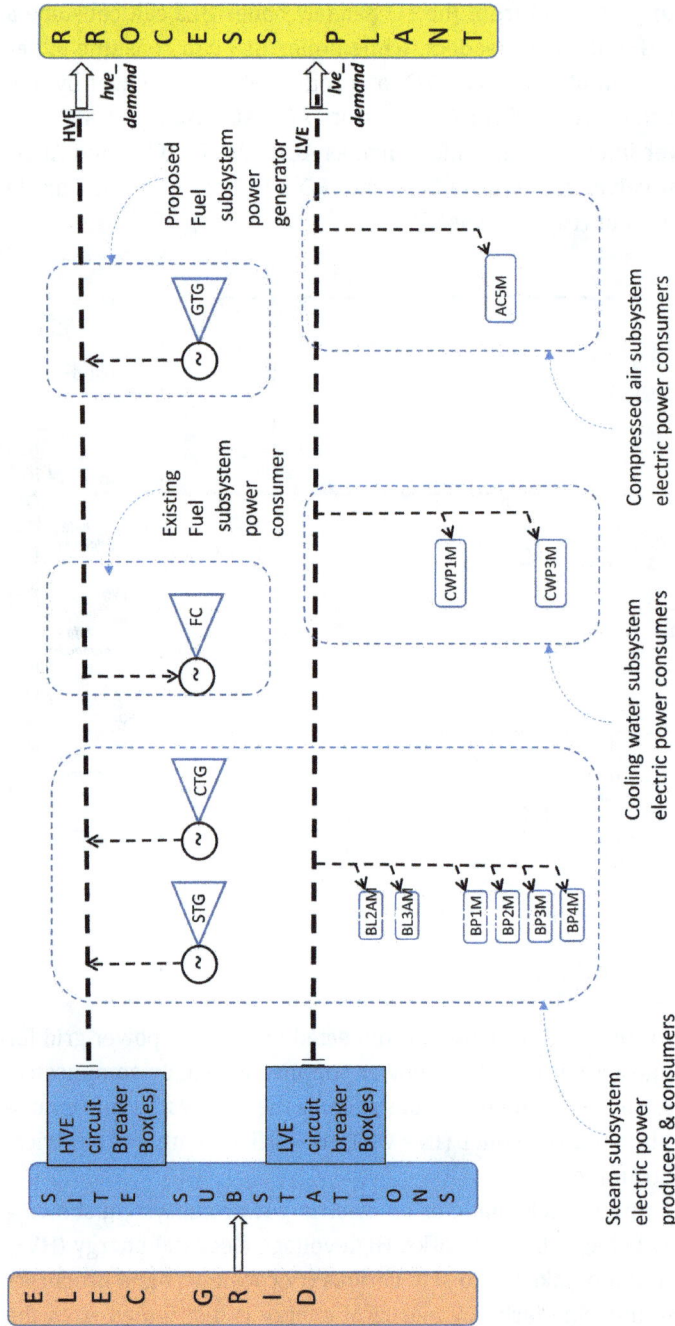

Fig. 1.5: Schematic of the sample electrical power subsystem.

an aggregate demand on the two electrical buses. Site substation(s) receive very high voltage (VHV) electrical energy from the local power grid, where energy is stepped down in transformers to high and low voltages for use at the site.

Modern process plants are adequately wired, so there usually are no limitations on the use and production of electrical power at various voltage levels. Consequently, we will use a simplified schematic comprising a single electrical bus connected to all the different producers and consumers of electrical energy, as shown in Fig. 1.6. In the subsequent model development, we will use the schematic of Fig. 1.6.

1.2.4 Cooling water subsystem

Most process plants require large amounts of cooling water for various cooling duties, such as in-process coolers and condensers. The cooling water system is usually a closed loop. Returned cooling water is treated and cooled in one or more cooling towers. The cooled water is pumped up to high pressure and distributed to the various consumers from cooling water supply header CW. The cooling water pumps are usually large and driven by turbine or motor drives.

Figure 1.7 is a simplified cooling water subsystem with three large cooling water pumps: CWP1, CWP2, and CWP3. Electrical motor CWP1M powers CWP1, HPS/MPS steam turbine driver CWP2T powers CWP2, and CWP3 is powered by a dual drive arrangement whereby it could either be powered by electrical motor CWP3M or by HPS/LPS steam turbine CWP3T.

1.2.5 Compressed air subsystem

Depending on the process needs, a process plant may have a centralized air compression subsystem for producing compressed air. Atmospheric air is filtered and compressed in one or more air compressors and distributed to consumers via high-pressure compressed air header CA.

Figure 1.8 is a simplified yet representative compressed air subsystem with five air compressors: AC1, AC2, AC3, AC4, and AC5. AC1, AC2, AC3, and AC4 are identical units powered by internal combustion engines AC1E, AC2E, AC3E, and AC4E consuming LP fuel gas. A large electric motor AC5M powers AC5.

1.3 The sample utility system and its characteristics

Although we have discussed individual utility subsystems, optimization is ideally performed for the entirety of the utility system. Figure 1.9 is a schematic of the overall sam-

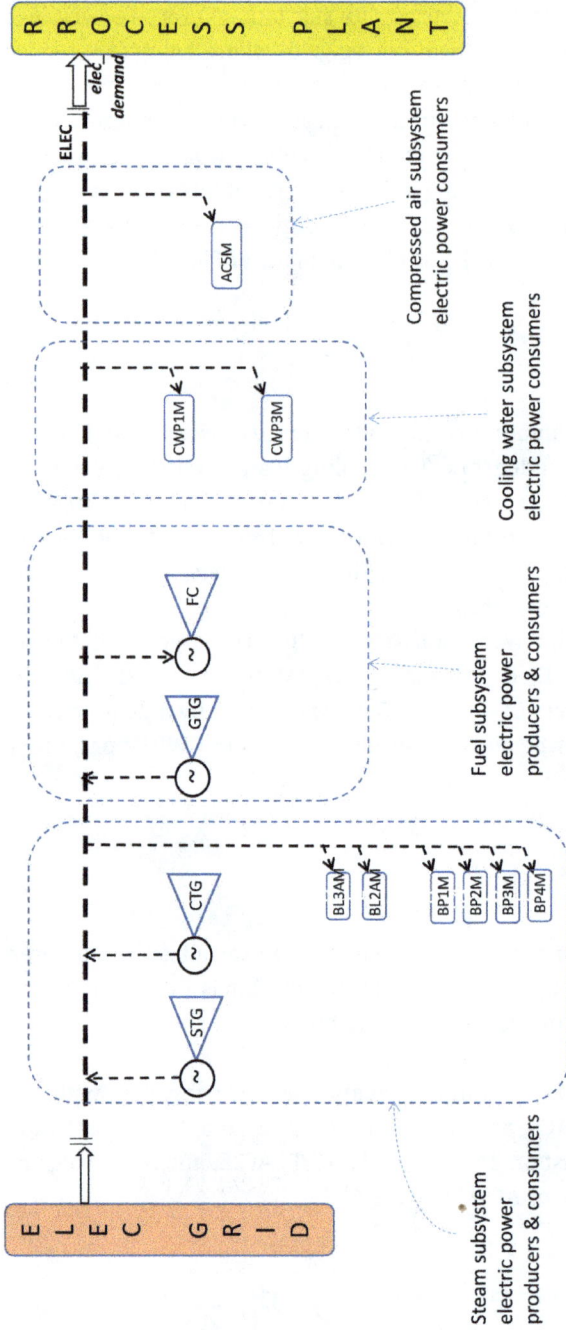

Fig. 1.6: Simplified schematic of the sample electrical power subsystem.

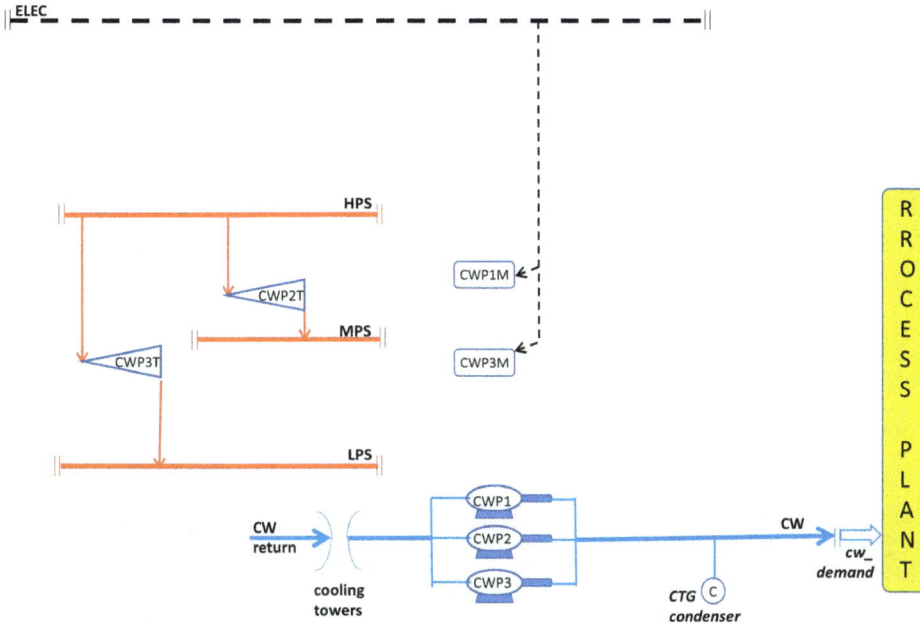

Fig. 1.7: Schematic of the sample cooling water subsystem.

ple utility system we will discuss in subsequent chapters. Certain characteristics of utility systems influence optimization system design; we will discuss these next.

Utility systems are designed to have large surplus capacity to provide increased utility needs during process start-ups, process shutdowns, and emergency conditions and to accommodate changes in process operating conditions. To fulfill this requirement, they invariably have surplus utility generation capacity. This surplus capacity is manifested as additional pieces of equipment. For example, a utility system may have four steam generators, whereas only two may be needed for normal process operation; similarly, it may also have spare pumps and compressors to accommodate significantly higher utility demands during abnormal process operations.

Some utility demand increases are planned activities for which surplus equipment could be started up in anticipation of the increased demand. However, some utility demand increases cannot be anticipated as they could be due to unplanned shutdowns or in response to process emergencies; for these occasions the utility system must have reserve capacity that could be brought online at very short notice. For example, increased steam demand could be met by maintaining a reserve capacity in the steam generators or by switching pump and compressor dual drives from the turbine to electric motors. Increased electric power demand could always be met by drawing more from the power grid; however, this may not be most cost-effective, especially if a new demand peak were to be created, and for such situations, surplus capacity could be maintained by keeping specific turbogenerators in slow roll mode

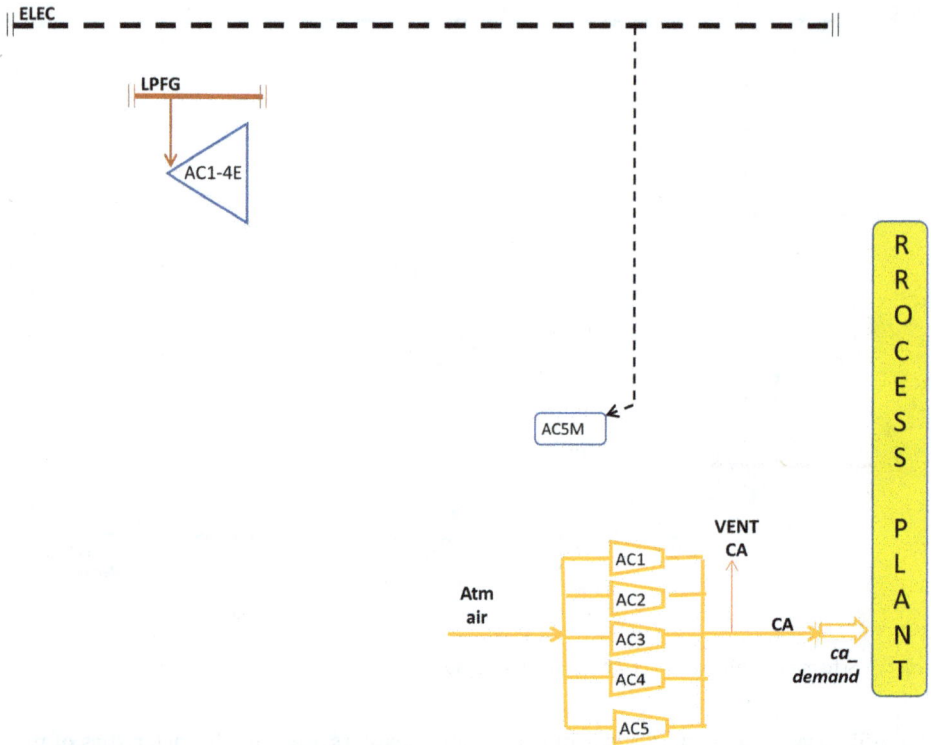

Fig. 1.8: Schematic of the sample air compression subsystem.

or by switching certain fans, pumps, and compressors from electric drives to steam turbine drives so they could respond almost instantaneously. These requirements imply that the optimizer must include additional constraints to model steam and power reserves and be able to make discrete on/off decisions.

Another characteristic of utility systems is that contracts with external energy suppliers are complex, so corporate entities or specialized business teams negotiate these contracts. Moreover, plant operations teams are typically organized by functional subsystems, so there may be an engineer for the steam subsystem, another for the fuel subsystem, and another for the power subsystem. Consequently, no person in the operations group fully understands all contractual agreements except in simple, basic terms. Still, the purchased energy costs heavily influence optimization results. Also, the current process demands, which change frequently, greatly influence the optimization results. In this dynamic environment, knowledge of past optimum policies is of little help in identifying low-cost operations. For these reasons, it is estimated that up to 35% of potential savings in utility system operations may remain unrealized in large refineries and petrochemical complexes [9, 10]. Thus, there is a solid economic case for using utility system optimizers to capture most, if not all, potential savings.

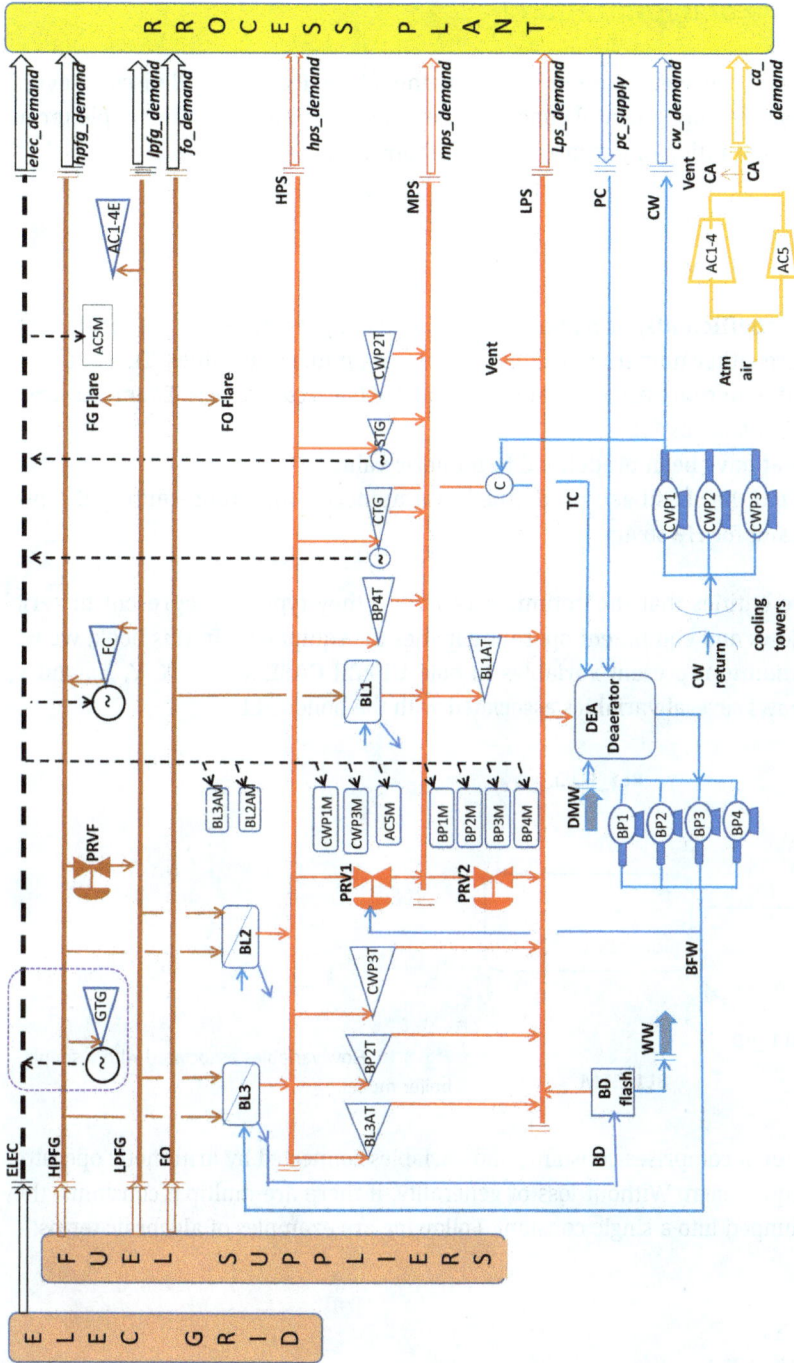

Fig. 1.9: Schematic of the sample utility system.

1.4 Overview of optimization

This section is an overview of optimization methodology. First, we will discuss several important basic yet fundamental concepts; then, define optimization in simple terms. A more formal definition of optimization will follow.

1.4.1 Basics

Constants, or **coefficients,** or **parameters,** have fixed predefined values for the optimizer. They are either numerical values or represent numerical values. This book will use a convention to represent constants in *italicized lowercase letters*. Examples are
– any number such as *1.23*
– k, k_i, k_1 that have been predefined to numeric values
– *hps_demand* that has been predefined to a numeric value representing the process demand for HP steam

Variables are entities that the optimizer can vary; they typically represent material and energy flows and equipment operating modes for equipment. In this book, we will use a convention to represent variables in bold **UPPER CASE** letters: **X, Y, X_i,** and **Y_i**. Figure 1.10 shows several variables associated with the boiler BL1.

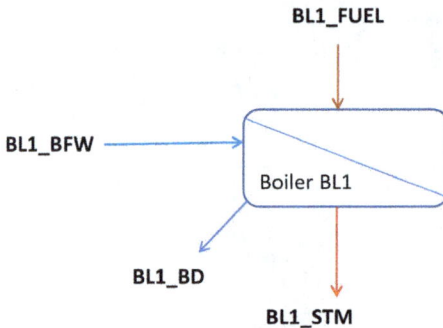

Fig. 1.10: Flow variables associated with a simple boiler model.

An algebraic **term** comprises constants and variables connected by arithmetic operators but not an "equal" sign. Without loss of generality, if there are multiple constants, they could all be lumped into a single constant. Following are examples of algebraic terms:
– $k_1 * \mathbf{X_1}$
– $\mathbf{X_2}$
– $k_2 * \mathbf{X_1} * \mathbf{X_2}$
– $k_1 * \mathbf{X_1} / (k_2 * \mathbf{X_2})$,
– $k_3 * \mathbf{X_1} / \mathbf{X_2}$, where $k_3 = k_1 / k_2$

where k_i is a known constant; $\mathbf{X}$ and $\mathbf{X_i}$ are variables.

An algebraic term is a **linear term** if it is of the form $k_i * \mathbf{X_i}$, i.e., a constant multiplied by a variable. An algebraic term is a **nonlinear term** if it is not of linear form. In the examples above, the first two terms are linear; all others are nonlinear.

An algebraic **expression** is a mathematical phrase; it comprises one or more "terms" connected by an addition or subtraction operator. Following are examples of the structure of an "algebraic expression":
–   Term1 + Term2 – Term3
–   –Term1 + Term2

An algebraic expression is a **linear expression** if it comprises only linear terms. Examples are:
– $k_1 * \mathbf{X_1} + k_2 * \mathbf{X_2}$
– $\sum k_i * \mathbf{X_i}$

An algebraic expression is a **nonlinear expression** if it is not a linear expression. An example follows:
– $k_1 * \mathbf{X_1} + k_2 * \mathbf{X_2} + k3 * \mathbf{X_1} / \mathbf{X_2}$

1.4.2 Optimization in simple terms

In simple terms, **optimization** refers to a search algorithm to find the best for the system under consideration, the best for the system as per a predefined criterion. The predefined criterion is called the **Objective Function**. Commonly used objective functions are "Profit Maximization" and "Cost Minimization." The system under consideration is defined by a set of **constraints** representing all relevant relationships among the various variables as well as minimum and maximum limits on the constraints.

The objective function, together with the constraints, defines the **optimization problem**. The best solution to the optimization problem is the **optimum solution** or the **optimum policy**.

The best solution found by an optimization algorithm may either be a **local optimum** or the **global optimum**. A local optimum is an optimum solution if there is no better solution in the immediate vicinity of the solution; there may be a better solution farther away, but that does not matter. The global optimum is the absolute best solution (as measured by the objective function) in the entire feasible space, i.e., there is no other solution with a better value of the objective function in the entire feasible space. Figure 1.11a shows the profit versus variable plot over the feasible range for a hypothetical single variable maximization problem; Fig. 1.11b shows a similar plot of

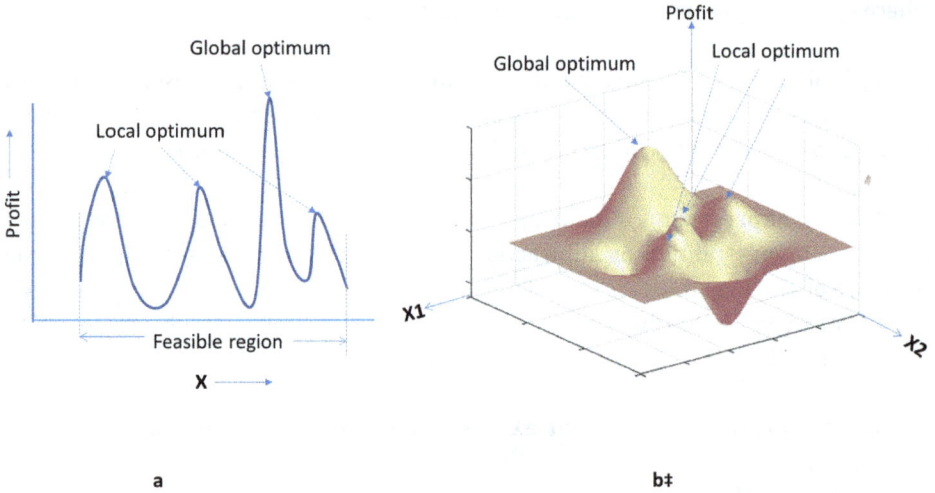

Fig. 1.11: Local and global optimums for a hypothetical profit maximization problem: (a) problem with one variable and (b) problem with two variables.

profit versus variables for an optimization problem in two variables; highlighted are the local optimums and the global optimum.

In the context of global optimization of utility system operations to minimize the operating cost, optimization is a search algorithm to determine the operations policy with the absolute least cost of operation that satisfies all process utility demands while respecting all constraints.

Almost always, and that will be the case here, an optimization problem is set up as a mathematical problem. The setup comprises defining an objective function and a set of constraints. The objective function is the criterion that evaluates to a numerical value and is a measure of the goodness of the solution. The constraints represent all the relationships among the various variables for the system under consideration.

The abovementioned definition of an optimization problem comprising an objective function and multiple constraints is rigorous and well-accepted. However, for completeness, I want to inform you of another nomenclature that, for historical reasons, is used by some in the industry; it is informal, and you are unlikely to find it in optimization textbooks. This nomenclature defines an optimization problem as comprising an objective function, a model, and constraints. In this framework, the constraints that describe the relationships among the variables are referred to as the **model**, and the minimum and maximum limits on variables are referred to as the constraints. The optimization algorithm moves the model variables within their prescribed limits until the optimum is found.

In optimization, the search for the optimum is usually iterative. The search starts from a starting point and moves towards the optimum in steps. Figure 1.12 conceptualizes optimization moves from the starting point P1 to the Global Optimum point P6 in five steps. In this figure, each point, P1 to P6, hits one or more constraints and represents a feasible operating point. By feasible, we mean that operations satisfy all constraints defined for the system. The entire space comprises two mutually exclusive regions: the feasible region, and the infeasible region that lies outside the feasible region. Figure 1.12 shows the feasible region as the shaded region enclosed within the dotted envelope; the region outside the dotted envelope is the infeasible region.

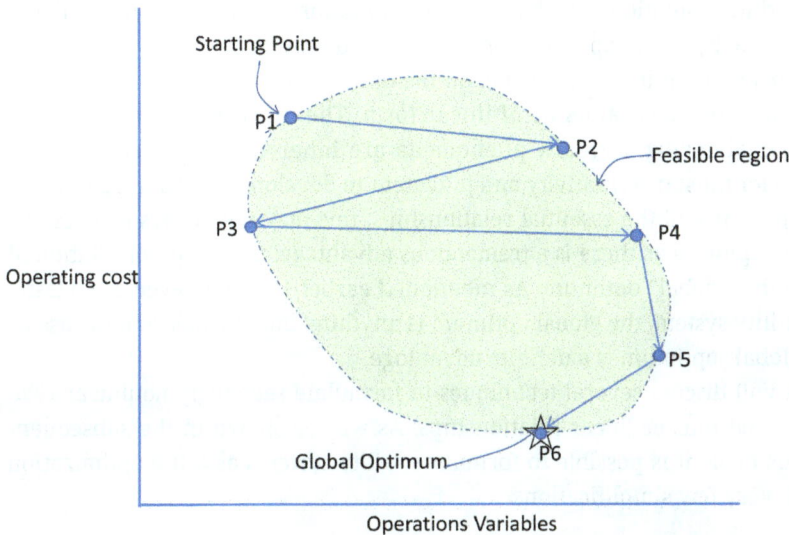

Fig. 1.12: Conceptual representation of optimization progression.

There are numerous optimization methods. They can be broadly classified into two categories: **Nonlinear Programming** (NLP) and **Linear Programming** (LP). Furthermore, if discrete or integer variables are present, they are referred to as **Mixed Integer Nonlinear Programming** (MINLP) and **Mixed Integer Linear Programming** (MILP).

Among the integer variables is a particular subclass called **Binary Variables**, which could be 0 or 1. Binary variables are very useful as they could represent the ON and OFF states of equipment; we will use such variables often. All others are called **General Integer Variables** and are not restricted to only 0 or 1.

The solution to optimization problems with discrete variables is usually carried out in two phases: in phase one, the problem is relaxed and discrete variables are treated as continuous variables; the optimum solution thus found forms a bound on the objective function. An efficient algorithm such as **Branch and Bound** determines the best discrete solution in phase two. Details of optimization methods and the branch and bound algorithm are beyond the scope of this book. Fortunately, the details of the optimization algorithm are not required to develop functional optimization models, and such would be our approach in this book. Interested readers are referred to an excellent textbook [11] on optimization.

Problem formulation in NLP can be more straightforward as it allows the use of general nonlinear functions. However, an NLP requires that the objective function and the constraints be continuous functions with continuous derivatives; this could be a severe limitation as some phenomena over the feasible range undergo abrupt transitions, which causes derivative and or function discontinuities, rendering NLP inapplicable. Another drawback of the NLP is that it only guarantees a "local" and not the "global" optimum solution, which is a severe limitation as the extent of additional benefits achievable by global optimization remains undiscovered.

Problem formulation in LP and MILP, on the other hand, requires that the objective function and the constraints be of linear form. The linearity requirement can sometimes be challenging, as many phenomena are inherently nonlinear. In such cases, the modeler must use creativity and judgment to develop a mixed integer linear model that represents all the essential relationships; this may not always be possible, but it is worth exploring as there is a tremendous advantage: the optimum solution of a MILP is also the "global" optimum. As mentioned earlier [9, 10], in even a modestly complicated utility system, the global optimum is anything but obvious, and the assurance of the "global" optimum is a definite advantage.

This book will discuss several techniques to formulate seemingly nonlinear relationships as mixed integer linear relationships. As will be shown in the subsequent chapters of this book, it is possible to formulate utility systems as MILP optimization problems with very few simplifications.

1.4.3 Optimization formal definition

A formal definition of optimization is: "optimization: noun: a method to determine the optimum solution to a mathematical optimization problem which comprises an objective function and a set of constraints." Figure 1.13 shows the structure of a generic optimization problem. Constraints can be organized into two categories:

Type I constraints, or simply constraints: represent the relationships among variables; these relationships can be equalities or inequalities. Constraints could be specified in an algebraic form or in standard form. In the standard form, all terms with variables are on the left side, and only constants appear on the right side. Following are two examples:

$$3 * X_1 - 2 * X_2 = 4 \qquad (1.1)$$

$$4 * X_1 + 0.5 * X_2 \le 8 \qquad (1.2)$$

Some Solvers, such as LINGO, also accept constraints in the familiar algebraic form. The algebraic form, however, is not unique. For example, the constraint in eq. (1.1) could be written as one of the following:

$$3 * X_1 = 2 * X_2 + 4 \qquad (1.3)$$

$$2 * X_2 = 3 * X_1 - 4 \qquad (1.4)$$

Converting from the algebraic form to the standard form is trivial, so we will feel free to use either form in this book.

Type II constraints or variable type specifications: By default, all variables in LP and MILP models are assumed to be continuous, so we need to specify only the exceptions. In our modeling, this means that binary and general integer variables must be explicitly specified.

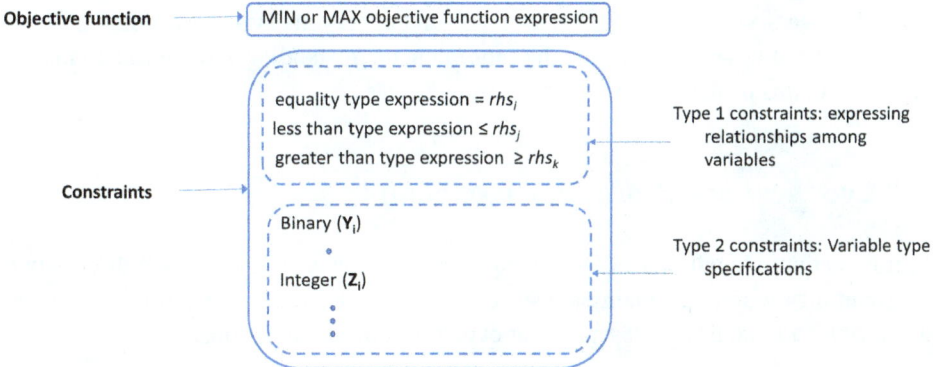

Fig. 1.13: Structure of a generic optimization problem.

Modeling or model building develops the optimization problem by defining the objective function and all the constraints. We will use MILP formulations for modeling of utility systems, so the objective function and the constraints will be required to be of linear form.

In this book, we will develop models using two commercially available modeling solvers: Solver in Microsoft Excel from Frontline Systems Inc. [15] and LINGO Solver from LINDO Systems Inc. [16] Although model development is similar, there are also differences that we will highlight as we go.

Defining the objective function is usually straightforward; in simple cases, it could be just the summation of unit cost of purchased energy multiplied by the amount of that purchased energy, i.e., $\sum cx_i * X_i$, where cx_i is the unit cost of the ith purchased energy and X_i is the amount of ith energy purchase. Note that not all objective functions can be modeled this simply; we will discuss modeling of more complicated cost functions in Chapter 8.

Modeling the Type I constraints that represent relationships among variables may require some skills. There is flexibility in developing these relationships; some may naturally be linear, while others may appear nonlinear but could easily be transformed into linear form. However, some constraints may be strictly nonlinear, in which case simplifaction may be required to develop a linear approximation. As we shall see in subsequent chapters, there is room for creativity when simplifying nonlinear relationships to mixed integer linear form.

Constraints have limits, and defining the bounds of constraints is usually straightforward. In LINGO, constraints are written in algebraic form, including equality and inequality symbols. However, Excel Solver requires each constraint to be specified by its type: equality, a lower bound ($\geq$ type) inequality, or an upper bound ($\leq$ type) inequality. We will simplify by specifying the lower and upper feasibility limits on all constraints; although this doubles the number of constraints, it simplifies model development in Excel. Note that in this framework, an equality constraint has the same value for the lower and the upper bound. By default, lower and upper limits of binary variables are 0 and 1; for general integers, we must specify the the lower and upper limits via constraints.

The following section discusses the conceptual understanding of optimization modeling. Part I of this book will go into the modeling details.

1.5 Overview of optimization modeling

In this section, we will discuss modeling in general terms. First, we will discuss two shortcomings when optimizing subsystems rather than the entire system. Next, we will discuss a modeling philosophy that overcomes these limitations.

1.5.1 Shortcomings of subsystem optimization

Utility systems have boundaries, usually the plant fences; what is outside is external to the system. Utility systems are seldom self-contained; they interact with external entities, such as external energy suppliers, that are shown as streams crossing the system boundaries.

In an optimization model, streams crossing the system boundary are of two types: fixed flow and variable flow. Fixed flow streams have specified flow rates (flow specs) and variable flow streams need cost structure (cost specs).

With this understanding, let us review the sample utility system shown in Fig. 1.9 by focusing only on the streams crossing the utility system boundary. Tab. 1.1 summarizes all such streams along with their specifications.

Fixed flow streams are the aggregate process demands; the process operation determines their flows, and the utility system must satisfy these demands. Variable flow streams broadly fall into two categories: external energy purchases and vent streams. Purchased energy streams have well-defined and legally binding contracts reflecting the true costs of these transactions. De-mineralized water, DMW, is internally produced; we will use an estimated unit cost. The cost of wastewater stream, WW, is usually tiny, so the error in assuming or ignoring its price is insignificant. The cost of vent streams, VENT, FGFLARE, and FOFLARE, are unnecessary as they are accounted for via increased DMW and purchased fuel costs. Therefore, when optimizing the utility system as a whole, the model will use true costs to calculate the objective function.

On the process utility demand side, there usually are measurements of significant flows, but not for smaller flows. We can also estimate process utility demands from the supply side. However, we must reconcile the measured and estimated flow discrepancies. We will discuss this topic thoroughly in Part II of this book.

Tab. 1.1: The sample utility system: specifications on streams crossing system boundaries.

Header ID	Stream id	Stream description	Stream type	UOM	Specification	Specfication basis
ELEC	BUY_ELEC	Electrical power from the grid	Variable	MW	Cost spec	Supply contract
HPFG	BUY_HPFG1	HPFG from supplier #1	Variable	MW	Cost spec	Supply contract
	BUY_HPFG2	HPFG from supplier #2	Variable	MW	Cost spec	Supply contract
LPFG	BUY_LPFG1	LPFG from supplier #1	Variable	MW	Cost spec	Supply contract
	BUY_LPFG2	LPFG from supplier #2	Variable	MW	Cost spec	Supply contract
FO	BUY_FO	FO from supplier	Variable	MW	Cost spec	Supply contract
DMW	BUY_DMW	Demin Water supply	Variable	t/h	Cost spec	Assumed unit cost
LPS	VENT	LP steam vent	Variable	t/h	Cost spec	cost ~ 0
WW	WW	Waste water to treatment	Variable	t/h	Cost spec	cost ~ 0
LPFG	FGFLARE	Fuel gas to flare	Variable	MW.th	Cost spec	cost ~ 0
FO	FOFLARE	Fuel oil to flare	Variable	MW.th	Cost spec	cost ~ 0
CA	VENTCA	Excess compressed air	Variable	kM3/h	Cost spec	cost ~ 0
ELEC	elec_demand	Elec power to process units	Fixed	MW	Flow spec	Aggregate process demand
HPFG	hpfg_demand	HPFG to process units	Fixed	MW.th	Flow spec	Aggregate process demand
LPFG	lpfg_demand	LPFG to process units	Fixed	MW.th	Flow spec	Aggregate process demand
FO	fo_demand	FO to process units	Fixed	MW.th	Flow spec	Aggregate process demand
HPS	hps_demand	HPS to process units	Fixed	t/h	Flow spec	Aggregate process demand
MPS	mps_demand	MPS to process units	Fixed	t/h	Flow spec	Aggregate process demand
LPS	lps_demand	LPS to process units	Fixed	t/h	Flow spec	Aggregate process demand
PC	pc_supply	Process condensate return	Fixed	t/h	Flow spec	Aggregate process demand
CW	cw_demand	Cooling water to process units	Fixed	kt/h	Flow spec	Aggregate process demand
CA	ca_demand	Compressed air to process units	Fixed	kM3/h	Flow spec	Aggregate process demand

Things are different when optimizing a subsystem; for example, let us consider the air compression subsystem, as shown in Fig. 1.8, and focus only on the streams crossing the subsystem boundary. Table 1.2 summarizes all such streams along with their specifications. This subsystem imports electric power and LP fuel gas and exports compressed air via the compress air header, CA. It also requires atmospheric air, which is just part of the environment. Compressed air goes to the process and has a flow spec. All other streams are variables that the optimizer determines; cost information will be required for these streams crossing the subsystem boundary, which will be problematic, as explained next.

Tab. 1.2: Sample compressed air subsystem: specification on streams crossing subsystem boundaries.

Header ID	Stream id	Stream description	Stream type	UOM	Specification	Specfication basis
ELEC	CA_ELEC	Air compressor power consumption	Variable	MW	Cost spec	Transfer price
LPFG	CA_LPFG	Air compressor fuel consumption	Variable	MW.th	Cost spec	Transfer price
CA	VENT_CA	Compressed air vent	Variable	kM3/h	Cost spec	Cost = 0
CA	ca_demand	Compressed air to process units	Fixed	kM3/h	Flow spec	Aggregate process demand

Consider the two external energy (fuel and electrical power) streams. Even though there are well-defined contracts with external energy suppliers, the contracts apply to the sitewide total energy import; it does not apply only to the portion of energy imported by the air compression subsystem: **CA_ELEC** and **CA_LPFG**. For example, LPFG fuel has two suppliers, each with a different contract and cost structure; it is unclear which of the two contractual costs to use for costing of **CA_LPFG**.

Defining an objective function for the air compression subsystem will require cost information for both purchased energy streams, which will be problematic. Those who have tried subsystem optimization have previously resolved this problem using "**transfer costs**" also known as "**transfer prices**" or **marginal costs** for these streams. Note, however, that a transfer or marginal cost is the incremental cost at the systemwide optimum point [12], which is yet to be determined and remains unknown. This problem could theoretically be resolved iteratively by initially assuming all incremental costs,

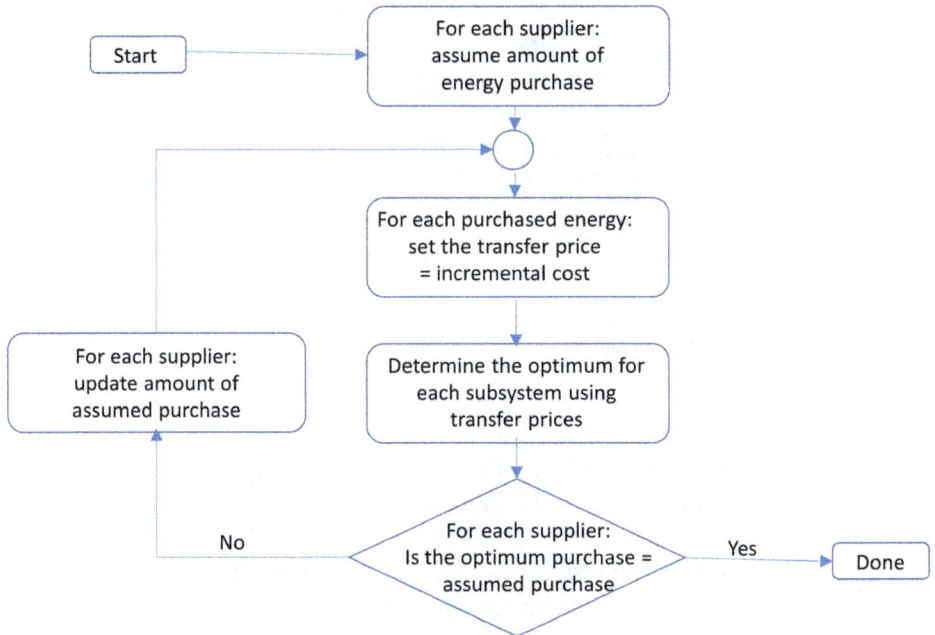

Fig. 1.14: Flowchart of a possible algorithm for system optimization using transfer prices.

optimizing each subsystem with the assumed costs, and iterating till convergence is achieved. Figure 1.14 is a flowchart of a possible convergence scheme. However, achieving this convergence would be tedious, time-consuming, and is seldom practiced. In practical terms, subsystem optimizations are carried out using somewhat arbitrary values for the transfer prices and with limited effort to achieve convergence. The downside of this approach is that the calculated optimization policy would likely not be the true optimum, and the expected benefits will likely not be realizable.

Because of this problem with transfer prices, we want to do away with subsystem optimizations and do a systemwide optimization with system boundary expanded to the plant fence where there is accurate pricing information in the form of purchased energy contracts.

1.5.2 Missed opportunities by suboptimizations

In the previous section, we made a strong case for doing systemwide optimization instead of multiple subsystem optimizations, as doing so obviates the need for iterating over transfer costs. In practice, transfer costs are somewhat arbitrarily chosen, and thus, the calculated optimum policies are somewhat fictitious and the calculated benefits are not realizable.

There is yet another reason for performing systemwide optimization. The reason is that when optimizing the system as a whole, additional opportunities emerge that do not appear when doing multiple subsystem optimizations. This can be best illustrated by considering an example. The electrical subsystem in Fig. 1.6 is a straightforward energy balance: **Electric Import + Internal power generation = internal power consumption +** *elec_demand*. This subsystem has no degrees of freedom, and hence, optimizing this subsystem in isolation calculates the cost of purchased power based on the electric power contract, and there are no alternatives to consider. Note that almost all power contracts impose a heavy penalty for establishing a new power demand peak; in such cases, it would make sense to take drastic steps to avoid making a new demand peak and thus avoid the cost penalty, perhaps by utilizing even inefficient internal power generation such as in the CTG in the steam subsystem or switching some of the dual drive services from the motor to steam turbine drives even if doing so results in venting of steam. The multiple suboptimization approach would miss the exploration of such possibilities.

1.5.3 Modeling philosophy

We will start with two quotes relevant to modeling in general and then outline a pragmatic modeling philosophy that we will use in modeling utility systems.

The first quote is by a British mathematician, George E. P Box of Box and Jenkins fame [13]:

All models are wrong, but some are useful.

This is a profound statement. The sentiment expressed here is that a model, an abstraction of reality, can never be perfect, as one could always find flaws. Reality is inherently complex, and no model can capture all nuances of reality. Hence, by rigorous comparison with reality, any model, especially of any complex system, will necessarily be wrong if one focuses on discrepancies. However, on the practical side, according to Box, the measure of goodness of a model must not be how faithfully it mimics reality but the extent of its usefulness. The author fully concurs with this sentiment.

For example, consider an industrial steam generator; it comprises hundreds of components and has a variety of control systems to produce desired amount of steam at the desired conditions.

On an overall level, the boiler could be in one of the three states: shut down, i.e., neither producing steam nor consuming any fuel; normal operation, i.e., consuming fuel and producing stream; and standby operation, i.e., not making any steam but consuming a small amount of fuel and be ready to make more steam on short notice. When producing steam, there is an active combustion process with radiant and convective heat transfer from the hot flue gas to the fluid in the steam drum and the boiler coils; boiling and superheating of steam; associated controls to produce steam at the desired temperature and pressure conditions; and blowdown flow to control the buildup of solids in the drum to reduce scaling in the boiler tubes and the drums. Each of these processes has complex science and technology behind it that could be described in terms of many sophisticated nonlinear equations. What if one takes the trouble to develop such a sophisticated model that closely mimics the boiler operations when producing steam, but the model cannot adequately model the standby operation? Such a model would not be very useful, as it would not allow the optimizer to consider the standby option when considering all possible options in the search for the global optimum. The model mentioned above, although very sophisticated, would fail the usefulness test.

The second quote is from one of the most famous scientists of our times, German-born physicist, Albert Einstein of the Theory of Relativity fame [14]:

Make everything as simple as possible, but not simpler.

This is another profound statement with practical implications regarding modeling for optimization.

Typically, engineers tend to start modeling rigorously in detailed and sophisticated terms, which is fine when modeling simpler systems such as a unit operation or a section of a unit. However, this approach becomes increasingly burdensome when modeling and later maintaining large-scale systems of some complexity, such as an entire utility system or production unit. Due to resource limitations, this approach often stops the modeling at the subsystems modeling level resulting in only subsystem level optimizations. As we discussed earlier, the problem with this approach is that a composite optimization of subsystems yields a smaller total benefit than a systemwide optimizer could produce. Here, Einstein's advice is invaluable; instead of starting with a complex model, start simplistically and add complexity, per Box's advice, only if it enhances the model's usefulness.

Considering the above, Fig. 1.15 conceptualizes expected optimization benefits versus optimization scope. A larger scope, preferably sitewide optimization, will give significantly higher optimization benefits, as you will deal with the true cost of purchased energy per contractual agreements with external suppliers and not arbitrarily assumed transfer costs, resulting in true and realizable benefits.

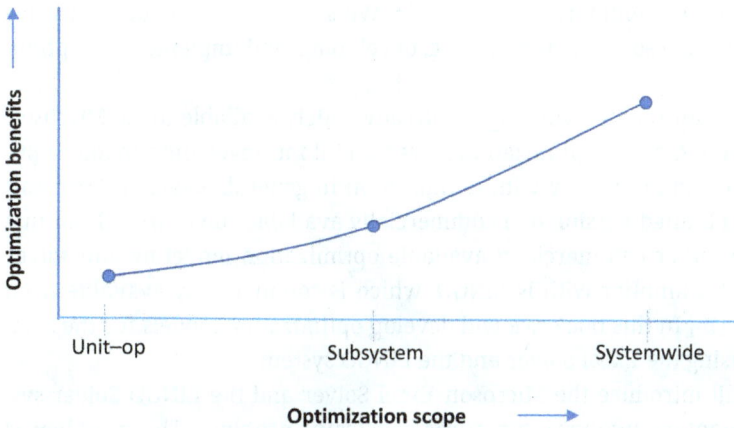

Fig. 1.15: Optimization benefits versus optimization scope.

Our recommendation is to start simple with the intention of developing a model for the entire utility system and add complexity only as needed.

Our modeling philosophy can be summarized as follows:

1. **Model scope**: Keep the scope of optimization large, as large as possible, preferably the entire system up to the plant boundary. Doing so, we will be using the true costs per legally binding contracts with all external entities, and there would be no need for "transfer costs." The calculated benefits would be true benefits realizable by implementing the optimal policy.

2. **Modeling simplicity**: Keep the models as simple as possible such that all the important relationships and features are captured. Such a model would be a complete and realistic representation of the actual operations without using undue amounts of development and maintenance resources. Start with a simple model and introduce complexity judiciously and only when it makes the model more useful. In practical terms, it means to start with simple input-output models, and as we encounter nonlinear behavior, be creative and seek ways to capture the essential fea-

tures and important relationships while still staying within the mixed integer linear programming framework.

In subsequent chapters, the philosophy mentioned above will guide us; we will develop simple yet useful models for the global optimization of utility system operations.

1.6 Overview of MILP solvers

MILP solution technology is mature; several robust and reasonably priced modeling and solver packages are commercially available. We should avail of such software packages and focus our resources on designing, developing, and implementing optimization systems.

Furthermore, optimization technology is already widely available to all. Microsoft Excel includes a version of the optimization solver useful for developing smaller optimization models to gain familiarity with optimization in general. Solver in Microsoft Excel, however, is a limited version of a commercially available solver from Frontline Systems Inc. [15] Another commercially available optimization modeling and solver package the author is familiar with is LINGO, which is commercially available from Lindo Systems Inc. [16] In this book, we will develop optimization models for the sample utility system using the Excel Solver and the LINGO system.

This section will introduce the Microsoft Excel Solver and the LINGO Solver system using an elementary two-variable, three-constraint problem. This problem is from the first chapter of LINGO User's Guide [17]. The problem description and formulation are as follows:

> *The Enginola Television Company produces two types of TV sets, the "Astro" and the "Cosmo." There are two production lines, one for each set. The Astro production line has a capacity of 60 sets per day, whereas the capacity for the Cosmo production line is only 50 sets per day. The labor requirements for the Astro set is 1 person-hour, whereas the Cosmo requires a full 2 person-hours of labor. Presently, there is a maximum of 120 man-hours of labor per day that can be assigned to production of the two types of sets. If the profit contributions are $20 and $30 for each Astro and Cosmo set, respectively, what should be the daily production to maximize profit?*

To define the problem in mathematical terms, first define variables, then an objective function, and then add constraints:

ASTRO = units of Astro TVs to be produced per day
COSMO = units of Cosmo TVs to be produced per day

Profit Objective function in dollars per day is:

$$\text{Maximize } 20^*\textbf{ASTRO} + 30^*\textbf{COSMO} \tag{1.5}$$

Subject to the daily labor and production capacity constraints:

$$\textbf{ASTRO} + 2\,{}^*\textbf{COSMO} \leq 120 \qquad (1.6)$$

$$\textbf{ASTRO} \leq 60 \qquad (1.7)$$

$$\textbf{COSMO} \leq 50 \qquad (1.8)$$

1.6.1 Using Excel Solver

If you wish to try this on your own, ensure that the Solver add-in is loaded in your Excel; if not, follow the instructions on the Microsoft support website [18] to load the Solver add-in.

In general terms, on an Excel Spreadsheet, we need to define spreadsheet cells for the two variables (ASTRO and COSMO), then define the objective function and the three constraints as formulas in four different cells, and lastly, specify the constraint limits in separate cells. Note that the optimum value of variables, objective function, and constraints are all populated by the optimizer.

Excel is a highly flexible system. There is a wide variety of ways to specify this optimization model. However, when embarking on a large development project, adopting a structure that is easy to use and yet scalable to large problems behooves.

Figure 1.16 shows the modeling structure we will use. It uses the freeze feature of Excel at cell G10 to create four regions on the spreadsheet. We reserve the top four lines for model ID, objective function, and all model input data; this effectively partitions the Excel spreadsheet into six areas. Figure 1.16 shows the six areas separated by dotted lines. Note that the cells highlighted in yellow are nominally for user input. Following is the description of each region:

Region ❶: The top left area, cells A1 to F4, are for top-level information. Rows 1 and 2 are for entering the problem and Case ID. Row 3 is the objective function: cell C3 shows the UOM and F3 is the objective function formula.

Region ❷: Rows 1 through 4, column G onwards, are to store all the inputs and parameters used in defining the optimization problem.

Region ❸: Rows 5 through 9, columns A through F, show headers for regions 4 and 5.

Region ❹: Rows 5 through 9, column G onwards, are dedicated to optimization variables. Row 5 shows the variable numbers, and Rows 6 and 7 are for variable ID and Units of Measure (UOM), respectively. Row 8 is for the variable cost coefficients, populated using the input values specified in Region 2. Note that variables that do not have cost value should be left blank; the solver treats them as zero. Row 9 is for the optimum values of the variables populated by the optimizer upon successful completion of the solver.

Region ❺: Columns A to F, rows 10 onwards, are dedicated to constraints.

Column A shows the constraint numbers; Columns B and C are for constraint ID and UOM, respectively; Columns D and E are for constraint low and high limit values, respectively. Column F shows the values of the constraints calculated using SUM-PRODUCT formulas, such as shown; it will show the optimum constraint values upon successful completion of the solver.

Region ❻: Rows 10 onwards and column G onwards are dedicated to the coefficients that define the constraints. We recommend that cells without coefficients be left blank, indicating the absence of coefficients; the solver treats them as zero.

This way of setting up the optimization problem is entirely scalable to problems of any size.

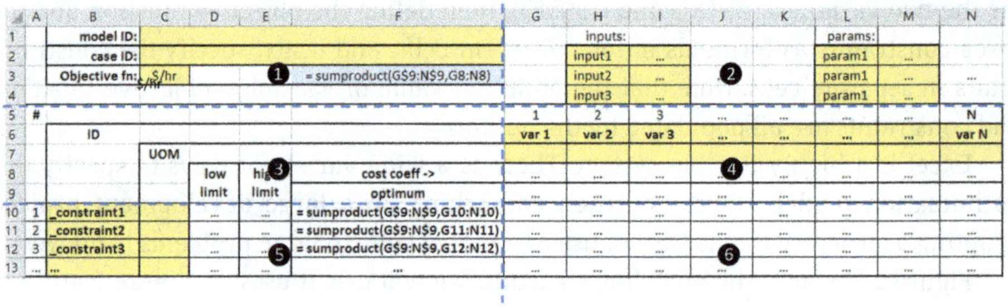

Fig. 1.16: Proposed MILP model structure in Microsoft Excel.

After the solver has solved, the optimum value of the objective function appears in cell F3; the optimum values of the variables are in row 9, starting column G onwards, and the optimum constraint values are in column F, starting row 10 onwards. Note that the spreadsheet input cells are highlighted in yellow background. For visual feedback, the constraint values in column F have conditional formatting; constraints at high limits are in blue background, and those at low limits are in pink background; equality constraints and those within the limits are not highlighted.

For the sake of consistency, the maximization objective function is highlighted in blue, and the minimization is in pink. As mentioned earlier, all constraints are specified in lower and upper limits. For equality constraints, both the upper and the lower limits will be the same value. For inequalities of the type "≥," the constraint RHS specifies the lower limit for the constraint, and the implied upper limit is +∞; however, computers being finite precision machines, specifying +∞ or very large numbers such as numbers in millions could cause numerical problems, so instead we will use a reasonably large number such as 999 as a proxy for +∞. For flexibility, we will designate this number by a parameter with the symbolic name "big," specified in the parameters section in area 2. For inequalities of the type "≤," the constraint RHS specifies the upper limit implied lower limit is −∞; again, we would use "-big" as a proxy for −∞. In instances where a constraint refers to a physical quantity, zero would be a better lower limit, such as the case in this example problem; so in this particular simple example, there is no need to use the "big" parameter, but later in this book you will see models that will use this parameter.

Figure 1.17a shows the Enginola Product Mix problem with inputs and parameters. Figure 1.17b shows the formulae: objective function calculation is in cell F3 and constraint calculations are in cells F10 through F12.

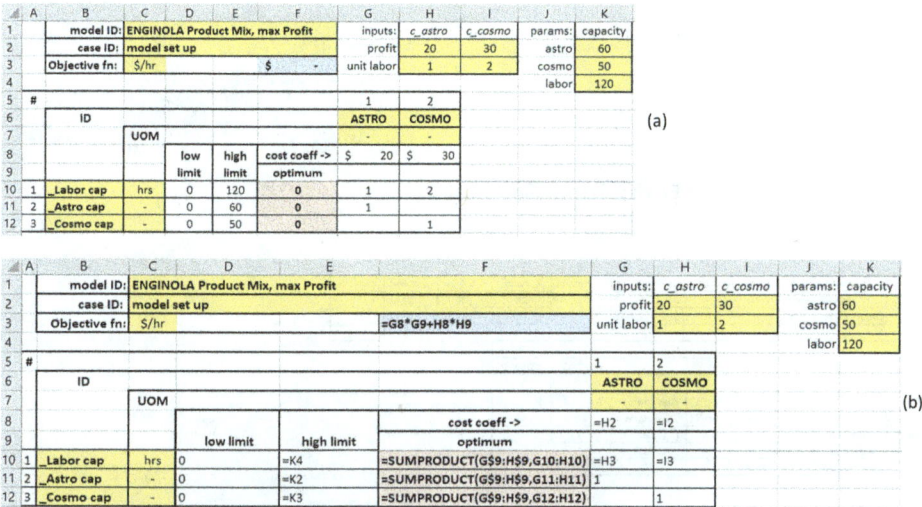

Fig. 1.17: Enginola product mix problem setup: (a) spreadsheet view and (b) formula view.

The optimizer in Excel Solver is configured on a pop-up window invoked by the Solver command found on the Data tab. Figure 1.18 shows the Solver configuration in the **Solver Parameters** pop-up window. Note that F3 is specified in the **Set Objective** box. The **Max** button is active, indicating maximization as the optimization direction. The **By Changing Variable Cells** box shows the optimization variables as cells G9 and H9. The **Subject to the Constraints** box shows the problem constraints. The optimization solution is obtained by pressing the **Solve** button on the pop-up window, as highlighted by a red box in Fig. 1.18.

The problem is solved instantly, and the **Solver Results** pop-up window opens; the global optimum solution is confirmed, as shown in the highlighted box in Fig. 1.19. Press **OK** to accept the solution; observe that the optimum policy is to produce 60 **ASTRO** and 30 **COSMO**. Note that both Labor and **ASTRO** capacities are at their high limits, as indicated by the light blue shading of the corresponding cells. The maximum profit is $2100/day.

This model file named **Enginola.xlsx** is available from the publisher's website.

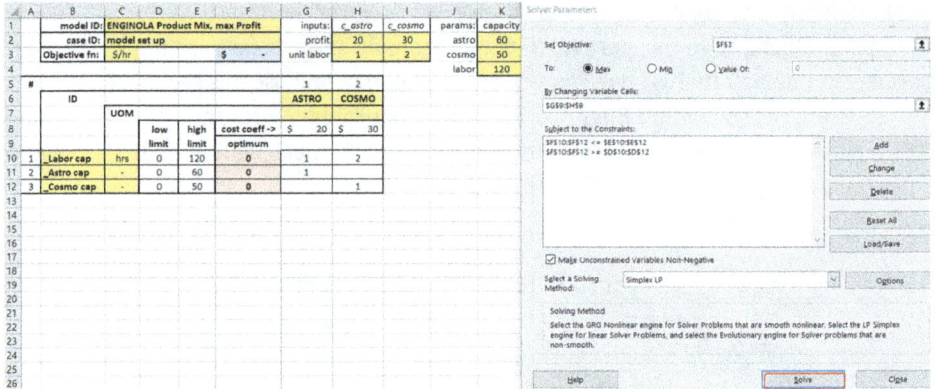

Fig. 1.18: Enginola model in Excel: Solver configuration.

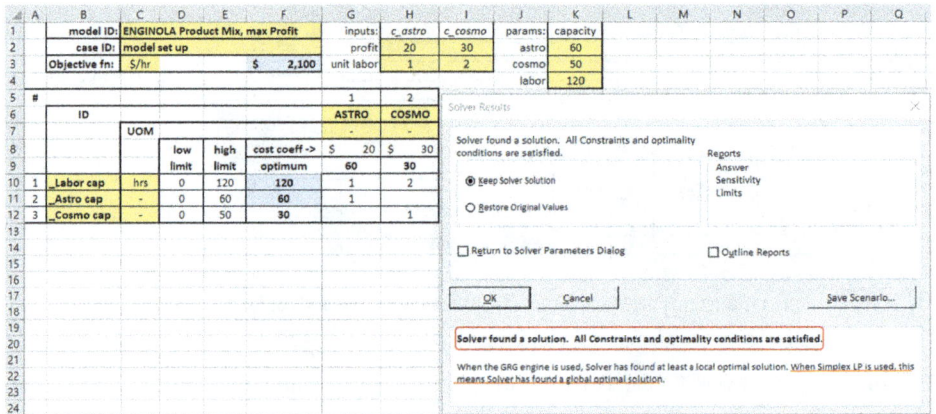

Fig. 1.19: Enginola problem solution in Microsoft Excel.

1.6.2 Using LINGO

If you wish to try this independently, ensure that LINGO software is installed on your computer. If not, a trial version of LINGO software is available from LINDO Systems' website [16]; follow the instructions to install it.

There are also several ways to set up the optimization problem in LINGO; the simplest way is to use it interactively by invoking LINGO and typing the problem formulation in algebraic form, as shown in Fig. 1.20; comments can be inserted anywhere in the text for documentation purposes. The main section starts with the keyword **MODEL:** and ends with the optional keyword **END MODEL**. Inputs and parameters for the problems are specified in **DATA:** section that ends with the keyword **END-**

DATA. The objective function is specified in the first line in the model section statement starting with **MAX** = followed by a linear expression followed by a ";" for our maximization problem. Constraints are written as algebraic equality or inequalities. Note that LINGO allows a shortcut in defining the constraints; you can use > to represent ≥ and < to represent ≤ . Comments start with a "**!**." All statements, including the comments, end with a ";."

Compared to Solver in Excel, LINGO does not require explicit declaration of the variables to be optimized; variables that are not specified values become the optimization variables. This is a very useful feature, especially when using the same model for multiple uses, as we will discuss in Chapter 9 on online implementation.

For the example problem, the production capacity constraints can be more efficiently defined using LINGO's **@BND** function to specify both the lower and the upper limits; however, we will not use this feature to maintain consistency with the Excel Solver model.

```
MODEL:
!
! Enginola Product Mix problem
! Ref: Linus Schrage, "Optimization Modeling with LINGO", 6th edition, p 1-2, (2021).
! ;

    DATA:
        !inputs;        c_astro = 20; c_cosmo = 30;         ! unit profit, $/unit;
                        labor_astro =  1; labor_cosmo = 2;   ! unit labor, hrs;
        !params;        cap_astro    = 60;                   ! ASTRO prduction caoacity ,units;
                        cap_cosmo = 50;                      ! COSMO production capacity, units;
                        cap_labor = 120;                     ! Labor acvailability, hours;
    ENDDATA

    ! Maximize profit;
    [_Enginola_Obj] MAX = c_astro * ASTRO + c_cosmo * COSMO;

    ! Capacity constraints;
    [_Labor_cap]    labor_astro * ASTRO + labor_cosmo * COSMO < cap_labor;
    [_Astro_cap]    ASTRO < cap_astro;
    [_Cosmo_cap]    COSMO < cap_cosmo;

END MODEL Eniginola PM
```

Fig. 1.20: Enginola model in LINGO.

To solve the problem, navigate to the Solver command on the **Solver** tab or alternately press the **Solve** button shown as the bull's eye icon highlighted in a red box in Fig. 1.20.

The solution is instantly produced, as shown in Fig. 1.21. Note that the solution obtained is the Global Optimum as indicated on the **Solver Status** window, and the solution values shown on the **Solver Report** window match those obtained in Excel.

Additional model statistics indicate two variables and four constraints. LINGO considers the objective function as constraint #1, the Labor constraint as constraint #2, and the production capacity constraints as constraints #3 and #4. In addition, LINGO provides sensitivity information such as reduced costs and dual prices. While sensitivity information is helpful for pure linear programming problems, they do not apply to MILP problems, so this book will not discuss them.

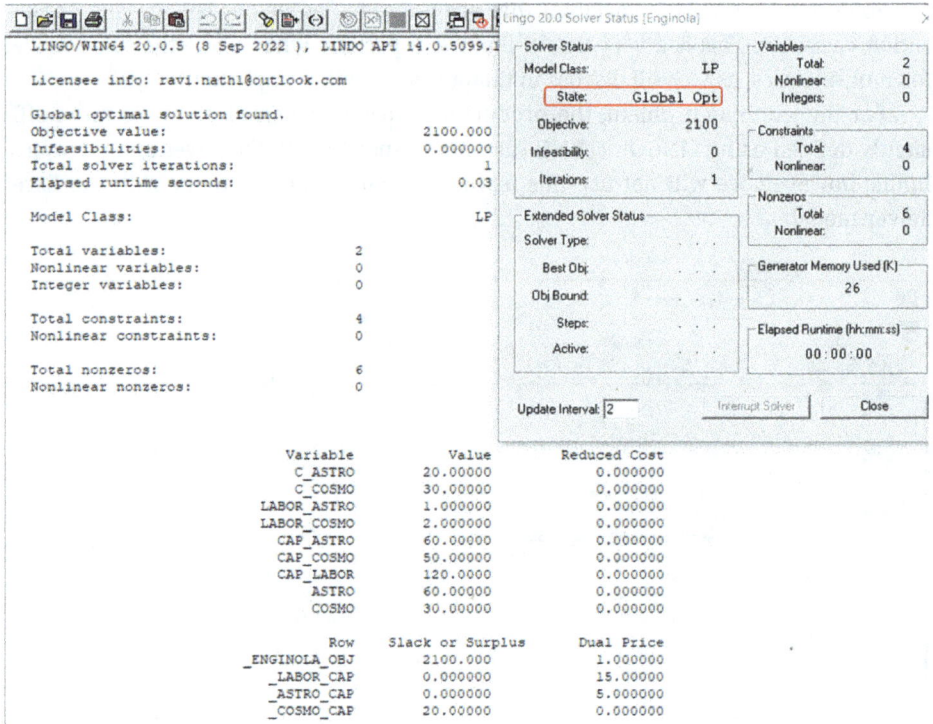

Fig. 1.21: Enginola problem solution in LINGO.

This model file named **Enginola.lg4** and Excel model file **Enginola.xlsx** are available from the publisher's website.

Summary

This chapter lays the foundation for this book. We presented essential concepts in optimization, optimization methods, shortcomings in subsystem optimizations, benefits of global optimization, and barriers to utility system optimization. We introduced a simple yet representative utility system and its subsystems in detail. We have outlined our modeling philosophy for use in the book. Finally, we presented two commercially available MILP solvers we will use to develop optimization models for the sample utility system and an example problem illustrating their interface.

Nomenclature

Symbol	UOM	Description
ACj	–	Air compressor #j
ACjE	–	Air compressor #j's Engine
ACjM	–	Air compressor #j's Motor
BD	–	Boiler Blow Down
BFW	–	Boiler Feed Water
BLj	–	Boiler #j
BLjAT	–	Boiler #j Auxiliaries' Steam Turbine
BLjAM	–	Boiler #j Auxiliaries' Motor
BPj	–	BFW Pump #j
BPjM	–	BFW Pump #j's Motor
BPjT	–	BFW Pump #j's Steam Turbine
CA	–	Compressed Air
CTG	–	Condensing TurboGenerator
CW	–	Cooling Water
CWPj	–	Cooling Water Pump #j
CWPjM	–	Cooling Water Pump #j's Motor
CWPjT	–	Cooling Water Pump #j's Steam Turbine
DEA	–	Deaerator Unit
DMW	–	Demineralized Water
ELEC	–	abbreviation for Electrical
FC	–	Fuel Compressor
FO	–	Fuel Oil
GTG	–	Gas TurboGenerator
HPFG	–	High-Pressure Fuel Gas
HPS	–	High-Pressure Steam
k, k$_i$	–	generic constant
LPFG	–	Low-Pressure Fuel Gas
LPS	–	Low-Pressure Steam
MPS	–	Medium-Pressure Steam
PC	–	Process Condensate
PRV	–	Pressure Reducing Station
STG	–	Steam back-pressure (noncondensing) TurboGenerator
X, X$_i$, Y, Y$_i$	–	generic variable

References

[1] "Petroleum Refining in the US", Wikipedia article, https://en.wikipedia.org/wiki/Petroleum_refining_in_the_United_States#:~:text=As%20of%20January%202019%2C%20there,US%2C%20distributed%20among%2030%20states.

[2] US Energy Information Administration, web site: https://www.eia.gov/dnav/pet/PET_PNP_CAPFUEL_DCU_NUS_A.htm

[3] "Tragedy of the Commons", Wikipedia article, https://en.wikipedia.org/wiki/Tragedy_of_the_commons

[4] "Nonlinear Programming", Wikipedia article, https://en.wikipedia.org/wiki/Nonlinear_programming

[5] "Integer Programming", Wikipedia article, https://en.wikipedia.org/wiki/Integer_programming

[6] "NP-Completeness", Wikipedia article, https://en.wikipedia.org/wiki/NP-completeness

[7] "Curse of Dimensionality", Wikipedia article, https://en.wikipedia.org/wiki/Curse_of_dimensionality

[8] Austin, G. T. "Shreve's Chemical Process Industries", 5th Edition, McGraw Hill, (1984).

[9] Rikhtegar, F. and S. Sadighi, "Optimization of energy consumption", Digital Refining, (Apriil 2015). https://www.digitalrefining.com/article/1001127/optimisation-of-energy-consumption

[10] Nath, R. et. al. "Experiences with Online STEAMPOP", Paper 28a, presented at the AIChE National Meeting New Orleans, (March 1988).

[11] Hillier, F. S. and G. J. Lieberman. "Introduction to Operations Research", 11th Edition, McGraw Hill, (2021).

[12] "Marginal Cost", Wikipedia article, https://en.wikipedia.org/wiki/Marginal_cost

[13] *Box, George EP*. http://www-sop.inria.fr/members/Ian.Jermyn/philosophy/writings/Boxonmaths.pdf *Science and statistics* http://www-sop.inria.fr/members/Ian.Jermyn/philosophy/writings/Boxonmaths.pdf *(PDF)* https://en.wikipedia.org/wiki/Journal_of_the_American_Statistical_Association *Journal of the American Statistical Association*, 71 (356): 791–799 (1976).

[14] Robinson, A, "Einstein said that – didn't he?" https://www.nature.com/articles/d41586-018-05004-4#:~:text=https://www.nature.com/articles/d41586-018-05004-4#:~:text=Everything%20should%20be%20made%20as,possible%20without%20having%20to%, (2018).

[15] Frontline Systems Inc, website: https://www.solver.com

[16] Lindo Systems Inc, website: https://www.lindo.com

[17] "Optimization Modeling with LINGO by Linus Schrage". available at: https://www.lindo.com/index.php/ls-downloads/user-manuals?id=112:lingo-documentation&catid=82

[18] Microsoft Support, web site: https://support.microsoft.com/en-us/office/load-the-solver-add-in-in-excel-612926fc-d53b-46b4-872c-e24772f078ca#:~:text=On%20the%20Tools%20menu%2C%20select,box%2C%20and%20then%20click%20OK.

Part I: **Model development**

Part I of the book discusses the development aspects of a utility system model with the aim of creating a global optimization model for a simple industrial process plant utility system. We assume no prior knowledge of or experience with modeling or optimization. Chapter 1 has already introduced the basic concepts of optimization and optimization philosophy.

We will start Part I with a general model development methodology. We will use this methodology throughout the book, first creating individual equipment models, then assembling the individual equipment models to create subsystem models and finally, integrating the subsystem models to develop a complete model for the sample utility system model presented in Chapter 1. The sample utility system model represents a simple process plant utility system.

All models will be developed in two commercially available mixed integer linear programming solver packages: the basic Excel solver that comes standard with Excel; but as the model grows, we will have to use the upgraded version called the Premium solver, which is fully compatible with the Basic solver, and another one called LINGO, which allows modeling in terms of algebraic equations. All model files are available from the publisher's website for this book.

Part II of the book will focus on the use and deployment aspects of the utility system optimization model.

https://doi.org/10.1515/9783111020679-002

2 Modeling basics

In general, a system can be viewed as a hierarchy. The system comprises smaller components, usually called subsystems, and each subsystem, in turn, has smaller subsystems until we reach the smallest modeling component. For a utility system, a three-level hierarchy is sufficient; it includes the subsystems, viz., steam, fuel, electric power, cooling water, and compressed air. Each of the subsystems comprises pieces of equipment or unit operations. For example, the steam subsystem includes boilers, turbo-generators, pressure-reducing valves, pumps, deaerators, blow-down flash units, and steam and water headers.

An equipment or an individual purchase contract would be our smallest modeling component. A subsystem model is created by merging its individual component models. Then, the system model is created by combining all the subsystem models, including models representing contracts with external suppliers.

In general, modeling is an iterative process of successive refinement. As mentioned in Chapter 1, we take a pragmatic approach to model development by starting simple and adding complexity only as needed and only when it enhances the model's usefulness. In general, modeling is also a creative process, so it is difficult to cast a rigid set of rules for such a creative process; nonetheless, we will provide some guidance. With such intent, we will discuss a systematic approach to modeling individual components in general terms in this chapter.

The following are the recommended steps for model development; however, do not be dogmatic about them; use them as guidelines:
1. Enumerate constants,
2. Enumerate operating modes,
3. Enumerate variables, and
4. Develop constraints, i.e., relationships among variables.

Initially, we will focus on developing the model for steam subsystem components as it has a wide variety of equipment that will form a solid basis for developing models for components in the other subsystems.

This chapter will discuss modeling in general terms; subsequent chapters will discuss the details of individual component modeling.

2.1 Enumerate constants

A good starting point is to start with an enumeration of the constants in the system. Although constants are values that do not change during an optimization run, most will change over time. Demands will usually change from one case to the next; electric power rates could change hourly; some fuel costs, such as spot prices, will vary daily;

https://doi.org/10.1515/9783111020679-003

others may change monthly or seasonally, while some others may have longer cycles. We will assign symbolic names to the constants whose values could change in the model as it eases the understanding of the model and allows for the possibility of changing the values systematically. We will start with constants for the steam subsystem, as that is our initial focus in modeling.

We will classify the constants into two categories: inputs and parameters. Inputs are values that change continuously from one optimization run to the next, such as the process utility demands. Parameters refer to values that change on a slower time scale, such as purchased energy costs and equipment performance coefficients. Table 2.1 shows an initial list.

Tab. 2.1: Initial list of constants for the steam subsystem.

Hedaer ID	Input or parameter	Symbolic ID	Sample value	UOM	Comments
Inputs: Process Utility demands:					
HPS	HPS process demand	hps_demand	100	t/h	Fixed at start of optimization time
MPS	MPS process demand	mps_demand	30	t/h	Fixed at start of optimization time
LPS	LPS process demand	lps_demand	30	t/h	Fixed at start of optimization time
PC	PC process supply	pc_supply	80	t/h	Fixed at start of optimization time
ELEC	ELEC process demand	elec_demand	30	MW	Fixed at start of optimization time
HPFG	HPFG process demand	hpfg_demand	30	MW.th	Fixed at start of optimization time
LPFG	LPFG process demand	lpfg_demand	30	MW.th	Fixed at start of optimization time
FO	FO process demand	fo_demand	30	MW.th	Fixed at start of optimization time
CW	CW process demand	cw_demand	15	kt/h	Fixed at start of optimization time
CA	CA process demand	ca_demand	30	kM3/hr	Fixed at start of optimization time
Parameters: Steam header related:					
	Pressure		44.0	bar(a)	Fixed operating condition
HPS	Temperature		395.0	°C	Fixed operating condition
	Enthapy	h_hps	3197.6	kJ/kg	From steam tables
	Pressure		18.5	bar(a)	Fixed operating condition
MPS	Temperature		275.0	°C	Fixed operating condition
	Enthapy	h_mps	2971.3	kJ/kg	From steam tables
	Pressure		4.9	bar(a)	Fixed operating condition
LPS	Temperature		155.0	°C	Fixed operating condition
	Enthapy	h_lps	2756.4	kJ/kg	From steam tables
	Pressure		44.0	bar(a)	Fixed operating condition
HPS BD	Temperature, saturation		256.0	°C	Fixed operating condition
	Enthapy	h_hpbd	1115.6	kJ/kg	From steam tables
	Pressure		18.5	bar(a)	Fixed operating condition
MPS BD	Temperature, saturation		208.5	°C	Fixed operating condition
	Enthapy	h_mpbd	890.9	kJ/kg; kWh/t	From steam tables
	Pressure		53.0	bar(a)	Fixed operating condition
BFW to Boilers	Temperature		100.8	°C	Fixed operating condition
	Enthapy	h_bfwHP	426.3	kJ/kg; kWh/t	From steam tables
	Pressure		1.0	bar(a)	Fixed operating condition
BFW ex Deaerator	Temperature		100.0	°C	Fixed operating condition
	Enthapy	h_bfwLP	419.1	kJ/kg	From steam tables
	Pressure		1.0	bar(a)	Fixed operating condition
Process Condensate	Temperature		70.0	°C	Assumed
	Enthapy	h_pc	293.0	kJ/kg	From steam tables
	Pressure		0.1	bar(a)	fixed at design time
Turbine Condensate	Temperature		49.4	°C	fixed at design time
	Enthapy	h_tc	206.9	kJ/kg	from steam tables

Note that enthalpy values in Tab. 2.1 are from the traditional steam tables [1]; using commercially available process simulation programs [2–4] may also be convenient.

This list presented in Tab. 2.1 is an initial list that we will expand as the model development progresses and as we add additional equipment models and additional constants emerge during the iterative modeling process.

2.2 Enumerate operating modes and degrees of freedom

Next, enumerate all possible operating modes of the equipment and the degrees of freedom associated with each mode of operation; this is valuable information to ensure that the model can operate in every operating mode.

An operating mode, also referred to as a "state" or simply a "mode," is an operational strategy in which equipment can operate sustainably.

For simple equipment, there would generally be just one operating mode, the normal mode of operation; however, for more complex equipment, multiple operating modes are the norm. For services with spare equipment, such as boilers, only some would operate most of the time, while others would be shut down. For such equipment, there would be at least two operating modes: the ON mode operation in which the equipment provides useful service and consumes energy and the OFF mode operation in which the equipment shuts down and neither consumes energy nor produces anything useful. Mathematically, a binary variable, 0 or 1, could represent these two modes. A value of 0 represents the OFF state and a value of 1 would mean the ON mode.

In OFF mode, all material and energy flows to and from the equipment must have zero values. In ON mode, the equipment operates within its specified operating limits while consuming energy.

Note that the startup process from OFF mode to ON mode for many equipment, especially the more complex ones, is seldom instantaneous, and the startup time may be too long to respond on time satisfactorily. For such equipment, we will need an additional operating mode; in this mode, the equipment would be ready to start on short notice, usually within minutes. This operating mode is referred to as "standby," "slow roll," or "warm" mode has low energy consumption and no useful output; its value is the ability to start producing quickly on demand. An example of equipment that may require this capability is a boiler; this is because the utility system is obligated to satisfy the process steam demand at all times, so it must be ready to respond quickly even to sudden and significant increases in process steam demand. This excess steam production capacity available on short notice is called the "steam reserve" or simply "reserve."

An equipment's degree of freedom (DOF) is a number; it refers to the minimum number of independent variables needed to specify a particular mode of operation completely.

For a boiler in the ON mode, the degree of freedom is 1, which means that one variable needs to be specified to specify the operations of the boiler; that variable usually is the amount of steam production from the boiler. But since variables are related, varying one also moves many other variables. Theoretically, one could specify the boiler feed water flow or fuel flow instead of steam production. However, traditionally, operations staff prefer to specify steam production.

Table 2.2 is an initial list of states for the equipment in the steam subsystem of the sample utility system, along with their degrees of freedom.

Tab. 2.2: Initial enumeration of expected operating modes of steam subsystem equipment.

Equipment type	Operating mode	Description	DOF
Boiler	ON	Boiler operating within specified operating limits	1
	OFF	Boiler is shut down	0
	Stand by	Boiler not producing steam but consuming some fuel, is in a state of readiness	0
Steam turbo generator	ON	Turbo-generator operating within specified operating limits	# of stages
	OFF	Turbo-generator is shut down	0
	Slow Roll	Turbo-generator is not producing but using some steam, is in a state of readiness	0
Pump	ON	Pump operating within specified operating limits	1
	OFF	Pump is shut down	0

2.3 Enumerate variables

Next, enumerate all variables associated with the equipment under consideration, as it would help develop equipment models. We will start with the modeling boiler #1. The variables on the flow diagram in Fig 1.10 were a good starting point for the expected variables; we will now update this list as additional variables have emerged.

Note that two binary variables are added to the initial list: one to indicate the ON/OFF status of the boiler and the other for the standby status. Table 2.3 is an updated list of variables for boiler #1 in the sample utility system.

Tab. 2.3: List of expected variables for Boiler #1.

Variable ID	Description	UOM
BL1_BFW	Boiler feed water flow	t/h
BL1_STM	Steam generation	t/h
BL1_BD	Blowdown flow	t/h
BL1_FUEL	Fuel consumption	MW.th
BL1_X	On/OFF status: 0=OFF, 1=ON	0/1
BL1_Y	Standby status: 1=standby, 0=otherwise	0/1

2.4 Develop relationships among variables

All variables represent physical or logical quantities and are related to each other by the underlying physical or logical phenomena. We must add these relationships via constraints so the model is aware of these relationships. Relationships among variables fall into the following three categories:

1. **Material balance relationships**

Material balance relationships are usually straightforward. They are mathematical representations of the conservation of mass.

If a piece of equipment has a single inlet and outlet stream, the material balance constraint is unnecessary; we could use a single variable to represent both the inlet and the outlet flows. For example, a pressure-reducing valve without de-superheating, such as PRV2 in the steam subsystem of the sample utility system, a single flow variable, **PRV2_STM**, characterizes it and implicitly specifies the material balance relationship, IN = OUT.

An overall material balance would be required if a piece of equipment has multiple input and output streams. If the inlet or outlet streams are also correlated, additional material balances will be required to specify the relationship between the correlated streams. In this context, a variable is independent if it can be specified independently of other variables; otherwise, it is dependent or correlated. The number of additional material balances would equal the number of correlated streams.

2. **Energy balance relationships**

If energy transfer occurs even indirectly, such as the mixing of streams with significantly different temperatures, or if there are external energy streams, then energy balance relationships are required.

Also, energy balance constraints will be necessary if energy conversion occurs, such as power generation in a turbogenerator. At times, rigorous energy balances are complicated due to the complexity of the physical phenomena or lack of necessary variables. In such cases, performance curves would be a valuable alternative. Manufacturers typically provide such performance curves or data; alternatively, plant tests could generate performance curves for the equipment at the site. Sometimes, the performance curves introduce new variables; in such cases, a careful examination or transformation of variables may obviate the need for adding new variables. At times, such relationships may be nonlinear, and in such cases, alternative formulations to keep the relationship linear or piecewise linear must be investigated.

3. **Logical Relationships**

Logic variables (binary and general integer variables) may be required to model multiple operating regions; if so, additional relationships would be necessary to bracket each operating region. We will discuss this situation extensively in the next chapter when modeling boilers.

If multiple pieces of equipment are logically connected, we could use a single logical variable for related equipment to keep them in sync. We will see examples of these in subsequent chapters.

Based on the above discussion, Tab. 2.4 is an initial list of constraint relationships for modeling a boiler.

Tab. 2.4: Initial list of constraint relationships for a generic boiler.

Equipment ID	Relationship	Comments
Boiler	Material balances	Overall mass balance, and empirical equation to define the blowdown flow
	Energy balalnce	Relationship is complex, we will use performance curve
	Logic relationship relating to On/Off and standby modes	To ensure that ON/OFF and Standby modes are mutually exclusive

2.5 Component model testing

After a preliminary model for a component has been created, it is essential to test the model to ascertain that all states are achievable and that performance relationships represent the actual operations of the component. If deficiencies are found, updating and retesting the model is imperative.

Summary

In this chapter, we discussed a general framework for modeling. Modeling entails representing the physical and logical reality in a mathematical form that can be solved using existing mathematical solvers. Physical reality is complex, and attempting to replicate that reality in mathematical form is usually impractical. Our emphasis, therefore, must be to create models that represent reality well enough and still be useful, making modeling a creative activity.

Nomenclature

Symbol	UOM	Description
CA	–	Compressed Air
CW	–	Cooling Water
ELEC	–	abbreviation for Electrical
FO	–	Fuel Oil
HPFG	–	High Pressure Fuel Gas
HPS	–	High Pressure Steam
LPFG	–	Low Pressure Fuel Gas
LPS	–	Low Pressure Steam
MPS	–	Medium Pressure Steam
PC	–	Process Condensate
BL1_BD	tph	Boiler #1: blowdown flow
BL1_BFW	tph	Boiler #1: boiler feed water flow
BL1_FUEL	MW.th	Boiler #1: fuel consumption
BL1_STM	tph	Boiler #1: steam production
BL1_X	Binary	Boiler #1: steam generation status: 1 = ON or producing steam, 0 otherwise
BL1_Y	Binary	Boiler #1: standby status: 1 = standby, 0 otherwise
PRV_STM	tph	Pressure-reducing Valve: steam flow

References

[1] Anonymous, "Steam Tables", available in print from ABB Combustion Engineering; or electronic format from https://che.k-state.edu/docs/imported/SteamTable.pdf.
[2] HYSIS, Process Simulation Software from Aspen Technology Inc.
[3] UniSim Design, Process Simulation Software from Honeywell International Inc.
[4] ChemCAD, Process Simulation Software from Chemstations Inc.

3 Modeling boilers

Steam is a widely used utility for process heating and to drive steam turbines that power rotating equipment such as fans, pumps, compressors to tranport fluids, and generators to produce electrical power. Typically, steam is produced and distributed at the site at various pressure levels.

Steam is produced in a boiler or a steam generator. Energy for steam production comes from fuel combustion with a slight excess of ambient air to ensure near-complete combustion. Typically, fire in the boiler is contained in the firebox at near-ambient pressure while water and steam are confined within the tubes and drums of the boiler maintained at the desired high pressure. Combustion products or flue gas comprise CO_2, water vapor, and nitrogen carried with the combustion air and the remaining oxygen. The flue gas attains a very high temperature in the combustion chamber and is progressively cooled down before being discharged into the atmosphere, usually in a tall chimney. Radiant heat in the combustion chamber and the flue gas cooling provide the thermal energy to the incoming boiler feed water (BFW) to produce steam at the desired temperature and pressure conditions. Modern boilers achieve high thermal efficiency by recovering as much energy as economical from the flue gas by preheating the incoming boiler feed water and combustion air.

Although the boiler feed water for steam generation is demineralized and high-purity, it still contains trace amounts of dissolved solids that build up in the boiler tubes and drums over time, causing scaling and reducing heat transfer efficiency. A low concentration of the dissolved solids is maintained to prevent excessive scaling by drawing a small amount of water from the boiler drums. This purge stream is called the boiler blowdown. Its flow is small, usually in the 1–2% range of the steam production flow.

The boiler manufacturer provides the normal operating range and efficiency versus the steam generation curve [1]. The efficiency curve is an upside-down parabolic shape with a peak. The recommended operating point is usually at or near the highest efficiency level.

This chapter will model all three boilers in the sample utility system. We will start with boiler #1. We will have considerable discussion as several new modeling techniques will be introduced. Subsequent boiler models would require much less discussion. Consistent with our modeling philosophy, we will start with a simple model and iteratively and judiciously add complexity till we get a realistic model of the boiler operation.

https://doi.org/10.1515/9783111020679-004

3.1 Boiler #1 modeling

In this section, we will start with a simple model and only consider the on and the off modes of boiler operation; later on, we will expand the model to include the standby mode.

For boiler #1, the nominal design point is 175 tph of steam production and the operational low and high limits are 70 and 210 tph, respectively, represented symbolically by *bl1_stm_ll* and *bl1_stm_hl*. Table 2.3 shows the four flow and two logical variables; however, this section will only consider the first five variables, the four flow variables, and the first logical variable, **BL1_X**. The four flow variables **BL1_BFW**, **BL1_STM**, **BL1_BD**, and **BL1_FUEL** correspond to the flows of boiler feed water, steam production, blowdown production, and fuel energy consumption, respectively. The boiler model comprises relationships among these variables over entire range and the two modes of boiler operation.

The boiler produces steam within its normal operating limits in the ON operating mode. In the OFF mode, the boiler is shutdown, with neither stream generation nor fuel consumption. We will use the binary variable **BL1_X** to model these two modes of operation, as discussed in the following section. We will develop the material and energy balance relationships that tie all the variables in the subsequent subsections and produce the first version of the MILP model for boiler #1.

3.1.1 Boiler #1: ON/OFF mode considerations

During normal operation, boiler #1 steam generation is between *70* and *210* tph. The following two simple constraints represent the steam production limits in the ON mode operation.

$$\textbf{BL1\_STM} \geq 70 \tag{3.1}$$

$$\textbf{BL1\_STM} \leq 210 \tag{3.2}$$

However, using the above two constraints precludes the possibility of OFF mode operation during which the steam generation must be zero because **BL1_STM** = 0 violates the lower limit constraint in eq. (3.1).

A vital question at this point is, can we modify the above two equations so they could represent the OFF mode operation? Let us explore by considering using the binary variable **BL1_X**, which can either be 0 or 1. Note that **BL1_X** is 0 when the boiler is in the OFF mode and 1 when in the ON mode. Let us multiply the RHS of the above two equations by **BL1_X** and use the symbolic name for steam generation limits; in doing so, we get:

$$\textbf{BL1\_STM} \geq bl1\_stm\_ll * \textbf{BL1\_X} \tag{3.3}$$

$$\textbf{BL1\_STM} \leq bl1\_stm\_hl * \textbf{BL1\_X} \tag{3.4}$$

Note that when **BL1_X** is 1, eq. (3.3) is identical to eq. (3.1), and eq. (3.4) is identical to the eq. (3.2); so, the above two equations represent the ON state of the boiler as before. And, when **BL1_X** is zero, eq. (3.3) and eq. (3.4) become:

$$BL1\_STM \geq 0 \qquad (3.5)$$

$$BL1\_STM \leq 0 \qquad (3.6)$$

Note that the above two equations can simultaneously be true only when **BL1_STM** = 0, representing the boiler OFF state.

Using the binary variable **BL1_X** in eqs. (3.3) and (3.4), we have found a set of equations capable of representing both the ON and the OFF operating modes of the boiler as far as steam production limits are concerned. That is exactly what we were attempting to do. In a later section, we will use **BL1_X** to ensure that fuel usage is also zero in the OFF mode. Note that both relationships are of a linear form, suitable for the MILP model; again, the constraints are:

$$BL1\_STM \geq bl1\_stm\_ll * BL1\_X \qquad (\_BL1\_STM\_LL)$$

$$BL1\_STM \leq bl1\_stm\_hl * BL1\_X \qquad (\_BL1\_STM\_HL)$$

3.1.2 Boiler #1: material balances

There are two outlet material streams, so an overall material balance would be required. The overall material balance relationship is simply a statement of the overall material balance IN = OUT. For boiler #1, it can be written as the following constraint:

$$BL1\_BFW = BL1\_STM + BL1\_BD \qquad (\_BL1\_MB1\}$$

In addition, as discussed earlier, the flow of the blowdown stream is proportional to steam production. Blowdown stream flow increases with increased steam production as increased steam production necessitates increased boiler feed water intake, which increases the inflow of dissolved solids in the boiler; these solids do not go out with the steam but build up in the boiler drum. So, with increased boiler feed water intake, the blowdown flow must also increase to maintain the desired solids concentration in the steam drum water. So, the blowdown stream is correlated to steam production and requires an additional material balance relationship.

The second material balance would relate boiler blowdown flow to steam generation flow. The blowdown stream is a small amount of water purged from the boiler drum to keep the solid particle concentration in the boiler water in the steam drum to an acceptable level; in practice, this control is achieved by maintaining the boiler water conductivity to an acceptable level.

Since boilers typically use demineralized water, the dissolved solids concentration in the incoming boiler feed water is low and is usually not measured. Blowdown flow is also relatively small, usually in the range of 1 to 2% of the steam production flow. Rigorous modeling of dissolved solid buildup in the boiler is impractical due to lack of measurements. As the blowdown flow is small, a simple correlation-based model, as written below, should suffice.

$$\textbf{BL1\_BD} = bl1\_bdr * \textbf{BL1\_STM} \qquad\qquad (\_BL1\_MB2)$$

where, $bl1\_bdr$ is the observed blowdown ratio, a typical value is 2% (or 0.02) of the steam production flow; it is an observed value and a known constant value at the start of optimization.

3.1.3 Boiler #1: energy balance

In the boiler, incoming boiler feed water absorbs radiant energy from fuel combustion and the cooling of hot flue gas. As expected, generating more steam requires more fuel; however, this relationship is complex and not completely linear. The boiler manufacturer usually provides this information as a "boiler efficiency curve" or "boiler efficiency versus steam" table, as shown in Fig. 3.1a and 3.1b. Note that the boiler efficiency curve is highly nonlinear; it has the shape of an upside-down parabola. As shown in Fig. 3.1a, the peak efficiency of around 80% occurs around its nominal design point of 175 tph steam generation. Note that efficiency drops as operation moves away from the design point in either direction, at first gradually and then more rapidly.

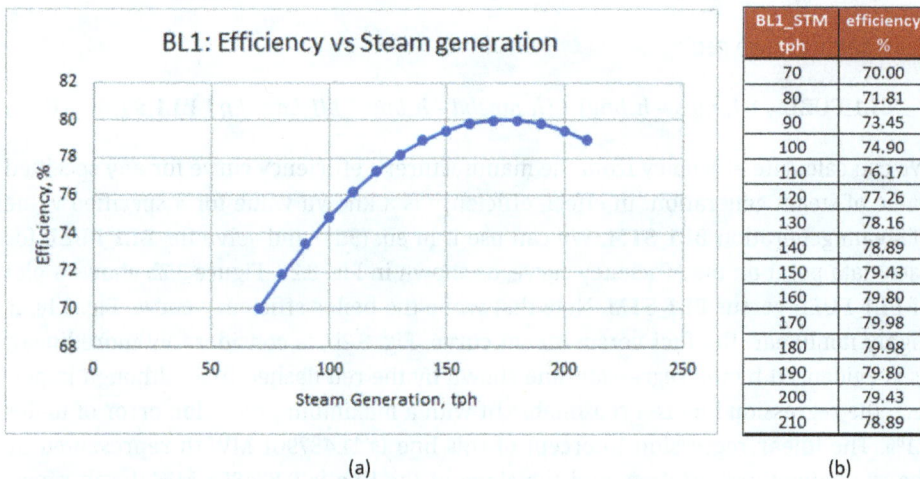

BL1_STM tph	efficiency %
70	70.00
80	71.81
90	73.45
100	74.90
110	76.17
120	77.26
130	78.16
140	78.89
150	79.43
160	79.80
170	79.98
180	79.98
190	79.80
200	79.43
210	78.89

(a)   (b)

Fig. 3.1: Boiler #1 efficiency vs. steam generation (a) in graphical form and (b) in tabular form.

We need to incorporate this boiler efficiency-related information in the boiler model. Note that the inclusion of efficiency as a variable in the optimization problem formulation is problematic as the efficiency relationship is highly nonlinear, and its inclusion will render the optimization a nonlinear problem and thus jeopardize the possibility of global optimization. Therefore, it is desirable to seek alternative model formulations that include the efficiency information but keep the model formulation linear or nearly linear. Let us see how this would be possible.

Realize that boiler efficiency, represented by the Greek symbol η, is a derived quantity. Essentially, it is the ratio of "energy absorbed to raise steam" to "total fuel energy supplied."

$$\eta = \text{total energy absorbed to generate steam / total boiler fuel energy} \qquad (3.7)$$

Let us look at this definition in energy terms: incoming boiler feed water (BFW) gets heated to the saturation temperature; then most of it is converted to steam, and the remainder goes out as blowdown. Associated energy flows are:

$$\text{Energy absorbed in steam} = (h\_mps - h\_bfw)^* \text{ } \textbf{BL1\_STM} \qquad (3.8)$$

$$\text{Energy absorbed in blow down} = (h\_mpsbd - h\_bfw)^* \text{ } \textbf{BL1\_BD} \qquad (3.9)$$

where h_mps, h_mpsbd and h_bfw are the specific enthalpies of MP steam, MP blowdown, and boiler feed water, respectively.

Combining the above three equations with the material balance eq. (_BL1-MB2) and substituting **BL1_FUEL** for the boiler fuel energy consumed, we get the following:

$$\eta = [(hmps - h\_bfw) + (h\_mpsbd - h\_bfw) * bl1\_bdr] * \textbf{BL1\_STM}/\textbf{BL1\_FUEL} \qquad (3.10)$$

Rearranging it, we get:

$$\textbf{BL1\_FUEL} = [(h\_mps - h\_bfw) + (h\_mpsbd - h\_bfw)^* bl1\_bdr] / \eta * \textbf{BL1\_STM} \qquad (3.11)$$

We can calculate efficiency from the manufacturer's efficiency curve for any specified value of steam generation. In effect, efficiency is a known value for a specified value of steam generation **BL1_STM**. We can use it in eq. (3.11) and solve for **BL1_FUEL** for each data point on the efficiency curve, as shown in Fig. 3.2a. Figure 3.2b shows a plot of **BL1_FUEL** versus **BL1_STM**. Note that while the boiler efficiency curve, Fig. 3.1a, is highly nonlinear, the fuel versus steam curve, Fig. 3.2b, is considerably more linear, as is evidenced by the regression line shown by the red dashed line. Although imperfect, the regression line is a reasonable fit with a maximum prediction error of under 2.1%. The linear regression intercept of this line is 11.487901 MW.th represented by the symbolic name *bl1_fg_a*, and the slope of the line is 0.824854 MW.th/tph represented by the symbolic name *bl1_fg_b*.

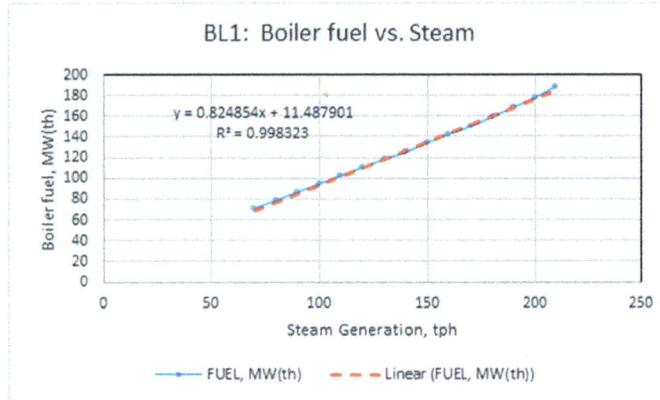

BL1_STM tph	efficiency %	FUEL, MW
70	70.00	70.7
80	71.81	78.8
90	73.45	86.6
100	74.90	94.4
110	76.17	102.1
120	77.26	109.8
130	78.16	117.6
140	78.89	125.5
150	79.43	133.5
160	79.80	141.7
170	79.98	150.3
180	79.98	159.1
190	79.80	168.3
200	79.43	178.0
210	78.89	188.2

(a)   (b)

Fig. 3.2: Boiler BL1 Fuel vs. steam generation (a) in tabular form and (b) in graphical form.

The following equation represents the boiler fuel vs. steam relationship in Fig. 3.2b:

$$\textbf{BL1\_FUEL} = bl1\_fg\_a + bl1\_fg\_b * \textbf{BL1\_STM} \qquad (3.12)$$

While the above equation adequately represents the ON mode operation, it does not accurately represent the OFF mode. Operation In OFF mode with 0 steam production, eq. (3.12) would predict $\textbf{BL1\_FUEL} = bl1\_fu\_a > 0$ instead of the correct value 0. How can we correct this shortcoming? Again, we can resolve this problem using the binary variable $\textbf{BL1\_X}$, similar to section 3.1.1. Doing so gives the following linear constraint:

$$\textbf{BL1\_FUEL} = bl1\_fg\_a * \textbf{BL1\_X} + bl1\_fg\_b * \textbf{BL1\_STM} \qquad (BL1\_EB1)$$

This constraint correctly represents both the ON and the OFF operating modes, which we were looking for.

3.1.4 Boiler #1 base model: BL1-v1

The material and energy balance relationships mentioned above tie all the boiler variables. These constraints and a cost objective function would form a starter model for the boiler. As a matter of principle, this boiler model must be tested and validated.

Note that at this stage, this model comprises a single boiler, BL1, so its boundary would be an envelope surrounding the boiler. Streams crossing the system boundary are steam $\textbf{BL1\_STM}$; blowdown, $\textbf{BL1\_BD}$; fuel, $\textbf{BL1\_FUEL}$; and boiler feed water, $\textbf{BL1\_BFW}$. MP steam is produced to meet the process demand, a known quantity represented by the symbol *mps_demand*. Since the generation of steam to meet the process demand is a requirement, its cost is immaterial, and the cost structure for this stream is unnecessary.

Fuel enters the subsystem and is from external suppliers, so it would require a cost structure consistent with the purchased fuel oil contract, but at this stage, for the sake of simplicity, we will assume a constant unit cost, c_fo. Boiler feed water entering the system would be a cost and would require a cost structure; however, since it is produced internally (within the utility system), there is no market price as such. At this stage, for the sake of simplicity, we will ignore this cost. The blowdown stream leaving the subsystem, would be a credit, and require a cost structure; however, it being an internal stream with no market value, we will again ignore its cost. Anyway, we will adequately account for these costs in the systemwide optimization.

$$\text{Objective function: Minimize } c\_fo * \textbf{BL1\_FUEL} \tag{3.13}$$

To test this model, we must specify MP steam demand; otherwise, the minimum cost solution would be the trivial zero cost-cost solution with the boiler turned OFF. So, for testing purposes, we will have to include an additional constraint such as the following:

$$\textbf{BL1\_STM} \geq mps\_demand \tag{\_MPS\_DMND}$$

Why is the constraint an inequality and not an equality? Good observation! We need the inequality to cover the test cases in which steam demand is low, i.e., below the boiler steam low limit, and in such cases, the boiler would have to produce more than the demanded steam. Anyway, the systemwide model would have a steam vent, and then we would use equality constraint, instead.

Figure 3.3 shows the complete boiler model, BL1-v1, implementation in LINGO.

All variables are physical quantities, so they must be nonnegative; however, that is the default in LP and MILP, so no additional constraints are necessary. All constraint relationships are labeled with labels starting with a "_" symbol to avoid confusion with any variable name.

We will implement this model both in Excel and in LINGO. We will test the model by specifying various values for *mps_demand* and then analyze the optimum policy for reasonableness; Tab. 3.1 shows the details of the four test cases and their expected results.

Figure 3.4 shows the complete model in Microsoft Excel. Figure 3.5 shows the formulae used in the spreadsheet: objective function is in cell F3, constraints are in cells F10 through F15, constraint limits are in columns D and E, cost coefficients are in row 8, and the model constraints matrix is in the range G10:K15. All the inputs and parameters used in the spreadsheet are in area ❷, and the MILP matrix is in area ❺ of the spreadsheet. Note that the optimizer populates the optimum value of variables in row 9, column G onwards, and the constraint values at the optimum in column F, rows ten onwards. The popup window in Fig. 3.4 shows the configuration of the Solver. Press-

```
MODEL:   TITLE BOILER #1 Model: BL1-v1;
!
! Nath, "Industrial Process Plants: Global Optimization of Utility Systems", Chap 3, (2023).;
! ;
  DATA:
      !inputs;     c_fo   = 18;            ! fuel cost, $/MWh;
                   mps_demand = 100;       ! mps demand, tph;
      !params;     bl1_bdr = 0.02;         ! blow down ratio;
                   bl1_fg_a = 11.4879;     ! fuel vs steam, intercept, MW(th);
                   bl1_fg_b = 0.824854;    ! fuel vs steam, slope, MW(th) / tph;
                   bl1_stm_ll = 70;        ! boiler operating limit, lower, tph;
                   bl1_stm_hl = 210;       ! boiler operating limit, higher, tph;
  ENDDATA

  ! Objective;     MIN = c_fo * BL1_FUEL;

  ! material balances;
  [_BL1_MB1]       BL1_BFW = BL1_STM + BL1_BD;
  [_BL1_MB2]       BL1_BD  = bl1_bdr * BL1_STM;

  ! energy balance;
  [_BL1_EB1]       BL1_FUEL = bl1_fg_a * BL1_X + bl1_fg_b * BL1_STM;

  ! generation limits;
  [_BL1_STM_LL]    BL1_STM > bl1_stm_ll * BL1_X;
  [_BL1_STM_HL]    BL1_STM < bl1_stm_hl * BL1_X;

  ! Binary variable specifications;
                   @BIN( BL1_X );

  ! demand constraint;
  [_MPS_DMND]      BL1_STM > mps_demand;

END MODEL BL1-v1
```

Fig. 3.3: Boiler #1, Base Model: BL1-v1 in LINGO.

Tab. 3.1: Boiler model: BL1-v1: test cases.

Case #	Attribute	Symbolic name	Value	UOM	Expected key results
1	MPS demand	mps_demand	100	tph	**BL1_STM** = 100, **BL1_X** = 1, obj fn > 0
2	MPS demand	mps_demand	50	tph	**BL1_STM** = 70, **BL1_X** = 1, obj fn < case 1 obj fn
3	MPS demand	mps_demand	0	tph	**BL1_STM** = 0, **BL1_X** = 0, obj fn = 0
4	MPS demand	mps_demand	220	tph	No feasible solution!

ing the Solve button on the popup solves the model and populates the solution. Figure 3.4 shows the case #1 solution.

Figure 3.6a shows the solution for test case #2; note that the boiler produces 70 tph, the production low limit, even though the demand is only 50 tph. Figure 3.6b shows the solution for test case #3; note that the optimizer shuts down the boiler, **BL1_X** = 0, as steam demand is 0. Figure 3.7 shows the infeasible solution for test case #4 as the demand exceeds the boiler capacity.

Figure 3.8 shows the test case #1 solution in LINGO. Note that LINGO output echoes inputs and parameter values in addition to the optimal solution; these variables appear in the same order as the model. As expected, LINGO produces results identical to those from the Excel Solver.

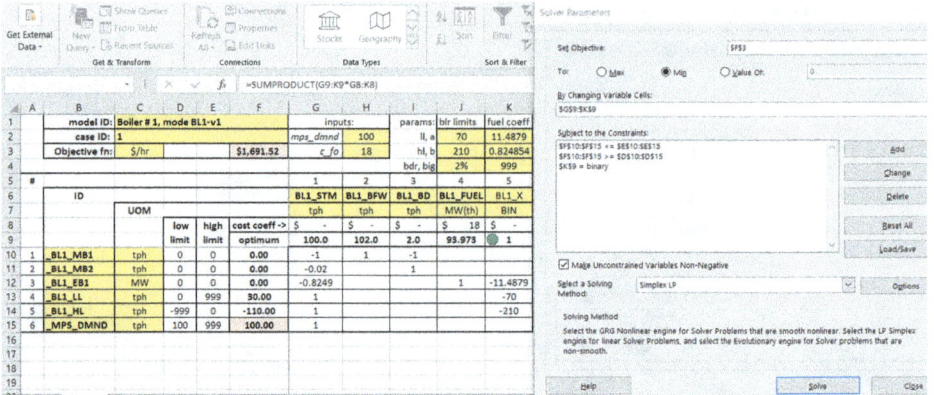

Fig. 3.4: Boiler #1 model: BL1-v1: Microsoft Excel set up.

	A	B	C	D	E	F	G	H	I	J	K
1		model ID:	Boiler # 1, mode BL1-v1				inputs:		params:	blr limits	fuel coeff
2		case ID:	1				*mps_dmnd*	100	*ll, a*	70	11.487901
3		Objective fn:	$/hr			=SUMPRODUCT(G9:K9*G8:K8)	*c_fo*	18	*hl, b*	210	0.824854
4									*bdr, big*	0.02	999
5	#						1	2	3	4	5
6		ID					BL1_STM	BL1_BFW	BL1_BD	BL1_FUEL	BL1_X
7			UOM				tph	tph	tph	MW(th)	BIN
8				low	high	cost coeff ->	0	0	0	=H3	0
9				limit	limit	optimum	100	102	2	93.973301	1
10	1	_BL1_MB1	tph	0	0	=SUMPRODUCT(G$9:K$9,G10:K10)	-1	1	-1		
11	2	_BL1_MB2	tph	0	0	=SUMPRODUCT(G$9:K$9,G11:K11)	=-J4		1		
12	3	_BL1_EB1	MW	0	0	=SUMPRODUCT(G$9:K$9,G12:K12)	=-K3			1	=-K2
13	4	_BL1_LL	tph	0	=K4	=SUMPRODUCT(G$9:K$9,G13:K13)	1				=-J2
14	5	_BL1_HL	tph	=-K4	0	=SUMPRODUCT(G$9:K$9,G14:K14)	1				=-J3
15	6	_MPS_DMND	tph	=H2	=K4	=SUMPRODUCT(G$9:K$9,G15:K15)	1				

Fig. 3.5: Boiler #1 Excel model: BL1-v1: display of formulae.

In the next subsection, we will modify this model to add the "standby" or "warm" mode operation capability.

3.1.5 Boiler #1: model with standby capability: BL1-v2

In this section, we will add the capability of standby mode for the boiler operation. In the standby mode, the boiler does not produce any steam but consumes a small amount of fuel energy, $bl1_fg_s$, to keep the boiler warm and ready to go instantaneously. In this mode **BL1_STM** = 0 but, **BL1_FUEL** = $bl1_fu_s > 0$.

Adding this capability to the model would require using the second binary variable, **BL1_Y**, which, when 1, would imply standby mode operation, and when 0, would indicate normal, on and off mode operations. Energy balance, **BL1_EB1**, will have to be modified to include fuel consumption in the standby model as shown below:

(a) test case #2 — case ID: 2

#	ID	UOM	low limit	high limit	cost coeff -> optimum	1 BL1_STM tph	2 BL1_BFW tph	3 BL1_BD tph	4 BL1_FUEL MW(th)	5 BL1_X BIN
					$ -	$ -	$ -	$ -	$ 18	$ -
					70.0	71.4	1.4	69.228		● 1
1	_BL1_MB1	tph	0	0	0.00	-1	1	-1		
2	_BL1_MB2	tph	0	0	0.00	-0.02		1		
3	_BL1_EB1	MW	0	0	0.00	-0.8249			1	-11.4879
4	_BL1_LL	tph	0	999	0.00	1			-70	
5	_BL1_HL	tph	-999	0	-140.00	1			-210	
6	_MPS_DMND	tph	50	999	70.00	1				

model ID: Boiler # 1, mode BL1-v1; inputs: mps_dmnd 50; Objective fn: $/hr $1,246.10; c_fo 18; params: blr limits ll, a 70 / hl, b 210 / bdr, big 2%; fuel coeff 11.4879 / 0.824854 / 999

(b) test case #3 — case ID: 3

#	ID	UOM	low limit	high limit	cost coeff -> optimum	1 BL1_STM tph	2 BL1_BFW tph	3 BL1_BD tph	4 BL1_FUEL MW(th)	5 BL1_X BIN
					$ -	$ -	$ -	$ -	$ 18	$ -
					0.0	0.0	0.0	0.000		● 0
1	_BL1_MB1	tph	0	0	0.00	-1	1	-1		
2	_BL1_MB2	tph	0	0	0.00	-0.02		1		
3	_BL1_EB1	MW	0	0	0.00	-0.8249			1	-11.4879
4	_BL1_LL	tph	0	999	0.00	1			-70	
5	_BL1_HL	tph	-999	0	0.00	1			-210	
6	_MPS_DMND	tph	0	999	0.00	1				

model ID: Boiler # 1, mode BL1-v1; inputs: mps_dmnd 0; Objective fn: $/hr $0.00; c_fo 18; params: blr limits ll, a 70 / hl, b 210 / bdr, big 2%; fuel coeff 11.4879 / 0.824854 / 999

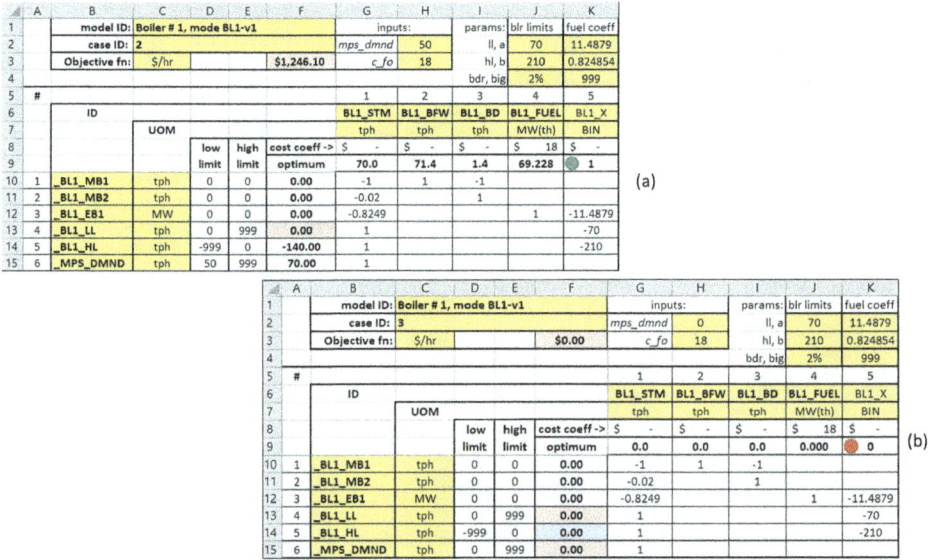

Fig. 3.6: Boiler #1, Excel model BL1-v1: (a) test case #2 solution and (b) test case #3 solution.

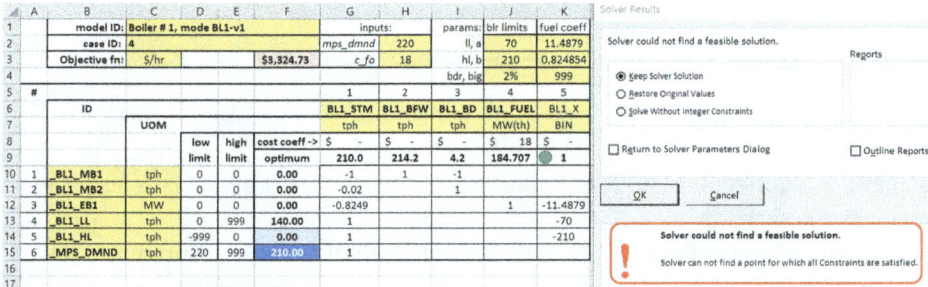

test case #4 — case ID: 4

#	ID	UOM	low limit	high limit	cost coeff -> optimum	1 BL1_STM tph	2 BL1_BFW tph	3 BL1_BD tph	4 BL1_FUEL MW(th)	5 BL1_X BIN
					$ -	$ -	$ -	$ -	$ 18	$ -
					210.0	214.2	4.2	184.707		● 1
1	_BL1_MB1	tph	0	0	0.00	-1	1	-1		
2	_BL1_MB2	tph	0	0	0.00	-0.02		1		
3	_BL1_EB1	MW	0	0	0.00	-0.8249			1	-11.4879
4	_BL1_LL	tph	0	999	140.00	1			-70	
5	_BL1_HL	tph	-999	0	0.00	1			-210	
6	_MPS_DMND	tph	220	999	210.00	1				

model ID: Boiler # 1, mode BL1-v1; inputs: mps_dmnd 220; Objective fn: $/hr $3,324.73; c_fo 18; params: blr limits ll, a 70 / hl, b 210 / bdr, big 2%; fuel coeff 11.4879 / 0.824854 / 999

Solver Results

Solver could not find a feasible solution.

Reports

● Keep Solver Solution
○ Restore Original Values
○ Solve Without Integer Constraints

☐ Return to Solver Parameters Dialog   ☐ Outline Reports

OK   Cancel

Solver could not find a feasible solution.
Solver can not find a point for which all Constraints are satisfied.

Fig. 3.7: Boiler #1, Excel model BL1-v1, test case #4: infeasible solution.

$$\textbf{BL1\_FUEL} = bl1\_fg\_a * \textbf{BL1\_X} + bl1\_fg\_b * \textbf{BL1\_STM} + bl1\_fg\_s * \textbf{BL1\_Y} \quad (\_BL1\_EB1S)$$

where **BL1_Y** is a binary variable corresponding to the standby mode and $bl1\_fu\_s$ is the energy consumption when in standby mode.

Furthermore, the standby mode and the normal modes are mutually exclusive. The following constraint expresses this fact.

$$\textbf{BL1\_X} + \textbf{BL1\_Y} \le 1 \quad\quad\quad (\_BL1\_LGC1)$$

Note that this constraint does allow both integer variables to be zero, which would represent the shutdown state, i.e., zero steam production as well as zero fuel consumption.

Standby mode is necessary as the utility system must be able to produce additional steam on short notice to accommodate changing process steam demands. The

```
LINGO/WIN64 20.0.21 (23 Aug 2023), LINDO API 14.0.5099.     Lingo 20.0 Solver Status [BL1-v1]                    ×

Licensee info: ravi.nathl@outlook.com                       ┌Solver Status───────────┐  ┌Variables────────┐
                                                            │ Model Class:      MILP │  │    Total:      5 │
Global optimal solution found.                              │                        │  │ Nonlinear:     0 │
Objective value:                       1691.519             │   State:     Global Opt│  │ Integers:      1 │
Objective bound:                       1691.519             │                        │  └─────────────────┘
Infeasibilities:                       0.000000             │ Objective:     1691.52 │  ┌Constraints──────┐
Extended solver steps:                        0             │                        │  │    Total:      7 │
Total solver iterations:                      0             │ Infeasibility: 3.55271e-15│ Nonlinear:    0 │
Elapsed runtime seconds:                   0.03             │ Iterations:          0 │  └─────────────────┘

Model Class:                             MILP               ┌Extended Solver Status──┐  ┌Nonzeros─────────┐
                                                            │                        │  │    Total:     14 │
Total variables:            5                               │ Solver Type:   B-and-B │  │ Nonlinear:     0 │
Nonlinear variables:        0                               │ Best Obj:      1691.52 │  ┌Generator Memory Used (K)┐
Integer variables:          1                               │ Obj Bound:     1691.52 │  │        28        │
                                                            │ Steps:               0 │  └─────────────────┘
Total constraints:          7                               │ Active:              0 │  ┌Elapsed Runtime (hh:mm:ss)┐
Nonlinear constraints:      0                               └────────────────────────┘  │     00:00:00     │

Total nonzeros:            14                               Update Interval: 2    [Interrupt Solver]   [Close]
Nonlinear nonzeros:         0


Model Title: BOILER #1 Model: BL1-v1

                             Variable          Value        Reduced Cost
            Inputs───────────   C_FO        18.00000           0.000000
                            MPS_DEMAND      100.0000           0.000000
                            BL1_BDR       0.2000000E-01        0.000000
                              BL1_FG_A     11.48790            0.000000
            Params──────────  BL1_FG_B      0.8248540          0.000000
                           BL1_STM_LL       70.00000           0.000000
                           BL1_STM_HL      210.0000            0.000000
                             BL1_FUEL       93.97330           0.000000
                              BL1_BFW      102.0000            0.000000
            Solution ───────   BL1_STM     100.0000            0.000000
                               BL1_BD        2.000000          0.000000
                                BL1_X        1.000000        206.7822

                              Row    Slack or Surplus    Dual Price
```

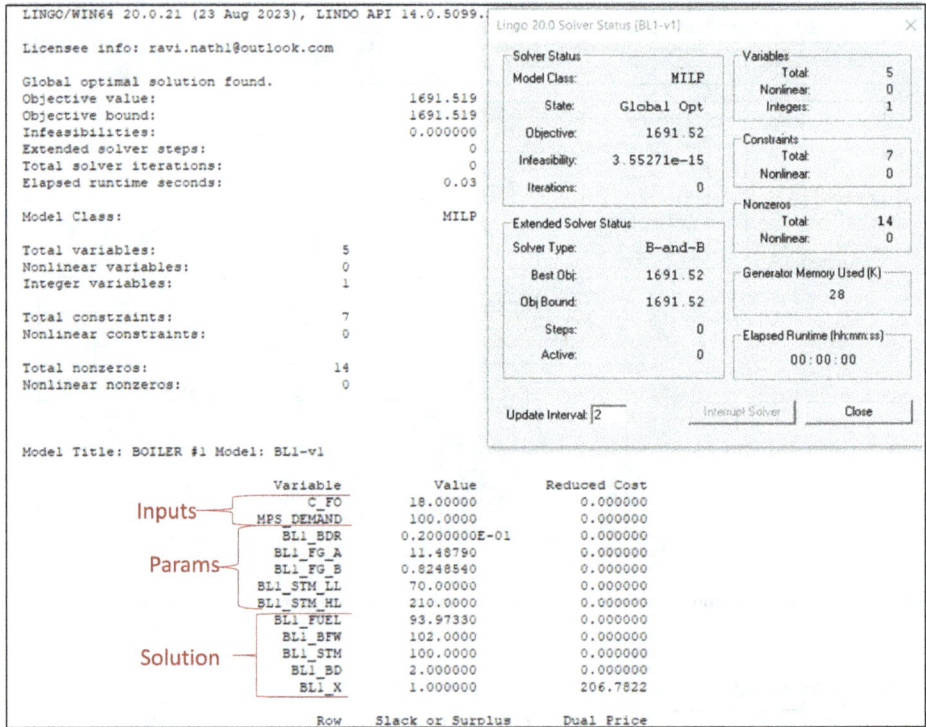

Fig. 3.8: Boiler #1 LINGO model: BL1-v1: test case #1 solution.

amount of additional steam that could be produced on short notice is referred to as the "steam reserve."

Steam reserve is the amount of additional steam that could be produced instantaneously by the utility system. Steam reserve is a system property, so the excess capacity of all boilers must be considered for systems with multiple boilers.

For the highest-pressure, the reserve would equal the unused capacity in the operating and the standby highest-pressure boilers. However, for lower-pressure steam, the reserve should also include excess capacity (above and beyond the higher-pressure steam reserve requirements) in the higher-pressure headers, as the higher-pressure steam can be let down to the lower-pressure header via connected pressure reducing valves or PRVs.

Let us consider boiler reserve for the current case of only one boiler in the system. In such a case, when the boiler is operating, the steam reserve, **MPS_RSV**, is the unused MPS boiler steam generation capacity, i.e.,

$$\mathbf{MPS\_RSV} = bl1\_stm\_hl * \mathbf{BL1\_X} - \mathbf{BL1\_STM} \tag{3.14}$$

When the boiler in the system is not producing but is in standby mode, the stream reserve is the total production capacity of the boiler.

$$\mathbf{MPS\_RSV} = bl1\_stm\_hl * \mathbf{Bl1\_Y} \tag{3.15}$$

Since the steam-producing mode and the standby mode are mutually exclusive, we can merge the above two equations to get a single equation that would cover both operating modes.

$$\mathbf{MPS\_RSV} = bl1\_stm\_hl * (\mathbf{Bl1\_X} + \mathbf{BL1\_Y}) - \mathbf{BL1\_STM} \tag{3.16}$$

Rearranging the above, we get the following constraint:

$$\mathbf{MPS\_RSV} + \mathbf{BL1\_STM} = bl1\_stm\_hl * (\mathbf{Bl1\_X} + \mathbf{BL1\_Y}) \tag{\_MPS\_RSV}$$

Since steam reserve is a potential cost, the optimizer would not select this mode of operations unnecessarily; therefore, we need to add a minimum steam reserve requirement constraint to the model for testing purposes. The steam reserve requirement constraint is:

$$\mathbf{MPS\_RSV} \geq mps\_reserve \tag{\_MPS\_RSRV}$$

where *mps_reserve* is the minimum steam reserve requirement.

The complete model with steam reserve requirement, **BL1-v2**, will be obtained by modifying the **BL1-v1** model with the constraints developed in this subsection. Figure 3.9 shows the complete model as implemented in LINGO. Figure 3.10 shows the implementation of the same model in Microsoft Excel.

For testing this model, we will use the same test cases as the previous version with an additional requirement of MPS reserve. Table 3.2 shows the test cases along with their expected results. Both LINGO and Excel Solver produce the expected results.

In the next section, we will further refine the boiler fuel versus steam production relationship by implementing a two-segment performance curve.

3.1.6 Boiler #1 model with segmented performance curve: BL1-v3

So far, we have used a linear approximation for boiler fuel versus steam performance curve; overall, it is a reasonable fit with a maximum prediction error of 2.1%. However, we can improve the fit further, as mentioned earlier, by approximating the performance curve with multiple linear segments. We will illustrate this technique with a two-segment approximation, bringing the maximum prediction error down to 0.4%, a significant improvement.

```
MODEL:  TITLE BOILER #1 Model: BL1-v2;
! Ref: Nath, "Industrial Process Plants: Global Optimization of Utility Systems", Chap. 3, (2023).;
! ;
  DATA:
      !inputs;      c_fo  = 18;                        ! fuel cost, $/MWh.th;
                    mps_demand = 100; mps_reserve = 50;  ! mps demand & reserve, tph;
      !params;      bl1_bdr = 0.02;                    ! blow down ratio;
                    bl1_fg_a = 11.4879; bl1_fg_b = 0.824854;  ! fuel vs steam, intercept & slope;
                    bl1_fg_s = 3;                      ! standby mode fuel consumption, MW.th;
                    bl1_stm_ll = 70;  bl1_stm_hl = 210;  ! boiler operating limits, tph;
  ENDDATA

  ! Objective;    MIN = c_fo * BL1_FUEL;

  ! material balances & reserve calculation;
  [_BL1_MB1]      BL1_BFW = BL1_STM + BL1_BD;
  [_BL1_MB2]      BL1_BD  = bl1_bdr * BL1_STM;
  [_BL1_RSV]      BL1_RSV + BL1_STM = bl1_stm_hl * (BL1_X  + BL1_Y);

  ! energy balances;
  [_BL1_EB1S]     BL1_FUEL = bl1_fg_a * BL1_X + bl1_fg_s * BL1_Y + bl1_fg_b * BL1_STM;

  ! Boiler steam generation limits;
  [_BL1_STM_LL]   BL1_STM > bl1_stm_ll * BL1_X;
  [_BL1_STM_HL]   BL1_STM < bl1_stm_hl * BL1_X;

  ! Logic constraint;
  [_BL1_LGC1]     BL1_X + BL1_Y < 1;

  ! Binary variable specifications;
                  @BIN(BL1_X);      @BIN(BL1_Y);

  ! Steam demand & reserve constraint;
  [_MPS_DMND]     BL1_STM > mps_demand;
  [_MPS_RSRV]     BL1_RSV > mps_reserve;

END MODEL BL1-v2
```

Fig. 3.9: Boiler #1, model with standby mode capability: BL1-v2: in LINGO.

	A	B	C	D	E	F	G	H	I	J	K	L	M
1		model ID:	Boiler # 1 MODEL BL1-v2				inputs:			params:	blr limits	fuel coeff	big
2		case ID:	1				mps_dmnd	100		ll, a	70	11.4879	999
3		Objective fn:	$/hr			$1,691.52	mps_rsrv	50		hl, b	210	0.82485	
4							c_fo	18		bdr, s	2.0%	3.00	
5	#						1	2	3	4	5	6	7
6		ID					BL1_STM	BL1_BFW	BL1_BD	MPS_RSV	BL1_FUEL	BL1_X	BL1_Y
7			UOM				tph	tph	tph	tph	MW.th	BIN	BIN
8				low	high	cost coeff ->	$ -	$ -	$ -	$ -	$ 18	$ -	$ -
9				limit	limit	optimum	100.0	102.0	2.0	110	93.973	1	0
10	1	_BL1_MB1	tph	0	0	0.00	-1	1	-1				
11	2	_BL1_MB2	tph	0	0	0.00	-0.02		1				
12	3	_MPS_RSV	tph	0	0	0.00	1			1		-210	-210
13	4	_BL1_EB1S	MW.th	0	999	0.00	-0.824854				1	-11.4879	-3.00
14	5	_BL1_STM_LL	tph	0	999	30.00	1					-70	
15	6	_BL1_STM_HL	tph	-999	0	-110.00	1					-210	
16	7	_MPS_DMND	tph	100	999	100.00	1						
17	8	_MPS_RSRV	tph	50	999	110.00				1			
18	9	_BL1_LGC1	tph	0	1	1.00						1	1

Fig. 3.10: Boiler #1, model with standby mode capability: BL1-v2: in Microsoft Excel.

A visual inspection of the fuel versus steam curve in Fig. 3.2 shows an upward bias from around 170 tph steam generation onwards. This observation makes sense as the boiler efficiency starts dropping from about 175 tph, as shown in Fig. 3.1; around 170 tph, the curve begins to steepen. Figure 3.11 shows approximation by two linear seg-

Tab. 3.2: Boiler #1 model: BL1-v2 test cases.

Case #	Attribute	Symbolic name	Value	UOM	Expected key results
all	MPS reserve	mps_resereve	50	tph	
1	MPS demand	mps_demand	100	tph	**BL1_STM** = 100, **BL1_X** = 1, obj fn > 0
2	MPS demand	mps_demand	50	tph	**BL1_STM** = 70, **BL1_X** = 1, obj fn < case 1 obj fn
3	MPS demand	mps_demand	0	tph	**BL1_STM** = 0, **BL1_Y** = 1, obj fn < case 2 obj fn
4	MPS demand	mps_demand	220	tph	No feasible solution!

ments: the blue line segment connecting the lower range values, corresponding to 70 to 170 tph steam generation, and the red line segment connecting the higher range values, corresponding to 170 to 210 tph steam generation. The dashed lines are extensions of these segments. Figure 3.10 shows the two regression lines, their extensions, and regression coefficients. Note that the regression coefficients of the blue line are named $bl1\_fg\_c$ and $bl1\_fg\_d$, and those of the red line are $bl1\_fg\_e$ and $bl1\_fg\_f$.

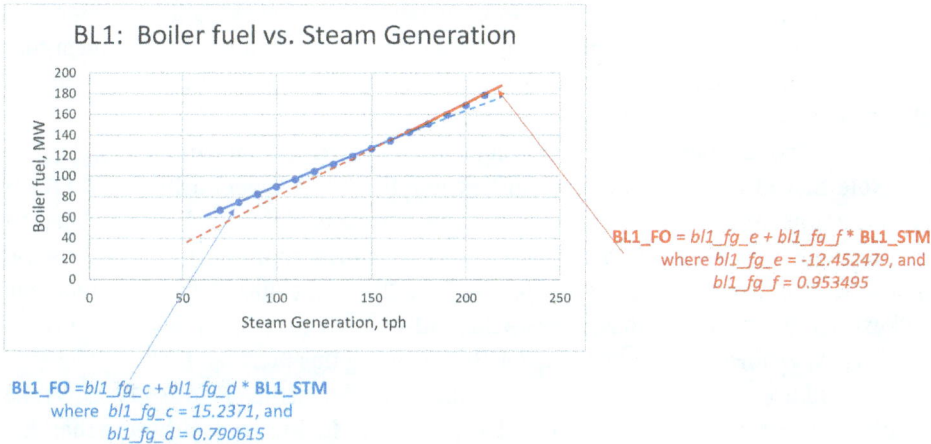

BL1_FO = $bl1\_fg\_e$ + $bl1\_fg\_f$ * **BL1_STM**
where $bl1\_fg\_e$ = -12.452479, and
$bl1\_fg\_f$ = 0.953495

BL1_FO = $bl1\_fg\_c$ + $bl1\_fg\_d$ * **BL1_STM**
where $bl1\_fg\_c$ = 15.2371, and
$bl1\_fg\_d$ = 0.790615

Fig. 3.11: Boiler #1: fuel vs. steam curve with two-segment approximation.

We are attempting to represent the fuel curve by the two linear segments – the blue linear segment in the lower range of 70–170 tph steam generation and the red linear segment in the upper range of 170–210 tph steam generation. How do we go about modeling these two linear segments in our model? First, let us make some observations. Note that in the lower production range, the extension of the red line, shown with the red dashed line, underestimates fuel consumption represented by the blue line; and in the upper range, the extension of the blue line, shown with the dashed blue line, underestimates the fuel consumption represented by the red line. Furthermore, fuel is a cost that the optimizer is trying to minimize. Let us exploit these two properties and modify the eq. (_BL1_EBS) to the following two constraints:

$$\textbf{BL1\_FUEL} \geq bl1\_fg\_c * \textbf{BL1\_X} + bl1\_fg\_d * \textbf{BL1\_STM} + bl1\_fg\_s * \textbf{BL1\_Y} \qquad (\_BL1\_EB1B)$$

$$\textbf{BL1\_FUEL} \geq bl1\_fg\_e * \textbf{BL1\_X} + bl1\_fg\_f * \textbf{BL1\_STM} + bl1\_fg\_s * \textbf{BL1\_Y} \qquad (\_BL1\_EB1R)$$

Let us analyze the behavior of these two constraints. When the boiler is down, i.e., **BL1_X** = 0, the two constraints will be identical and become similar to the previous constraint in eq. (_BL1_EB1S). When the boiler is producing steam, i.e., **BL1_X** = 1, in the lower range, the blue line eq. (_BL1_EB1B) would predict higher fuel than that predicted by the extension of red line, constraint eq. (_BL1_EB1R); so eq. (_BL1_EB_1B) would be more constraining and will dominate. Similarly, in the upper range, the constraint in eq. (_BL1_EB1R) would predict higher fuel than that predicted by the extension of blue line eq. (_BL1_EB1B), and the constraint in eq. (_BL1_EB_1 R) would dominate – precisely the behavior we wanted.

The boiler model with a two-segment fuel performance curve will be the same as in the previous subsection when the objective function is minimizing fuel, except that the constraint in eq. (_BL1_EB1S) will be replaced by the two constraints mentioned above, eq. (_BL1_EB1B) and eq. (_BL1_EB1R).

The complete model, BL1-v3, as implemented in LINGO, is shown in Fig. 3.12.

Note that LINGO allows the use of "include files" in a model using the **@FILE()** interface function [2]; we will use this function. Once a model is final, we will keep all the associated constraints in a separate file; doing so would make the model integration process efficient and less error-prone. Now that the boiler #1 model is final, we will extract its constraints to a separate file, BL1-v4.lng. Figure 3.13 shows the "include file." Fig. 3.14 shows the complete model, BL1-v4, using this included file.

We will use the same test cases as for BL1–v2 model testing. Figure 3.15 shows the solution for test case #1 in the Excel Solver; note that fuel usage is different than before and more accurate as it is using the two-segment performance curve. Note that both Excel Solver and LINGO Solver give the expected solutions.

In this section, we have approximated the fuel vs. steam curve by two linear segments and reduced the maximum approximation error by 80% (from 2.07% to 0.42%) by adding just one additional constraint; this was because the curve was getting steeper with increasing steam production as is usually the case with such curves. Two remarks:

(1) If the curve was not steepening, then an additional binary integer variable would be required for defining the two regions; such a technique will be discussed in Chapter 8 when we model external purchased energy contracts involving tier pricing.

(2) It is possible to improve approximation accuracy to any desired degree by increasing the number of linear segments; however, two or three segments are usually sufficient in practice.

```
MODEL:   TITLE BOILER #1 Model: BL1-v3;
! Ref: R Nath, "Industrial Process Plants: Global Optimization of Utility Systems", Chap. 3, (2023).;
! ;
 DATA:
  !inputs;  c_fo  = 18;                              ! fuel cost, $/MWh;
            mps_demand = 100; mps_reserve = 50;      ! mps demand & reserve, tph;
  !params;  bll_bdr = 0.02;                          ! blow down ratio;
            bll_fg_c = 15.2371; bll_fg_d = 0.790615; ! fuel vs steam, seg 1 coefficients;
            bll_fg_e =-12.4525; bll_fg_f = 0.953495; ! fuel vs steam, seg 2, ccefficients;
            bll_fg_s = 3;                            ! standby fuel consumption , MW(th);
            bll_stm_11 = 70;  bll_stm_h1 = 210;      ! boiler operating limits, tph;
  ENDDATA

  ! Objective;    MIN = c_FO * BL1_FUEL;

  ! material balances & reserve calculations;
  [_BL1_MB1]     BL1_BFW = BL1_STM + BL1_BD;
  [_BL1_MB2]     BL1_BD = bll_bdr * BL1_STM;
  [_BL1_RSV]     BL1_RSV + BL1_STM = bll_stm_h1 * (BL1_X  + BL1_Y);

  ! energy balances;
  [_BL1_EB1B]     BL1_FUEL > bll_fg_c * BL1_X + bll_fg_s * BL1_Y + bll_fg_d * BL1_STM;
  [_BL1_EB1R]     BL1_FUEL > bll_fg_e * BL1_X + bll_fg_s * BL1_Y + bll_fg_f * BL1_STM;

  ! Boiler steam generation limits;
  [_BL1_STM_LL]  BL1_STM > bll_stm_11 * BL1_X;
  [_BL1_STM_HL]  BL1_STM < bll_stm_h1 * BL1_X;

  ! Logic constraint;
  [_BL1_LGC1]    BL1_X + BL1_Y < 1;

  ! Binary variable specifications;
                 @BIN( BL1_X );    @BIN( BL1_Y );

  ! Steam demand & reserve constraint;
  [_MPS_DMND]    BL1_STM > mps_demand;
  [_MPS_RSRV]    BL1_RSV > mps_reserve;

END MODEL BL1-v3
```

Fig. 3.12: Boiler #1, model with a two-segment performance curve: BL1-v3: in LINGO.

```
! BL1 constraints file: BL1-v4.lng;

  ! material balances & reserve calc;
  [_BL1_MB1]     BL1_BFW  = BL1_STM + BL1_BD;
  [_BL1_MB2]     BL1_BD   = bll_bdr * BL1_STM;
  [_BL1_RSV]     BL1_RSV  + BL1_STM = bll_stm_h1*(BL1_X  + BL1_Y);

  ! energy balances;
  [_BL1_EB1B]     BL1_FUEL > bll_fg_c*BL1_X + bll_fg_s*BL1_Y + bll_fg_d*BL1_STM;
  [_BL1_EB1R]     BL1_FUEL > bll_fg_e*BL1_X + bll_fg_s*BL1_Y + bll_fg_f*BL1_STM;

  ! generation limits;
  [_BL1_STM_LL]  BL1_STM  > bll_stm_11 * BL1_X;
  [_BL1_STM_HL]  BL1_STM  < bll_stm_h1 * BL1_X;

  ! logic constraint & variable specs;
  [_BL1_LGC1]    BL1_X + BL1_Y < 1;
                 @BIN(BL1_X); @BIN(BL1_Y);
!;
```

Fig. 3.13: Boiler #1, model constraints file: BL1-v4.lng.

```
MODEL:  TITLE BOILER #1 Model: BL1-v4;
! ;
! Ref: R Nath, "Industrial Process Plants: Global Optimization of Utility Systems", Chapter 3, (2023);
! ;
  DATA:
   !inputs;   c_fo  = 18; mps_demand = 100; mps_reserve = 50;      ! mps demand & reserve, tph;
   !params;   bl1_bdr = 0.02;          bl1_fg_s = 3;               ! bd ratio, standby fuel [MW.th];
              bl1_fg_c = 15.2371;      bl1_fg_d = 0.790615;        ! fuel [MW.th] vs steam [tph], seg 1;
              bl1_fg_e =-12.4525;      bl1_fg_f = 0.953495;        ! fuel [MW.th] vs steam [tph], seg 2;
              bl1_stm_ll = 70;         bl1_stm_hl = 210;           ! boiler operating limits, tph;
  ENDDATA

   ! Objective;    MIN = c_FO * BL1_FUEL;

   ! constraints;   @FILE('C:\BookDG\LingoModels\BL1-v4.lng');

   ! Demand & reserve;
   [_MPS_DMND]    BL1_STM > mps_demand;
   [_MPS_RSRV]    BL1_RSV > mps_reserve;

END MODEL BL1-v4
```

Fig. 3.14: Boiler #1 model using "include file" BL1-v4.lng in LINGO.

	A	B	C	D	E	F	G	H	I	J	K	L	M
1		model ID:	Boiler # 1 MODEL BL1-v3				inputs:		params:	blr limits	fuel coeff	Seg 2 coeff	big
2		case ID:	1				mps_dmnd	100	ll, a	70	15.2371	-12.4525	999
3		Objective fn:	$/hr			$1,697.37	mps_rsrv	50	ul, b	210	0.790615	0.953495	
4							c_fo	18	bdr, s	2%	3.00		
5	#						1	2	3	4	5	6	7
6		ID					BL1_STM	BL1_BFW	BL1_BD	BL1_RSV	BL1_FUEL	BL1_X	BL1_Y
7			UOM				tph	tph	tph	tph	MW.th	BIN	BIN
8				low	high	cost coeff ->	$ -	$ -	$ -	$ -	$ 18	$ -	$ -
9				limit	limit	optimum	100.0	102.0	2.0	110.0	94.299	● 1	● 0
10	1	_BL1_MB1	tph	0	0	0.00	-1	1	-1				
11	2	_BL1_MB2	tph	0	0	0.00	-0.02	0	1				
12	3	_BL1_RSV	tph	0	0	0.00	1			1		-210	-210
13	4	_BL1_EB1B	MW.th	0	999	0.00	-0.790615				1	-15.2371	-3
14	5	_BL1_EB1R	MW.th	0	999	11.40	-0.953495				1	12.4525	-3
15	6	_BL1_STM_LL	tph	0	999	30.00	1					-70	
16	7	_BL1_STM_HL	tph	-999	0	-110.00	1					-210	
17	8	_MPS_DMND	tph	100	999	100.00	1						
18	9	_MPS_RSRV	tph	50	999	110.00				1			
19	10	_BL1_LGC1	tph	0	1	1.00						1	1

Fig. 3.15: Boiler #1 model: BL1-v3: Excel Solver solution for test case #1.

This concludes the modeling of boiler #1. We will continue this chapter by modeling the other two boilers in the sample utility system; however, we will skip the details for brevity unless new concepts are involved.

3.2 Boiler #2 modeling

Boiler #2 is similar to Boiler #1 except that it produces HP steam and can consume a mix of HPFG and LPFG. The model will be identical to boiler #1 model, except for adding two variables corresponding HPFG and LPFG fuel consumption. HPS demand and reserve requirements will replace MPS demand and reserve requirements. It will also have different operating range and fuel versus steam regression coefficients.

Boiler #2 operating range is between 80 tph and 240 tph, and the design production rate is 200 tph. Figure 3.16 shows the efficiency curve and Fig. 3.17 shows the two-segment fuel curve and regression coefficients.

BL2: Efficiency vs. Steam Generation

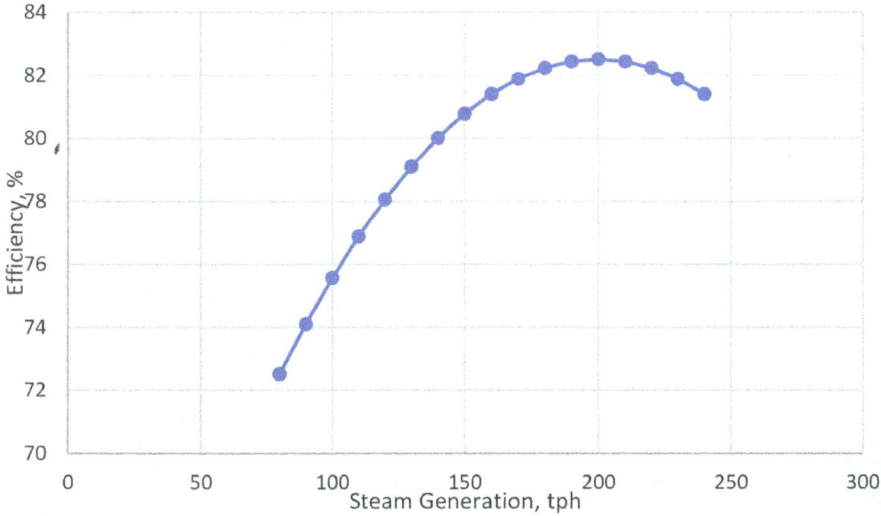

Fig. 3.16: Boiler #2: efficiency curve.

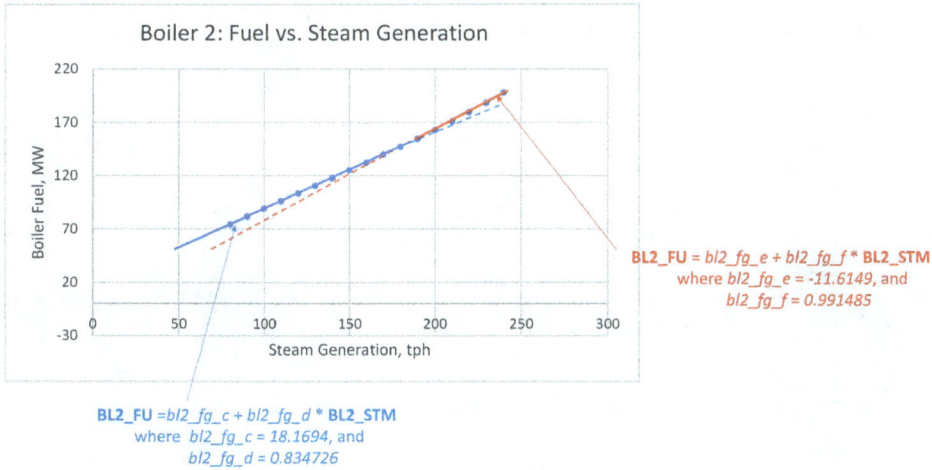

Fig. 3.17: Boiler #2: fuel vs. steam curve with two-segment approximation.

3.2.1 Boiler #2 model consuming two fuels in any proportion: BL2-v1

This model would be relatively straightforward; we will use the boiler #1 model files, replace all occurrences of BL1 with BL2, add a constraint equating **BL2_FUEL** to the sum of **BL2_HPFG** and **BL2_LPFG**, and replace MPS demand and reserve terms by corresponding HPS terms. We will modify the objective function to include both HPFG and LPFG terms.

Figure 3.18 shows the complete model in LINGO, and Fig. 3.19 shows the boiler #2 constraints file, BL2-v1.lng.

```
MODEL:  TITLE BOILER #2 Model BL2-v1;
!
! Ref: R Nath, "Industrial Process Plants: Global Optimization of Utility Systems",
!           Chapter 3, (2023).;
 DATA:
    !inputs;    c_lpfg      = 19;  c_hpfg     = 20;      ! fuel costs, $/MWh;
                hps_demand = 100; hps_reserve = 50;      ! hps demand & reserve, tph;
    ! params;   b12_bdr = 0.02;                          ! blow down ratio;
                b12_fg_c = 18.1694; b12_fg_d = 0.834726; ! fuel vs steam, seg 1 coefficients;
                b12_fg_e =-11.6149; b12_fg_f = 0.991485; ! fuel vs steam, seg 2, coefficients;
                b12_fg_s = 4;                            ! standby mode fuel consumption , MW(th);
                b12_stm_ll = 80;  b12_stm_hl = 240;      ! boiler operating limits, tph;
    ENDDATA

 ! Objective;
 [_BL2_V1_OBJ]   MIN = c_lpfg * BL2_LPFG + c_hpfg * BL2_HPFG;

 ! constraints from include file;
                 @FILE('C:\BookDG\LingoModels\BL2-v1.lng');

 ! Demand & reserve constraint;
 [_MPS_DMND]     BL2_STM > hps_demand;
 [_MPS_RSRV]     BL2_RSV > hps_reserve;


END MODEL BL2-v1
```

Fig. 3.18: Boiler #2, LINGO model with two-segment performance curve: BL2-v1.

```
! BL2 constraints file: BL2-v1.lng;

 ! material balances & reserve calc;
 [_BL2_MB1]       BL2_BFW = BL2_STM + BL2_BD;
 [_BL2_MB2]       BL2_BD = b12_bdr * BL2_STM;
 [_BL2_RSV]       BL2_RSV + BL2_STM = b12_stm_hl * (BL2_X  + BL2_Y);

 ! energy balances;
 [_BL2_EB1]       BL2_FUEL = BL2_LPFG + BL2_HPFG;
 [_BL2_EB1B]      BL2_FUEL > b12_fg_c * BL2_X + b12_fg_s * BL2_Y + b12_fg_d * BL2_STM;
 [_BL2_EB1R]      BL2_FUEL > b12_fg_e * BL2_X + b12_fg_s * BL2_Y + b12_fg_f * BL2_STM;

 ! generation limits;
 [_BL2_STM_LL]    BL2_STM > b12_stm_ll * BL2_X;
 [_BL2_STM_HL]    BL2_STM < b12_stm_hl * BL2_X;

 ! logic constraint & variable specs;
 [_BL2_LGC1]      BL2_X + BL2_Y < 1;
                  @BIN( BL2_X );     @BIN( BL2_Y );
!;
```

Fig. 3.19: Boiler #2, LINGO model constraints file: BL2-v1.lng.

Although the changes are relatively straightforward, validating the model to confirm that no errors have crept in is a good practice. We will use the same four test cases to validate the BL1-v3 model except for replacing MP steam with HP steam. Figure 3.20 shows the solutions for test case #1; note that Excel Solver produces the expected results.

			inputs:				params:	blr limits	fuel coeff	Seg 2 coeff	
model ID:	Boiler # 2 Model BL2-v1										
case ID:	1		hps_dmnd	100	c_lpfg	19	LL, a	80	18.1694	-11.6149	
Objective fn:	$/hr		$1,931.20	hps_rsrv	50	c_hpfg	20	UL, b	240	0.834726	0.991485
							bdr, s, big	2%	4.00	999	

#						1	2	3	4	5	6	7	8	9
	ID					BL2_STM	BL2_BFW	BL2_BD	BL2_RSV	BL2_FUEL	BL2_LPFG	BL2_HPFG	BL2_X	BL2_Y
		UOM				tph	tph	tph	tph	MW.th	MW.th	MW.th	BIN	BIN
			low limit	high limit	cost coeff -> optimum	$ -	$ -	$ -	$ -	$ -	$ 19	$ 20	$ -	$ -
						100	102	2	140	101.642	101.642	0	1	0
1	_BL2_MB1	tph	0	0	0.00	-1	1	-1						
2	_BL2_MB2	tph	0	0	0.00	-0.02		1						
3	_BL2_RSV	tph	0	0	0.00	1			1				-240	-240
4	_BL2_EB1	MW.th	0	0	0.00					1	-1	-1		
5	_BL2_EB1B	MW.th	0	999	0.00	-0.83473					1	1	-18.1694	-4.00
6	_BL2_EB1R	MW.th	0	999	14.11	-0.99149					1	1	11.6149	-4.00
7	_BL2_STM_LL	tph	0	999	20.00	1							-80	
8	_BL2_STM_UL	tph	-999	0	-140.00	1							-240	
9	_HPS_DMND	tph	100	999	100.00	1								
10	_HPS_RSRV	tph	50	999	140.00				1					
11	_BL2_LGC1	tph	0	1	1.00								1	1

Fig. 3.20: Boiler #2, model with two-segment performance curve: BL2-v1: in Microsoft Excel.

3.3 Boiler #3 modeling

Boiler #3 is similar to Boiler #2, except that it can consume only one fuel at a time, either HPFG or LPFG. Since the model will have to decide between the two fuels, new integer variable(s) and additional logic constraints will be required, as discussed below.

Boiler #3 operating range is 100–300 tph steam generation with a design production rate of 250 tph. Figure 3.21 shows the efficiency curve, and Fig. 3.22 shows the two-segment fuel vs. steam curve and regression coefficients.

3.3.1 Boiler #3 model: selecting between two possible fuels: BL3-v1

For simplicity, let us start with two binary variables; later, we can use just one binary integer variable, resulting in a more efficient model. The new binary variables are **BL3_W** and **BL3_Z**: **BL3_W** = 1 to indicate the use of LPFG and **BL3_Z** to indicate the use of HPFG. We will use these binary variables for setting up the corresponding fuel consumption upper limit constraint:

$$\text{BL3\_LPFG} \leq big * \text{BL3\_W} \qquad (\_\text{BL3\_LPF\_HL})$$

$$\text{BL3\_HPFG} \leq big * \text{BL3\_Z} \qquad (\_\text{BL3\_HPF\_HL})$$

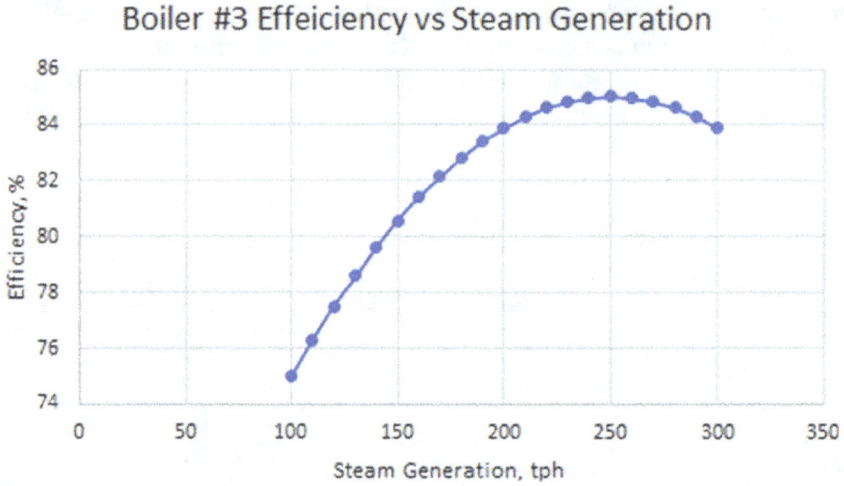

Fig. 3.21: Boiler #3: efficiency curve.

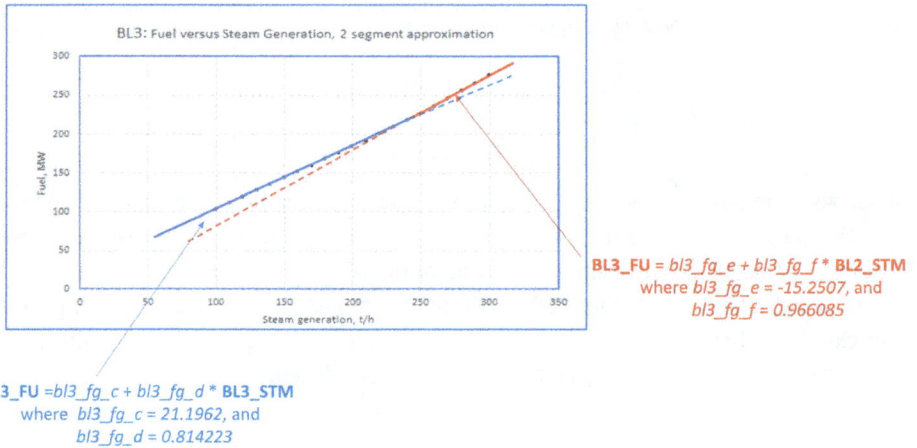

BL3_FU = *bl3_fg_e* + *bl3_fg_f* * **BL2_STM**
where *bl3_fg_e* = -15.2507, and
bl3_fg_f = 0.966085

BL3_FU = *bl3_fg_c* + *bl3_fg_d* * **BL3_STM**
where *bl3_fg_c* = 21.1962, and
bl3_fg_d = 0.814223

Fig. 3.22: Boiler #3: fuel vs. steam generation: two-segment curve.

where *big* is a reasonably large number, such as 999. The above two constraints ensure that LPFG is unavailable when **BL3_W** =0 and HPFG is unavailable when **BL3_Z** is 0.

Since only one fuel can be used, both **BL3_W** and **BL3_Z** cannot be simultaneously 1; this could be specified using the following logic constraint:

$$BL3\_W + BL3\_Z \leq 1 \tag{3.17}$$

However, fuel would be required only when either the boiler is generating stream, i.e., **BL3_X** = 1, or the boiler is in standby mode, i.e., **BL3_Y** = 1. Using this information, we come up with a tighter and more efficient formulation for the above constraint:

$$\textbf{BL3\_W} + \textbf{BL3\_Z} = \textbf{BL3\_X} + \textbf{BL3\_Y} \qquad\qquad (\_\textbf{BL3\_LGC2})$$

Note another interesting property of the above constraint. Since it is equality, if three of the four variables are binaries, the fourth will naturally become a binary. So, one of the variables need not be specified as a binary, leading to an even more efficient formulation.

The rest of the boiler #3 model is similar to the boiler #2 model. Figures 3.23 and 3.24 show boiler #3, constraints file BL3-v1.lng and the complete boiler #3 model, respectively.

```
! BL3 constraints file: BL3-v1.lng;

  ! material balances & reserve calculations;
  [_BL3_MB1]      BL3_BFW = BL3_STM + BL3_BD;
  [_BL3_MB2]      BL3_BD  = bl3_bdr * BL3_STM;
  [_BL3_RSRV]     BL3_RSV + BL3_STM = bl3_stm_hl * (BL3_X + BL3_Y);

  ! energy balances;
  [_BL3_EB1]      BL3_FUEL = BL3_LPFG + BL3_HPFG;
  [_BL3_EB1B]     BL3_FUEL > bl3_fg_c * BL3_X + bl3_fg_s * BL3_Y + bl3_fg_d * BL3_STM;
  [_BL3_EB1R]     BL3_FUEL > bl3_fg_e * BL3_X + bl3_fg_s * BL3_Y + bl3_fg_f * BL3_STM;

  ! generation & fuel consumption constraints;
  [_BL3_STM_LL]   BL3_STM  > bl3_stm_ll * BL3_X;
  [_BL3_STM_HL]   BL3_STM  < bl3_stm_hl * BL3_X;
  [_BL3_LPF_HL]   BL3_LPFG < big * BL3_W;
  [_BL3_HPF_HL]   BL3_HPFG < big * BL3_Z;

  ! Logic constraint & variable specs;
  [_BL3_LGC1]     BL3_X + BL3_Y < 1;
  [_BL3_LGC2]     BL3_W + BL3_Z = BL3_X + BL3_Y;
                  @BIN( BL3_X );    @BIN( BL3_Y );    @BIN( BL3_Z );
!;
```

Fig. 3.23: Boiler #3 model LINGO constraints file: BL3-v1.lng.

We will use the same four test cases that we used to validate the boiler #2 model. Figure 3.25 shows the Microsoft Excel implantation and solution for test case #1; note that Excel Solver produces the expected results.

Now that we have models for the three boilers, we can put them together to create a model for the subsystem of the three boilers in the next section.

```
MODEL: TITLE BOILER #3 Model: BL3-v1;
! Ref: R Nath, "Industrial Process Plants: Global Optimization of Utility Systems", Chap. 3, (2023).;
! ;
  DATA:
    !inputs;      c_lpfg     = 19;  c_hpfg     = 20;        ! fuel costs, $/MWh(th);
                  hps_demand = 100; hps_reserve = 50;       ! hps demand & reserve, tph;
    !params;      b13_bdr = 0.02; big = 999;               ! blow down ratio, big number;
                  b13_fg_c = 21.1962; b13_fg_d = 0.814223;  ! fuel vs steam, seg 1, coefficients;
                  b13_fg_e =-15.2507; b13_fg_f = 0.966085;  ! fuel vs steam, seg 2, coefficients;
                  b13_fg_s = 5;                             ! standby mode fuel usage, MW.th;
                  b13_stm_11 = 100; b13_stm_h1 = 300;       ! boiler operating limits, tph;
  ENDDATA

  ! Objective;   MIN = c_lpfg * BL3_LPFG + c_hpfg * BL3_HPFG;

  ! constraints from include file;
                  @FILE('C:\BookDG\LingoModels\BL3-v1.lng');

  ! demand & reserve constraint;
  [_HPS_DMND]    BL3_STM > hps_demand;
  [_HPS_RSRV]    BL3_RSV > hps_reserve;

END MODEL BL3-v1
```

Fig. 3.24: Boiler #3 model: BL3-v1: in LINGO.

	A	B	C	D	E	F	G	H	I	J	K	L	M	N	O	P	Q
1		model ID:	Boiler #3 model: BL3-v1				inputs:		c_lpfg		BL3 param:	blr limits	seg1 coeff	Seg 2 coeff			
2		case ID:	1				hps_dmnd	100	19		ll, a	100	21.1962	-15.2507			
3		Objective fn:	$/hr			$1,949.8	hps_rsrv	50	c_hpfg		hl, b	300	0.814223	0.966085			
4									20		bdr, s, big	2%	5	999			
5	#						1	2	3	4	5	6	7	8	9	10	11
6		ID					BL3_STM	BL3_BFW	BL3_BD	BL3_FUEL	BL3_LPFG	BL3_HPFG	BL3_RSV	BL3_W	BL3_X	BL3_Y	BL3_Z
7			UOM				tph	tph	tph	MW.th	MW.th	MW.th		0/1	BIN	BIN	BIN
8				low	high	cost coeff ->					$ 19	$ 20					
9				limit	limit	optimum	100.0	102.0	2.0	102.619	102.619	0.000	200.0	1.00	1.00	0.00	0.00
10	1	BL3_MB1	tph	0	0	0.00	-1	1	-1								
11	2	BL3_MB2	tph	0	0	0.00	-0.02		1								
12	3	BL3_RSV	tph	0	0	0.00	1						1		-300	-300	
13	4	BL3_EB1	MW.th	0	0	0.00				1	-1	-1					
14	5	BL3_EB1B	MW.th	0	999	0.00	-0.814223			1					-21.1962	-5.0	
15	6	BL3_EB1R	MW.th	0	999	21.26	-0.966085			1					15.2507	-5.0	
16	7	BL3_STM_LL	tph	0	999	0.00	1								-100.0		
17	8	BL3_STM_HL	tph	-999	0	-200.00	1								-300.0		
18	9	BL3_LPF_HL	tph	-999	0	-896.38					1				-999		
19	10	BL3_HPF_HL	tph	-999	0	0.00						1					-999
20	11	HPS_DMND	tph	100	999	100.00	1										
21	12	HPS_RSRV	tph	50	999	200.00							1				
22	13	BL3_LGC1	tph	0	1	1.00									1	1	
23	14	BL3_LGC2	tph	0	0	0.00								1	-1	-1	1

Fig. 3.25: Boiler #3: model BL3-v1: in Microsoft Excel showing test case #1 results.

3.4 Boiler subsystem model: BLSS-v1

As discussed earlier, some constraints, such as the steam demand and steam reserve constraints, apply to the system as a whole. However, in the individual boiler models, we had no choice but to use these constraints for individual boilers. Now that we have models for the individual boilers, we can aggregate them and apply the system-wide constraints more appropriately to the three boilers system.

The boiler subsystem model is created by aggregating the individual boiler models and adjusting for the systemwide constraints, such as those relating to steam reserve and demands. HPS reserve calculation and HPS demand and reserve constraints will be modified, as only one HPS reserve calculation will be required. Similarly, we will aggregate the objective function. Test cases are created by aggregating the previous test cases, as shown in Tab. 3.3.

Tab. 3.3: Boilers subsystem model: BLSS-v1 test cases.

Case #	Attribute	Symbolic name	Value	UOM	Expected key results
1	MPS demand	*mps_demand*	60	tph	**BL1_STM** = min, **BL2_STM** = 100, **MPS_RSV** > 50, **HPS_RSV** > 50, **BL__X** = 1, **BL__Y** = 0, obj fn > 0
	HPS demand	*hps_demand*	100	tph	
	MPS reserve	*mps_reserve*	50	tph	
	HPS reserve	*hps_reserve*	50	tph	
2	MPS demand	*mps_demand*	30	tph	**BL1_STM** = min, **BL2_STM** = min, **MPS_RSV** > 50, **HPS_RSV** > 50, **BL__X** = 1, **BL__Y** = 0, obj fn < case 1
	HPS demand	*hps_demand*	50	tph	
	MPS reserve	*mps_reserve*	50	tph	
	HPS reserve	*hps_reserve*	50	tph	
3	MPS demand	*mps_demand*	0	tph	**BL1_STM** = **BL2_STM** = 0, **MPS_RSV** > 50, **HPS_RSV** > 50, **BL__X** = 0, **BL__Y** = 1, obj fn < case 2
	HPS demand	*hps_demand*	0	tph	
	MPS reserve	*mps_reserve*	50	tph	
	HPS reserve	*hps_reserve*	50	tph	
4	MPS demand	*mps_demand*	90	tph	**BL1_STM** = 90, **BL2_STM** = 150, **MPS_RSV** > 50, **HPS_RSV** > 50, **BL__X** = 1, **BL__Y** = 0, obj fn > case 1
	HPS demand	*hps_demand*	150	tph	
	MPS reserve	*mps_reserve*	50	tph	
	HPS reserve	*hps_reserve*	50	tph	

Figure 3.26 shows the complete model as implemented in LINGO using the three include files created earlier. Figure 3.27 shows the Microsoft Excel Solver model; note that the coefficient matrix has a block diagonal structure; the individual boiler models are along the diagonal of the matrix, and the boiler reserve and demand models that span

```
MODEL:  TITLE BOILERS Subsystem Model: BLSS-v1;
! Nath, "Industrial Process Plants: Global Optimization of Utility Systems", Chapter 3, (2023).;
!;
DATA: ! inputs:  c_fo = 18; c_lpfg = 19; c_hpfg = 20;                              ! costs;
                 mps_demand = 60; mps_reserve = 50; hps_demand = 100; hps_reserve = 50;!demand, reserve
        ! params;
        bl1_fg_c = 15.2371; bl1_fg_d = 0.790615; bl1_fg_e =-12.4525; bl1_fg_f = 0.953495;   ! BL1 coeffs;
        bl1_bdr=0.02; bl1_fg_s = 3; bl1_stm_ll = 70; bl1_stm_hl = 210;
        bl2_fg_c = 18.1694; bl2_fg_d = 0.834726; bl2_fg_e =-11.6149; bl2_fg_f = 0.991485;   ! BL2 coeffs;
        bl2_bdr=0.02; bl2_fg_s = 4; bl2_stm_ll = 80; bl2_stm_hl = 240;
        bl3_fg_c = 21.1962; bl3_fg_d = 0.814223; bl3_fg_e =-15.2507; bl3_fg_f = 0.966085;   ! BL3 coeffs;
        bl3_bdr=0.02; bl3_fg_s = 5; bl3_stm_ll = 100; bl3_stm_hl = 300; big = 999;
   ENDDATA
! Objective;
  [_BLSS_V1_OBJ]  MIN = c_fo*BL1_FUEL + c_lpfg*(BL2_LPFG+BL3_LPFG) + c_hpfg*(BL2_HPFG+BL3_HPFG);

  ! constraints from include file;
                  @FILE('C:\BookDG\LingoModels\BL1-v4.lng');
                  @FILE('C:\BookDG\LingoModels\BL2-v1.lng');
                  @FILE('C:\BookDG\LingoModels\BL3-v1.lng');

! Steam demand & reserve constraint;
  [_MPS_RSV]      MPS_RSV = BL1_RSV;
  [_HPS_RSV]      HPS_RSV = BL2_RSV + BL3_RSV;
  [_MPS_DMND]     BL1_STM > mps_demand;
  [_MPS_RSRV]     MPS_RSV > mps_reserve;
  [_HPS_DMND]     BL2_STM + BL3_STM > hps_demand;
  [_HPS_RSRV]     HPS_RSV > hps_reserve;

END MODEL BLSS-v1
```

Fig. 3.26: Boilers subsystem model: BLSS-v1: in LINGO.

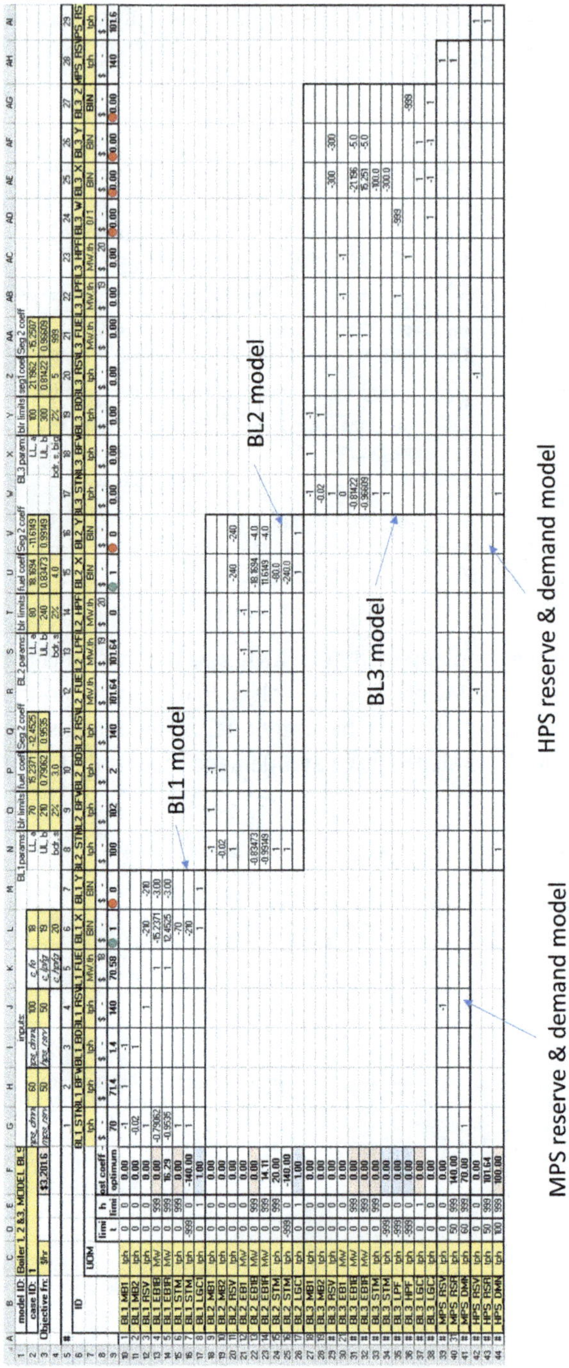

Fig. 3.27: Boilers subsystem model: BLSS–v1: in Microsoft Excel.

multiple boilers models have been placed near the end of the matrix, as highlighted on the figure. Table 3.4 shows the details of the solution for test case #1.

Tab. 3.4: Boilers subsystem model: BLSS-v1: Test case #1 optimal solution.

			Boilers Subsystem Model: BLSS-v1		
			Global optimal solution found.		
			Objective value:		$ 3,201.6
#	Variable ID	UOM	cost coeff	Opt. value	Cost
1	BL1_STM	tph	$ -	70	$ -
2	BL1_BFW	tph	$ -	71.4	$ -
3	BL1_BD	tph	$ -	1.4	$ -
4	BL1_RSV	tph	$ -	140	$ -
5	BL1_FUEL	MW.th	$ 18	70.58	$ 1,270.4
6	BL1_X	BIN	$ -	1	$ -
7	BL1_Y	BIN	$ -	0	$ -
8	BL2_STM	tph	$ -	100.0	$ -
9	BL2_BFW	tph	$ -	102.0	$ -
10	BL2_BD	tph	$ -	2.0	$ -
11	BL2_RSV	tph	$ -	140.0	$ -
12	BL2_FUEL	MW.th	$ -	101.64	$ -
13	BL2_LPFG	MW.th	$ 19	101.64	$ 1,931.2
14	BL2_HPFG	MW.th	$ 20	0.00	$ -
15	BL2_X	BIN	$ -	1	$ -
16	BL2_Y	BIN	$ -	0	$ -
17	BL3_STM	tph	$ -	0.0	$ -
18	BL3_BFW	tph	$ -	0.0	$ -
19	BL3_BD	tph	$ -	0.0	$ -
20	BL3_RSV	tph	$ -	0.0	$ -
21	BL3_FUEL	MW.th	$ -	0.00	$ -
22	BL3_LPFG	MW.th	$ 19	0.00	$ 0.0
23	BL3_HPFG	MW.th	$ 20	0.00	$ -
24	BL3_W	0 / 1	$ -	0	$ -
25	BL3_X	BIN	$ -	0	$ -
26	BL3_Y	BIN	$ -	0	$ -
27	BL3_Z	BIN	$ -	0	$ -
28	MPS_RSV	tph	$ -	140.0	$ -
29	HPS_RSV	tph	$ -	101.6	$ -

Summary

This chapter discusses the modeling of boilers in general and the three boilers in the sample utility system in particular. We started with a simple boiler model and refined it till it modeled all three operating modes: ON, OFF, and standby. We refined the model to enhance the boiler fuel versus steam relationship accuracy using a two-segment boiler performance curve. We concluded the chapter by aggregating the individual boiler models and creating a model for a subsystem of the three boilers, BLSS-v1. We tested all models with several test problems.

Nomenclature

Symbol	UOM	Description
blj_bdr	%	Boiler #j: blowdown ratio
blj_stm_hl	tph	Boiler #j: steam production: high limit
blj_stm_ll	tph	Boiler #j: steam production: low limit
blj_fg_s	MW.th	Boiler #j: fuel consumption in standby mode
blj_fg_a	MW.th	Boiler #j: fuel versus steam regression: intercept
blj_fg_b	MW.th/tph	Boiler #j: fuel versus steam regression: slope
blj_fg_c	MW.th	Boiler #j: fuel versus steam regression, lower range: intercept
blj_fg_d	MW.th/tph	Boiler: fuel versus steam regression, lower range: slope
blj_fg_e	MW.th	Boiler #j: fuel versus steam regression, upper range: intercept
blj_fg_f	MW.th/tph	Boiler #j: fuel versus steam regression, upper range: slope
c_fo	$/MWh.th	Fuel oil cost
CA	–	Compressed Air
CW	–	Cooling Water
ELEC	–	abbreviation for Electrical
FO	–	Fuel Oil
h_bfw	MW.th/t	enthalpy of boiler feed water steam
h_mps	MW.th/t	enthalpy of MP steam
h_mpsbd	MW.th/t	enthalpy of MP blowdown
HPFG	–	High Pressure Fuel Gas
HPS	–	High Pressure Steam
HPS	–	High Pressure Steam
hps_demand	tph	HP steam demand
hps_reserve	tph	HP reserve requirement
LPFG	–	Low Pressure Fuel Gas
LPS	–	Low Pressure Steam
MPS	–	Medium Pressure Steam
mps_demand	tph	MP steam demand
mps_reserve	tph	MP reserve requirement
PC	–	Process Condensate
BLj_BD	tph	Boiler #j: blowdown flow
BLj_BFW	tph	Boiler #j: boiler feed water flow
BLj_FUEL	MW.th	Boiler #j: fuel consumption
BLj_HPFG	MW.th	Boiler #j: HPFG consumption
BLj_LPFG	MW.th	Boiler #j: LPFG consumption
BLj_STM	tph	Boiler #j: steam production
BLj_W	Binary	Boiler #j: LPFG usage status: 1 = using LPFG, 0 otherwise
BLj_X	Binary	Boiler #j: operational status: 0 = OFF, 1 = ON

(continued)

Symbol	UOM	Description
BLj_Y	Binary	Boiler #j: standby status: 1 = standby, 0 otherwise
BLj_Z	Binary	Boiler #j: HPFG usage status: 1 = using HLPFG, 0 otherwise
HPS_RSV	tph	HP steam reserve
MPS_RSV	tph	MP steam reserve
η	–	generic efficiency

References

[1]   Boiler Energy Efficiency Measures, http://www.machineryspaces.com/boiler.html, (2016).
[2]   Anonymous, "LINGO The Modeling Language and Optimizer", p. 485, Lindo Systems Inc, (2020).

4 Modeling rotating equipment

Utility systems use a variety of rotating equipment. We can classify them into shaft power producers and shaft power consumers. We will model commonly used equipment of both kinds. The next chapter will discuss another type of equipment called turbogenerator, which combines a shaft work producer with a dedicated electrical generator.

Commonly used shaft work consumers are:
- Pumps,
- Fans (blowers), and
- Compressors.

Commonly used shaft work producers are:
- Electric motors,
- Steam turbines,
- Internal combustion engines, and
- Gas turbines.

Electric motors and steam turbines are the most commonly used among shaft work-producing devices. An electric motor is an electromagnetic device that converts electrical energy into mechanical energy, and a turbine is a mechanical device that converts the kinetic energy of an expanding fluid into mechanical energy. In a steam turbine, the expansion of high-pressure steam to lower pressure is the source of the kinetic energy.

In process plant utility systems internal combustion engines are mostly for specialized services, and gas turbines are used almost exclusively for electrical power generation. An internal combustion engine uses the expansion of the combustion product produced by fuel combustion, usually with atmospheric air, in an enclosed combustion chamber to move a piston to produce mechanical energy. In a gas turbine, direct combustion of fuel at high pressure in an enclosed chamber generates the expanding flue gas that drives the turbine.

The mechanical energy produced by motors, turbines, and engines is in the form of a rotating shaft that powers an energy-consuming device at the other end of the rotating shaft, i.e., a shaft work consumer is paired with a shaft work producer to achieve the process objective: moving a fluid against pressure or over a distance.

Among the shaft power-consuming devices in a utility system, pumps are used to transport liquids such as boiler feed water and cooling water; fans and blowers are used for transporting gases, such as boiler combustion air, across smaller distances; and compressors are used for transporting gases, such as compressed air, across more significant distances and or across larger pressure differentials.

A utility system has numerous types of pumps that come in various sizes, big and small; the smaller ones may not be worthy of individual models and may be modeled

https://doi.org/10.1515/9783111020679-005

as a fixed load to keep the optimization model simple, with practically no impact on the overall operating cost. What is small, however, is a judgment call on the part of the modeling team.

This chapter will discuss the modeling of the individual components mentioned above. We will discuss gas turbines in the next chapter.

4.1 Modeling electric motors

Electric motors are commonly used to supply shaft power because of easy availability, high reliability, and longevity. Larger motors enjoy high efficiency over the normal operating range, as shown in Fig. 4.1 [1, 2]. They, indeed, are the workhorse of the industry.

We will model a generic motor, M01, that will be used as a template for all motors in the sample utility system model. This model has two continuous variables and a binary variable. Continuous variables are **M01_ELE** and **M01_SP,** representing the input electrical and the output shaft power respectively. A binary variable **M01_X** represents its ON/OFF status. Note that motor efficiency, *m01_eta*, is the ratio of the output power to input power, as shown below:

$$m01\_eta = \textbf{M01\_SP} / \textbf{M01\_ELE} \tag{4.1}$$

Since the motor efficiency, especially for large industrial motors, is nearly constant over the normal operating range, we will keep things simple by assuming a constant, *m01_eta*. As shown below, we will rearrange equation (4.1) above to a linear constraint, relating input **M01_ELE** and output power **M01_SP**.

$$\textbf{M01\_SP} = m01\_eta * \textbf{M01\_ELE} \tag{\_M01\_EB1}$$

In addition, a motor has capacity limits; we will model these limits similar to those for boiler models.

To test this model, we must specify a demand for shaft work; otherwise, the minimum cost solution would be the trivial zero-cost solution, with the motor turned OFF. We will add the following constraint to add a shaft power demand.

$$\textbf{M01\_SP} = sp\_demand \tag{\_SP\_DMND}$$

Note that there is a unique solution to this model for a specified shaft power demand; so this is not an optimization problem as such. Nonetheless, we will formulate it as an optimization problem for easier integration into larger optimization models.

Since the problem formulated has a unique solution, any arbitrary objective function could be used; we will minimize the electrical power consumption, **M01_ELE**. Figure 4.2 is a complete LINGO model for a generic motor, M01.

All variables are physical quantities; they will be constrained to be nonnegative by default.

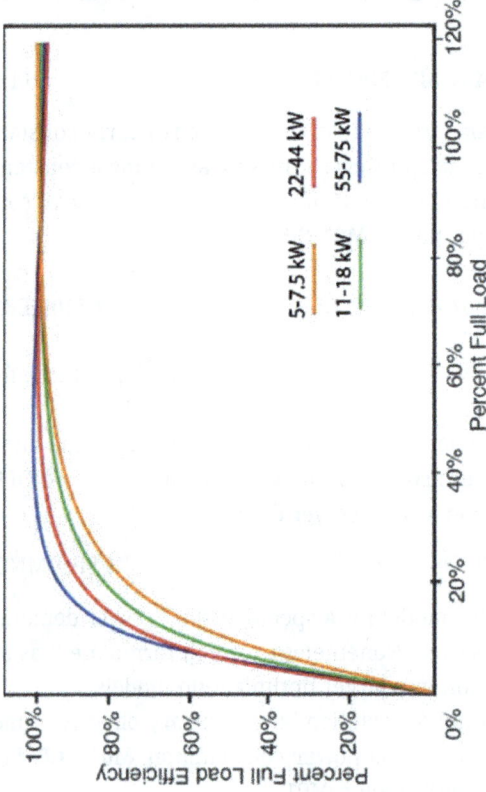

(a)

courtesy: www.naemotors.com

(b)

Fig. 4.1: Industrial motors: (a) motor efficiency as a function of load; (b) sample motor nameplate information.

```
MODEL:    TITLE Generic Motor model: M01-v1;
!
!Ref: Nath,"Industrial Process Plants: Global Optimization of Utility Systems", Chap. 4, (2023)
! ;

DATA:!inputs;     sp_demand = 1;                    ! shaft power demand, MW;
     !params;     m01_eta = 0.95;                   ! motor efficiency, fraction;
                  m01_sp_ll = 0.5;  m01_sp_hl = 1.5;  ! operating limits, MW;
  ENDDATA

  ! Objective;     MIN = M01_ELEC;

  ! energy balance;
  [_M01_EB1]      M01_SP = m01_eta * M01_ELEC;

  ! Capacity constraints;
  [_M01_SP_LL]    M01_SP > m01_sp_ll * M01_X;
  [_M01_SP_HL]    M01_SP < m01_sp_hl * M01_X;

  ! Integers;     @BIN( M01_X );

  ! Demand constraint;
  [_SP_DMND]      M01_SP = sp_demand;

END MODEL M01-v1
```

Fig. 4.2: Generic motor, M01-v1: in LINGO.

This model, M01-v1, has been implemented in Excel and LINGO. This is a simple model; so extensive testing is optional. We will test it with a single test case to ensure no apparent errors in the model formulation.

Figure 4.3 shows the complete generic motor model in Microsoft Excel, showing the solution to the test case problem. Note that the optimizer populates the optimum value of variables in row 9 and the value of constraints at the optimum level in column F. Note, both the Excel solver and LINGO produce identical results.

▲	A	B	C	D	E	F	G	H	I	J
1		model ID:	generic motor M01-v1				inputs:		params:	
2		case ID:	base case				sp_demand	1	m01_sp_ll	0.5
3		Objective fn:	MW			1.05			m01_sp_hl	1.5
4							big	999	m01_eta	95%
5	#						1	2	3	
6		ID					M01_SP	M01_ELE	M01_X	
7			UOM				MW.sp	MW	BIN	
8				low	high	cost coeff ->		$ 1.0		
9				limit	limit	optimum	1	1.053	⬤ 1	
10	1	_M01_SP_LL	MW.sp	0	999	0.50	1		-0.5	
11	2	_M01_SP_HL	MW.sp	-999	0	-0.50	1		-1.5	
12	3	_M01_EB1	MW	0	0	0.00	1	-0.95		
13	4	_SP_DMND	MW.sp	1	1	1.00	1			

Fig. 4.3: Generic motor, M01-v1: in Microsoft Excel.

4.2 Modeling steam turbines

Steam turbines are also commonly used to produce shaft power; their advantage is using self-generated steam and very high efficiency, especially when the turbine exhaust balances the steam load on the steam headers.

Modeling of steam turbines is relatively straightforward; a simple input-output model is sufficient. We will model a generic turbine, T01, and use it as a template for all turbines in the sample utility system model. It has two continuous variables and a binary variable. The continuous variables are **T01_STM** and **T01_SP**, representing the inlet throttle steam flow and the output shaft power. A binary variable **T01_X** represents its ON/OFF status.

We will have a performance curve constraint relating the input steam flow to the output shaft power and two constraints for capacity limits in terms of high and low steam flows. The performance curve relating shaft power to throttle flow is usually obtained by performing a plant test or using an equipment manufacturer-supplied performance curve. The relationship between the two variables is fairly linear; coefficients of linear relationship can be obtained by regression. Figure 4.4 shows sample turbine curves.

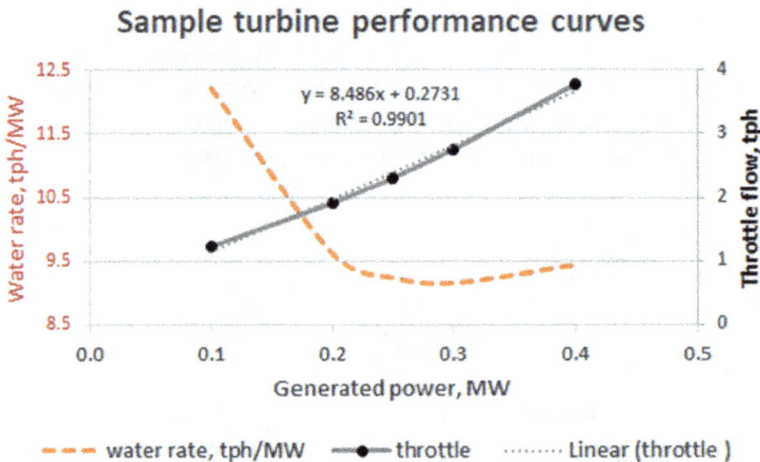

Fig. 4.4: Sample steam turbine, T01, performance curves.

The turbine model is simple. There is a unique solution for a specified shaft power demand, so this is not an optimization problem as such. Nonetheless, we will formulate it as an optimization problem and use an optimizer to solve it. We could use any arbitrary variable for the objective function; we will minimize the steam usage, **T01_STM**. Figure 4.5 is a complete model of a generic steam turbine model, T01-v1, as configured in LINGO.

```
MODEL:    TITLE Generic Turbine Model: T01-v1;
!
!Ref: Nath, "Industrial Process Plants: Global Optimization of Utility Systems", Chapt. 4, (2023);
! ;
  DATA:
       !inputs;     sp_demand = 0.25;        ! shaft power demand, MW;

       !params;     t01_stm_a = 0.27306;     ! throttle vs shaft power, intercept, tph;
                    t01_stm_b = 8.48596;     ! throttle vs shaft power, slope, tph/MW.sp;
                    t01_stm_ll = 1;          ! throttle limit, lower, tph;
                    t01_stm_hl = 5;          ! throttle limit, higher, tph;
  ENDDATA

  ! Objective;
  [_T01_OBJ]     MIN = t01_STM;

  ! Energy balance;
  [_T01_EB1]     T01_STM = t01_stm_a * T01_X + t01_stm_b * T01_SP;

  ! Capacity constraints;
  [_T01_STM_LL]  T01_STM > t01_stm_ll * T01_X;
  [_T01_STM_HL]  T01_STM < t01_stm_hl * T01_X;

  ! Binary variable specification;
                 @BIN( T01_X );

  ! Demand constraint;
  [_SP_DMND]     T01_SP = sp_demand;

END MODEL T01-v1
```

Fig. 4.5: Generic steam turbine model: T01-v1: in LINGO.

This model, T01-v1, has been implemented in Excel and LINGO. It is a simple model; so extensive testing is unnecessary. We will test it with a single test case to ensure no apparent errors in the model formulation.

Figure 4.6 shows the complete generic turbine model in Microsoft Excel and the test case problem #1 solution. Note, both the Excel solver and LINGO produce identical results.

	A	B	C	D	E	F	G	H	I	J	K
1		model ID:	Generic turbine single stage: T01-v1				inputs:	params:			
2		case ID:	1				sp_dmnd	t01_stm_ll	1	t01_stm_a	0.27306
3		Objective fn:	tph			2.39	0.25	t01_stm_hl	5	t01_stm_b	8.48596
4								big	999		
5	#						1	2	3		
6		ID					T01_SP	T01_STM	T01_X		
7			UOM				MW.sp	tph	BIN		
8				low	high	cost coeff ->	$ -	$ 1.00	$ -		
9				limit	limit	optimum	0.250	2.4	1		
10	1	_T01_STM_LL	tph	0	999	1.39		1	-1.00		
11	2	_T01_STM_HL	tph	-999	0	-2.61		1	-5.0		
12	3	_T01_EB1	MW	0	0	0.00	-8.49	1	-0.27		
13	4	_SP_DMND	MW.sp	0.25	0.25	0.25	1				

Fig. 4.6: Generic steam turbine model: T01-v1: in Microsoft Excel.

4.3 Modeling internal combustion engines

Using internal combustion engines (ICE) to generate shaft power is uncommon, but it does happen in some situations.

ICE modeling is relatively straightforward; a simple input-output model is adequate. We will model a generic ICE that will serve as a template for all ICEs in the sample utility system model. It has two continuous variables and a binary variable. Continuous variables are **E01_FG** and **E01_SP,** representing fuel consumption by the engine and the shaft power produced. A binary variable **E01_X** represents its ON/OFF status.

The performance curve is often available as load versus efficiency; this relationship is usually nonlinear, which is unsurprising. By eliminating efficiency and transforming the data to load vs. input energy, one can see that the relationship between the output power and input fuel is relatively linear; this may sound familiar, as we have seen similar behavior earlier when modeling boilers in Chapter 3. Figure 4.7 is a generic gas engine performance curve [3].

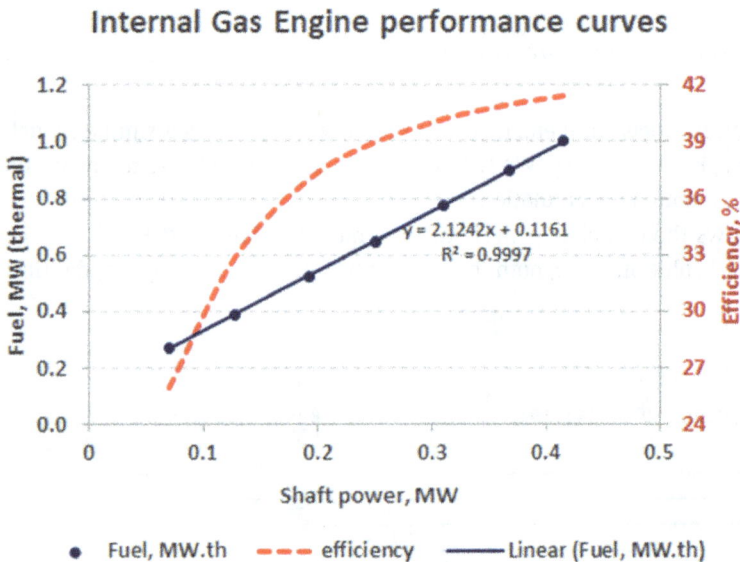

Internal Gas Engine performance curves

$$y = 2.1242x + 0.1161$$
$$R^2 = 0.9997$$

Fig. 4.7: Performance curves for a sample internal combustion engine.

We will have a constraint relating the input fuel consumption to the output shaft power and two constraints for the high and low capacity limits. The performance curve relating shaft power to fuel consumption is usually obtained by performing a plant test or using data provided by the equipment manufacturer; regressing this data gives a linear correlation.

There would be a unique solution for a specified shaft power demand; so this is not an optimization problem as such. Nonetheless, we will formulate it as an optimization problem to be able to use an optimization solver. Since there is a unique solution, we could use any arbitrary variable for its objective function; we will minimize the fuel consumption, **E01_FG**.

The model is simple; so extensive testing is unnecessary. We test it with a single test case to ensure no apparent errors in the model formulation.

This model, E01-v1, has been implemented in Excel and LINGO. Figure 4.8 is the complete model as configured in LINGO. Figure 4.9 shows the complete generic turbine model in Microsoft Excel, populated with the results for the test problem #1 solution. Both the Excel Solver and LINGO produce identical results.

```
MODEL:    TITLE Generic Internal Combustion Engine Model: E01-v1;
!
!Ref: Nath, "Industrial Process Plants: Global Optimization of Utility Systems",
!              Chapter 4, (2023).;

  DATA:
      ! inputs;    sp_demand = 0.36;        ! shaft power demand, MW;

      ! params;    e01_fg_a = 0.1161;       ! fuel vs shaft power, intercept, MW.th;
                   e01_fg_b = 2.1242;       ! fuel vs shaft power, slope, MW.th/MW;
                   e01_sp_ll = 0.1;         ! SP limit, lower, MW;
                   e01_sp_hl = 0.5;         ! SP limit, higher, MW;
  ENDDATA

  ! Objective;
  [_E01_OBJ]      MIN = E01_FG;

  ! Energy balance;
  [_E01_EB1]      E01_FG = e01_fg_a * E01_X + e01_fg_b * E01_SP;

  ! Capacity constraints;
  [_E01_SP_LL]    E01_SP > e01_sp_ll * E01_X;
  [_E01_SP_HL]    E01_SP < e01_sp_hl * E01_X;

  ! Binary variable specifications;
                  @BIN( E01_X );

  ! Demand constraint;
  [_SP_DMND]      E01_SP = sp_demand;

END MODEL E01-v1
```

Fig. 4.8: Generic ICE model: E01-v1: in LINGO.

4.4 Modeling pumps

Pumps are ubiquitous in the utility system and transport liquid streams such as BFW and cooling water. Pumps are consumers of shaft power, usually supplied by a motor or a steam turbine. Although numerous types of pumps exist, centrifugal pumps are

◢	A	B	C	D	E	F	G	H	I	J
1		model ID:	E01-v1: generic internal comb. Engine				inputs:	params:		
2		case ID:	1				sp_dmnd	e01_sp_ll, hl	0.1	0.5
3		Objective fn:	MW.th			**0.88**	0.36	e01_fg_a, b	0.1161	2.1242
4								big	999	
5	#						1	2	3	
6		ID					E01_SP	E01_FG	E01_X	
7			UOM				MW.sp	MW.th	BIN	
8				low	high	cost coeff ->		$ 1.00		
9				limit	limit	optimum	0.360	0.881	⬤ 1	
10	1	_E01_EB1	MW.th	0	0	0.00	-2.12	1	-0.12	
11	2	_E01_SP_LL	MW.sp	0	999	0.26	1		-0.10	
12	3	_E01_SP_HL	MW.sp	-999	0	-0.14	1		-0.5	
13	4	_SP_DMND	MW.sp	0.36	0.36	0.36	1			

Fig. 4.9: Generic ICE model: E01-v1: in Microsoft Excel, showing the test case solution.

the most commonly used for water transport because of their versatility and durability. They will be the subject of discussion below.

Modeling of pumps is relatively straightforward; a simple input–output model will suffice. We will model a generic pump that will serve as a template for all pumps in the sample utility system model. It has two continuous variables and a binary variable. The continuous variables are **P01_FLO** and **P01_SP**, representing the pump inlet liquid flow and the required shaft power. A binary variable **P01_X** will represent its ON/OFF status.

The performance curve relating the shaft power to inlet liquid flow is usually obtained by performing a plant test or from the manufacturer. Figure 4.10 shows sample curves comprising the pump head, efficiency, and shaft power requirement as a function of the pump inlet liquid flow.

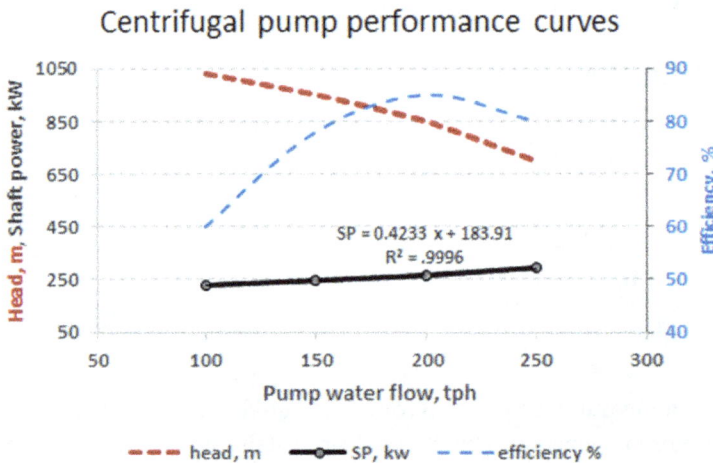

Fig. 4.10: Sample pump: performance curves.

Note that the pump head is very high at low inlet liquid flow rates, below the design minimum, which could cause pump vibration and stability problems. For this reason, a minimum bypass flow line is installed on industrial centrifugal pumps [4], as shown in Fig. 4.11a. The net effect of this arrangement is that the pump can accommodate very low flow rates. Still, its energy consumption does not drop below the level corresponding to the minimum design flow, as shown by the solid blue line in Fig. 4.11b. The performance curve comprises the solid blue line in the lower range and the solid red line in the upper range. The dashed blue line is an extension of the solid blue line to the upper range, and the dashed red line is an extension of the solid red line to the lower range.

Observe that the dashed lines underestimate the solid lines. We have encountered a similar situation before when modeling a boiler with a segmented performance curve in Section 3.1.6, and we will use a similar model approach to the pump performance curve in Fig. 4.11b using two greater than ($\geq$) type inequalities. Doing so eliminates the need for a binary variable and makes the optimization model more efficient. Doing so also eliminates the liquid flow lower limit constraint and gives a more realistic pump model.

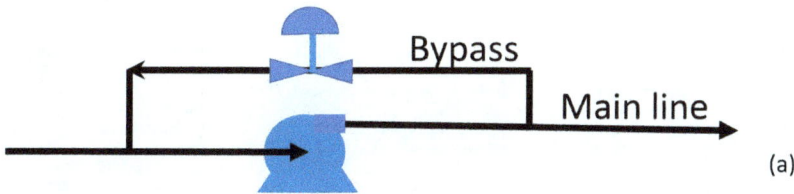

(a)

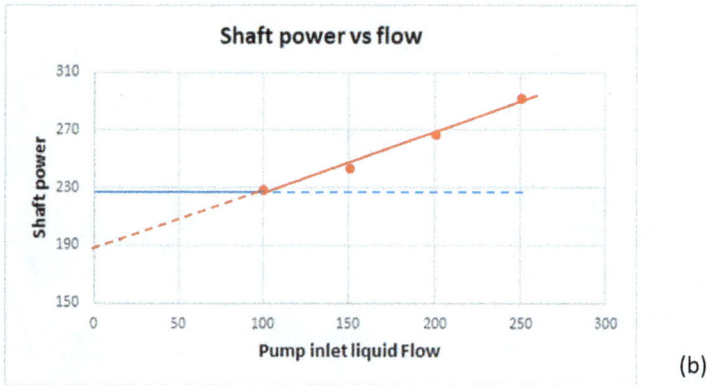

(b)

Fig. 4.11: Centrifugal pump: (a) minimum flow bypass line configuration. (b) actual shaft power vs. inlet liquid flow (in solid blue and red lines).

For testing purposes, the objective function will be to minimize the shaft power, **P01_SP**, by specifying the cost coefficient of 1 to **P01_SP**. We will test this model for four different values of inlet liquid flow demand, *flo_demand*. Tab. 4.1 shows the 4 test cases and the expected results.

Tab. 4.1: Generic pump model: P01-v1: test cases.

#	Attribute	Symbolic name	Value	UOM	Expected key results
1	flow demand	*flo_demand*	100	tph	**P01_FLO** = 100, **P01_X** = 1, obj fn > 0
2	flow demand	*flo_demand*	50	tph	**P01_FLO** = 50, **P01_X** = P01M_X =1, obj fn same as case 1
3	flow demand	*flo_demand*	0	tph	**P01_FLO** = 0, **P01_X** = 0, obj fn = 0
4	flow demand	*flo_demand*	150	tph	**P01_FLO** = 150, **P01_X** = 1, obj fn > for case 1

Figure 4.12 is the complete pump model in LINGO. Fig. 4.13 shows the same model and solution to test case #1 problem in Microsoft Excel. The Excel Solver populates the optimum value of variables in row 9 and the constraint values at the optimum point in column F. The Excel Solver and the LINGO Solver produce the expected results.

```
MODEL:    TITLE Generic Pump Model: P01-v1;
!
!Ref: Nath, "Industrial Process Plants: Global Optimization of Utility Systems", Chap. 4, (2023)
! ;
 DATA:!inputs;       flo_demand = 100;         ! pump inlet flow demand, tph;

       !params;      p01_sp_a = 0.183910;      ! inlet flow vs shaft power, intercept, MW.sp;
                     p01_sp_b = 0.000423;      ! throttle vs shaft power, slope, MW.sp/ tph;
                     p01_des_11 = 100;         ! inlet flow design low limit, tph;
                     p01_flo_h1 = 250;         ! inlet flow high limit, tph;
 ENDDATA

 ! Objective;     MIN = P01_SP;

 ! Performance curves;
 [_P01_EB1B]       P01_SP > (p01_sp_a + p01_sp_b * p01_des_11) * P01_X;
 [_P01_EB1R]       P01_SP > p01_sp_a * P01_X + p01_sp_b * P01_FLO;

 ! Capacity constraints;
 [_P01_FLO_HL]    P01_FLO < p01_flo_h1 * P01_X;

 ! Binary spec;   @BIN( P01_X );

 ! Inlet flow requirement;
 [_FLO_DMND]      P01_FLO = flo_demand;

END MODEL P01-v1
```

Fig. 4.12: Generic Pump model: P01-v1: in LINGO.

In the next subsection, we will add a motor drive to this pump model.

⊿	A	B	C	D	E	F	G	H	I	J
1		model ID:	P01-v1: generic pump				inputs:	params:	*big*	999
2		case ID:	1				*liq_dmnd*	*p01_des_ll, hl*	100	250
3		Objective fn:	MW.sp			0.226	100	*p01_sp_a, b*	0.183910	0.000423
4										
5	#						1	2	3	
6		ID					P01_LIQ	P01_SP	P01_X	
7			UOM				tph	MW.sp	BIN	
8				low	high	cost coeff ->		$	1.00	
9				limit	limit	optimum	100	0.226	1	
10	1	_P01_EB1B	MW.sp	0	999	0.00		1	-0.2262	
11	2	_P01_EB1R	MW.sp	0	999	0.00	-0.0004	1	-0.1839	
12	3	_P01_LIQ_HL	tph	-999	0	-150.00	1		-250.0	
13	4	_LIQ_DMND	tph	100	100	100.00	1			

Fig. 4.13: Generic pump model: P01-v1: in Microsoft Excel, showing test case #1 solution.

4.5 Modeling a pump with a motor driver

Pumps consume shaft power, usually provided by either a motor driver or a steam turbine driver; however, some large pumps may have dual drivers, i.e., the pump is connected to both a motor driver and a steam turbine driver, and either could power the pump. Pumps with dual drives provide an additional degree of freedom that the optimizer exploits to obtain a better fit with the site steam balance and purchased energy considerations.

In this subsection, we will model a generic pump, driven by a motor; subsequent subsections will model the other two possibilities.

Modeling a pump with a motor drive entails merging the pump model, P01-v1, with the motor model, M01-v1, eliminating duplication of variables, and rationalizing the relationships among the resulting variables to simplify the combined model. Make two observations: (1) the shaft power required by the pump, **P01_SP**, must equal the shaft power produced by the motor, **M01_SP**, and (2) the motor binary, **M01_X**, must equal the pump binary, **P01_X**. With these observations in mind, we could eliminate **M01_SP** and use **P01_SP**. Similarly, we could eliminate **M01_X** and use **P01_X**, instead.

Furthermore, for an adequately matched pump and motor pair, the motor should not be limiting, and in such cases, we could eliminate **M01_SP's** high limit and low limit constraints. However, there is a subtlety: note that the pump shaft power has a two-segment performance curve with two ≥ type constraints. If **P01_SP** (or a related variable) is not in the objective function and the pump is OFF, **P01_SP** could take on an arbitrary nonzero value without affecting the objective function, which would be confusing. We will eliminate such a possibility by retaining the **P01_SP** high limit constraint in the model.

As we may encounter pumps with a motor drive when creating a larger optimization model, we will save the model constraints to an external file called "P01MD-v1.lng" that would be included in larger LINGO models, as was done for the boilers in Chapter 3.

Figures 4.14 and 4.15 constitute the complete model of a generic pump model with a motor drive, as configured in LINGO. Figure 4.16 shows the full generic turbine model with motor drive in Microsoft Excel, showing the solution of the test case # 1 problem. The Excel and LINGO Solver produce identical and expected results for all test cases.

```
! Pump P01 with motor drive P01M: model constraints file: P01MD-v1.lng;

  ! P01 constraints;
  [_P01_EB1B]      P01_SP  > (p01_sp_a + p01_sp_b * p01_des_11) * P01_X;
  [_P01_EB1R]      P01_SP  >  p01_sp_a * P01_X + p01_sp_b * P01_FLO;
  [_P01_SP_HL]     P01_SP  < (p01_sp_a + p01_sp_b * p01_flo_hl) * P01_X;
  [_P01_FLO_HL]    P01_FLO <  p01_flo_hl * P01_X;

  ! P01M constraints;
  [_P01M_EB1]      P01_SP  = p01m_eta * P01M_ELE;

  ! Integers;       @BIN( P01_X );
!;
```

Fig. 4.14: Generic pump with motor drive model: P01MD-v1: constraints file: P01MD-v1.lng.

```
MODEL:   TITLE Generic Pump with motor drive model: P01MD-v1;
!
! Nath, "Industrial Process Plants: Global Optimization of Utility Systems", Chap. 4, (2023)
! ;
 DATA:!inputs;     flo_demand = 100;        ! flow demand, tph;

      !params;     p01_sp_a = 0.183910; p01_sp_b = 0.000423;  ! inlet flow vs shaft power;
                   p01_des_11 = 100;    p01_flo_hl = 250;     ! inlet flow limits, tph;
                   p01m_eta = 0.95;                           ! motor efficiency, fraction;
 ENDDATA

  ! Objective;    MIN = P01M_ELE;

  ! constraints;  @FILE('C:\BookDG\LingoModels\P01MD-v1.lng');

  ! Flow demand constraint;
  [_FLO_DMND]      P01_FLO = flo_demand;

END MODEL P01MD-v1
```

Fig. 4.15: Generic pump with motor drive model: P01MD-v1: in LINGO.

In the next subsection, we will model the pump with a dedicated turbine drive.

4.6 Modeling a pump with turbine drive

The model of a pump with a steam turbine drive is similar to that of a pump with a motor drive, except the motor performance equation curve is replaced by the turbine

A	B	C	D	E	F	G	H	I	J
1	model ID:	P01MD-v1: pump w motor drive				inputs:	params:	big	999
2	case ID:	1				liq_dmnd	p01_des_ll, hl	100	250
3	Objective fn:	MW			0.238	100	p01_sp_a, b	0.183910	0.000423
4							p01m_eta		95%
5	#						1	2	3	4
6		ID					P01_LIQ	P01_SP	P01_X	M01_ELE
7			UOM				tph	MW.sp	BIN	MW
8				low	high	cost coeff ->				$ 1.0
9				limit	limit	optimum	100	0.226	1	0.238
10	1	_P01_EB1B	MW.sp	0	999	0.00		1	-0.2262	
11	2	_P01_EB1R	MW.sp	0	999	0.00	-0.0004	1	-0.1839	
12	3	_P01_SP_HL	MW.sp	-999	0	-0.06		1	-0.2897	
13	4	_P01_LIQ_HL	tph	-999	0	-150.00	1		-250.0	
14	5	_P01M_EB1	MW	0	0	0.00		1		-0.95
15	6	_LIQ_DMND	tph	100	100	100.00	1			

Fig. 4.16: Generic pump with motor drive model: P01MD-v1: in Excel, showing test case #1 solution.

performance curve. We will use the same 4 test cases to validate this pump model, as shown in Tab. 4.1.

Figures 4.17 and 4.18 show the generic pump with a turbine drive model configured in LINGO. Figure 4.19 shows the same model in Microsoft Excel, showing the test case # 1 problem solution. Both the Excel Solver and the LINGO Solver produce identical and expected results.

```
! Pump P01 with turbine drive P01T: model constraints file: P01TD-v1.lng;

  ! Pump P01 constraints;
  [_P01_EB1B]      P01_SP  > (p01_sp_a + p01_sp_b * p01_des_ll) * P01_X;
  [_P01_EB1R]      P01_SP  >  p01_sp_a * P01_X + p01_sp_b * P01_FLO;
  [_P01_SP_HL]     P01_SP  < (p01_sp_a + p01_sp_b * p01_flo_hl) * P01_X;
  [_P01_FLO_HL]    P01_FLO <  p01_flo_hl * P01_X;

  ! Turbine P01T constraints;
  [_P01T_EB1]      P01T_STM = p01t_stm_a * P01_X + p01t_stm_b * P01_SP;

  ! Integers;     @BIN( P01_X );
!;
```

Fig. 4.17: Generic pump with turbine drive model constraints file: P01TD-v1.lng.

In the next subsection, we will add a dual drive capability to the pump model.

4.7 Modeling a pump with dual drives

In this section, we will model a generic pump with dual drives. In this configuration, the pump could be powered either by a motor or a turbine. Pumps with dual drives

```
MODEL:    TITLE Generic Pump P01 with turbine drive P01T model: P01TD-v1;
! ;
! Nath, "Industrial Process Plants: Global Optimization of Utility Systems", Chap. 4, (2023)
! ;
 DATA:!inputs;     flo_demand = 100;                          ! flow demand, tph;

      !params;;    p01_sp_a = 0.183910;    p01_sp_b = 0.000423;    ! P01: SP vs inlet flow ;
                   p01_des_ll = 100;       p01_flo_hl = 250;       ! P01: inlet flow limits;
                   p01t_stm_a = 0.27306;   p01t_stm_b = 8.48596;   ! P01T: steam vs SP;
   ENDDATA

  ! Objective;    MIN = P01T_STM;

  ! constraints;  @FILE('C:\BookDG\LingoModels\P01TD-v1.lng');

  ! Flow demand constraint;
  [_FLO_DMND]     P01_FLO = flo_demand;

END MODEL P01TD-v1
```

Fig. 4.18: Generic pump with turbine drive model: P01TD-v1: in LINGO.

	A	B	C	D	E	F	G	H	I	J	K
1		model ID:	P01TD-v1: pump w turbine drive				inputs:	params:		p01_sp_a	0.183910
2		case ID:	1				flo_dmnd	p01_des_ll	100	p01_sp_b	0.000423
3		Objective fn:	tph			2.19	100	p01_flow_hl	250	p01t_stm_a	0.273060
4								big	999	p01t_stm_b	8.485960
5	#						1	2	3	4	
6		ID					P01_FLO	P01_SP	P01_X	P01T_STM	
7			UOM				tph	MW.sp	BIN	TPH	
8				low	high	cost coeff ->				$	1.0
9				limit	limit	optimum	100.0	0.226	1	2.19	
10	1	_P01_EB1B	MW.sp	0	999	0.00		1	-0.2262		
11	2	_P01_EB1R	MW.sp	0	999	0.00	-0.0004	1	-0.1839		
12	3	_P01_SP_HL	MW.sp	-999	0	-0.06		1	-0.2897		
13	4	_P01_FLO_HL	tph	-999	0	-150.00	1		-250.0		
14	5	_P01T_EB1	tph	0	0	0.00		-8.4860	-0.2731	1	
15	6	_FLO_DMND	tph	100	100	100.00	1				

Fig. 4.19: Generic pump with turbine drive model: P01TD–v1: in Excel, showing test case #1 solution.

have an additional degree of freedom to choose the drive, which the optimizer would exploit to obtain a more economical operations policy.

Modeling a pump with dual drives entails merging the models of the pump, a motor, and a turbine. In addition, we need to add constraints that represent the relationships between the pump and its drivers and ensure that only one drive powers the pump. We will modify the objective function to a cost function to account for both possibilities.

Each driver would need a binary ON/OFF variable, and we also tie these two variables to the binary variable of the pump. As these two drive operations are mutually exclusive, we will add a constraint that equates **P01_X** to the sum of **P01M_X** and **P01T_X**, as shown by the logical constraint in the model.

We will use the 4 test cases as shown in Tab. 4.2 to test this model.

Figures 4.20 and 4.21 show the complete generic pump with turbine drive model P01DD-v1 configured in LINGO. Figure 4.22 shows the same model in Microsoft Excel,

Tab. 4.2: Generic pump model with dual drives: P01DD-v1: Test Cases.

#	Attribute	Symbolic name	Value	UOM	Expected key results
all	elec cost	c_elec	60	$/MWh	
1	flow demand	flo_demand	100	tph	P01_FLO = 100, P01_X = P01M_X = 1, obj fn > 0
	steam cost	c_stm	10	$/ton	
2	flow demand	flo_demand	50	tph	P01_FLO = 50, P01_X = P01M_X =1, obj fn same as case 1
	steam cost	c_stm	10	$/ton	
3	flow demand	flo_demand	0	tph	P01_FLO = 0, P01_X = P01M_X = P01T_X = 0, obj fn = 0
	steam cost	c_stm	10	$/ton	
4	flow demand	flo_demand	100	tph	P01_FLO = 150, P01_X = P01T_X = 1, obj fn < for case 1
	steam cost	c_stm	5	$/ton	

showing the test problem #1 solution. Both the Excel solver and LINGO produce identical and expected results.

```
! Pump P01 with dual drives P01M & P01T: model constraints file: P01DD-v1.lng;

! P01;
 [_P01_EB1B]       P01_SP  > (p01_sp_a + p01_sp_b * p01_des_11) * P01_X;
 [_P01_EB1R]       P01_SP  > p01_sp_a * P01_X + p01_sp_b * P01_FLO;
 [_P01_FLO_HL]     P01_FLO < p01_flo_hl * P01_X;

! P01M;
 [_P01M_EB1]       P01M_SP = p01m_eta * P01M_ELE;
 [_P01M_SP_HL]     P01M_SP < (p01_sp_a + p01_sp_b * p01_flo_hl) * P01M_X;

! P01T;
 [_P01T_EB1]       P01T_STM = p01t_stm_a * P01T_X + p01t_stm_b * P01T_SP;
 [_P01T_SP_HL]     P01T_SP  < (p01_sp_a + p01_sp_b * p01_flo_hl) * P01T_X;

! Logic;
 [_P01_SP]         P01_SP = P01M_SP + P01T_SP;
 [_P01_LGC1]       P01_X  = P01M_X + P01T_X;

! Integers;        @BIN( P01_X );      @BIN( P01M_X );
!;
```

Fig. 4.20: Generic pump with dual drives model constraints file: P01DD-v1.lng.

In the following section, we will model fans and blowers; these devices are used to transport gases, such as combustion air, over short distances, into the boiler combustion chamber.

4.8 Modeling boiler auxiliary fans and blowers

Depending on the design of the combustion system of a boiler, it may have a natural draft design, in which case, combustion air and flue gas would move without the assistance of a fan or a blower. However, many industrial boilers have dedicated fans or

```
MODEL:    TITLE Generic Pump P01 with dual drives P01M & P01T model: P01DD-v1: Base Case;
! ;
! Ref: Nath, "Industrial Process Plants: Global Optimization of Utility Systems", Chap. 4, (2023)
! ;
DATA:!inputs;    c_elec = 60;      c_stm = 10; flo_demand = 100; ! costs, flow demand;
     !params;    p01_sp_a = 0.183910;    p01_sp_b = 0.000423;    ! P01: SP vs inlet flow;
                 p01_des_ll = 100;       p01_flo_hl = 250;       ! P01: inlet flow limits, tph;
                 p01m_eta = 0.95;                                ! P01M: efficiency, fraction;
                 p01t_stm_a = 0.27306;   p01t_stm_b = 8.48596;   ! P01T: steam vs SP;
ENDDATA

  ! Objective;    MIN = c_elec * P01M_ELE + c_stm * P01T_STM;

  ! constraints;  @FILE('C:\BookDG\LingoModels\P01DD-v1.lng');

  ! Flow demand constraint;
  [_FLO_DMND]     P01_FLO = flo_demand;

END MODEL P01DD-v1
```

Fig. 4.21: Generic pump with dual drives model: P01DD-v1: in LINGO.

blowers to transport combustion air or flue gas. These devices are sometimes called boiler auxiliaries; we will use this term in this book. These devices are not small energy consumers; but they have relatively constant energy usage over the normal operating range of the boiler; so, modeling them as a constant load is usually adequate. If not, they could be modeled like the pumps discussed earlier in this chapter. Similar to pumps, these devices may be powered by dedicated motors, dedicated steam turbines, or dual drives. We will next discuss the modeling of each case.

In the sample utility system, boiler #1 has boiler auxiliaries powered by the steam turbine, BL1AT. We will assume a constant load on the steam turbine, which means a constant MP steam throttle flow, $bl1at\_c$ tph. The following constraint expresses this relationship;

$$\textbf{BL1AT\_STM} = bl1at\_c * \textbf{BL1\_X} \qquad (\_BL2AT\_EB1)$$

Boiler #2 auxiliaries in the sample utility system are powered by a motor, BL2AM. We will assume a constant load on the motor with a constant electrical power requirement of $bl2am\_c$. The following constraint expresses this relationship;

$$\textbf{BL2AM\_ELE} = bl2am\_c * \textbf{BL2\_X} \qquad (\_BL2AM\_EB1)$$

Boiler #3 has dual drives, powered by motor BL3AM or turbine BL3AT. BL3AM motor will consume electrical power, $bl3am\_c$, and BL3AT will require a constant flow of HP steam, $bl3at\_c$. The following constraints will model these relationships:

$$\textbf{BL3AM\_ELE} = bl3am\_c * \textbf{BL3AM\_X} \qquad (\_BL3AM\_EB1)$$

$$\textbf{BLA3T\_STM} = bl3at\_c * \textbf{BL3AT\_X} \qquad (\_BL3AT\_EB1)$$

$$\textbf{BL3\_X} = \textbf{BL3AM\_X} + \textbf{BL3AT\_X} \qquad (\_BL3A\_LGC)$$

The model will also need a binary specification for either **BL3AM_X** or **BL3AT_X**.

model ID: generic pump with dual drives
case ID: 1
Objective fn: $/hr = $ 14.29

inputs:

	value
flo_dmnd	100
c_stm	10
c_elec	60

params:

	value
p01_des_ll	100
p01_flo_hl	250
m01_eta	95%
p01_sp_a	0.183910
p01_sp_b	0.000423
t01_stm_a	0.273060
t01_stm_b	8.485960
big	999

Decision variables (optimum solution):

#	1	2	3	4	5	6	7	8	9
ID	P01_FLO	P01_SP	P01_X	P01M_SP	P01M_ELE	P01M_X	P01T_SP	P01T_STM	P01T_X
UOM	tph	MW.sp	BIN	MW.sp	MW	BIN	MW.sp	tph	0/1
optimum	100.0	0.226	1	0.226	0.238	1	0.000	0.0	0

Constraints (cost coeff -> optimum):

#	ID	UOM	low limit	high limit	optimum	P01_FLO	P01_SP	P01_X	P01M_SP	P01M_ELE	P01M_X	P01T_SP	P01T_STM	P01T_X
1	_P01_EB1B	MW.sp	0	999	0.00		1	-0.2262						
2	_P01_EB1R	MW.sp	0	999	0.00	-0.00042	1	-0.1839						
3	_P01_FLO_HL	tph	-999	0	-150.00	1		-250.0						
4	_M01_EB1	MW	0	0	0.00				1	-0.95				
5	_P01M_SP_HL	MW.sp	-999	0	-0.06				1		-0.2897			
6	_T01_EB1	tph	0	0	0.00							-8.485960	1	-0.273
7	_P01M_SP_HL	MW.sp	-999	0	0.00							1		-0.2897
8	_P01_SP	MW.sp	0	0	0.00		1		-1			-1		
9	_P01_LGC1	-	0	0	0.00			1			-1			-1
10	_FLO_DMND	tph	100	100	100.00	1								

Fig. 4.22: Generic pump with dual drives model: P01DD-v1: in Excel, showing test case #1 solution.

We will implement these constraints in the combined boiler subsystem model, BLSS-v2. For the sake of simplicity, we will ignore the cost of steam used by the turbine drives as later in Chapter 6, we will develop the model for the entire steam subsystem, and then the cost of steam streams would not be necessary. We will combine the model of the boiler subsystem model BLSS-v1 and the constraints mentioned above for boiler auxiliaries. We will also update the objective function to include additional terms relating to the auxiliaries.

We will test this model for the same test cases as for BLSS-v1, as shown in Tab. 3.3.

The complete model of the boiler subsystem with auxiliaries, BLSS-v2, is shown in Fig. 4.23 as configured in LINGO. Note that we are using the constraint files for the three boilers and adding the additional constraints relating to the auxiliaries. Figure 4.24 shows the same model in Microsoft Excel; however, the figure is significantly reduced in size to fit the printed page, which unfortunately makes the details difficult to visualize; however, we can discern the model structure. Tab. 4.3 shows the optimum results for the base case. LINGO and Excel Solver give identical and expected results for all test cases.

```
MODEL: TITLE BOILER Subsystem with Auxiliaries Model: BLSS-v2: Base Case;
! ;
! Nath, "Industrial Process Plants: Global Optimization of Utility Systems", Chapter 4, (2023)
! ;
DATA:!inputs;         c_fo =18; c_lpfg =19; c_hpfg =20; c_elec =60;   ! energy costs, $/MWh;
                      mps_demand = 100; mps_reserve = 50;             ! MPS demand & reserve, tph;
                      hps_demand = 100; hps_reserve = 50;             ! HPS demand & reserve, tph;

   ! BL1 params;      bl1_bdr = 0.02;    bl1_fg_s=3; bl1_stm_l1 =70;   bl1_stm_hl =210;
                      bl1at_c = 8;       bl1_fg_c = 14.438801;         bl1_fg_d = 0.749099;
                      big = 999;         bl1_fg_e =-11.80915;          bl1_fg_f = 0.903498;
   ! BL2 params;      bl2_bdr = 0.02;    bl2_fg_s=4; bl2_stm_l1 =80;   bl2_stm_hl =240;
                      bl2am_c = 0.5;     bl2_fg_c = 15.809680;         bl2_fg_d = 0.7262998;
                                         bl2_fg_e =-10.107546;         bl2_fg_f = 0.8627063;
   ! BL3 params;      bl3_bdr = 0.02;    bl3_fg_s=5; bl3_stm_l1 =100;  bl3_stm_hl =300;
                      bl3am_c = 0.5;     bl3_fg_c = 19.19346044;       bl3_fg_d = 0.7372921;
                      bl3at_c = 5;       bl3_fg_e =-13.80971;          bl3_fg_f = 0.874805;
   ENDDATA

! Objective;          MIN = c_fo * BL1_FUEL + c_lpfg * (BL2_LPFG+BL3_LPFG)
                      + c_hpfg *(BL2_HPFG+BL3_HPFG) + c_elec * (bl2am_ele+bl3am_ele);

   ! BL1;             @FILE('C:\BookDG\LingoModels\BL1-v4.lng');
   [_BL1AT_EB1]       BL1AT_STM = bl1at_c * BL1_X;

   ! BL2;             @FILE('C:\BookDG\LingoModels\BL2-v1.lng');
   [_BL2AM_EB1]       BL2AM_ELE = bl2am_c * BL2_X;

   ! BL3;             @FILE('C:\BookDG\LingoModels\BL3-v1.lng');
   [_BL3AT_EB1]       BL3AT_STM = bl3at_c * BL3AT_X;
   [_BL3AM_EB1]       BL3AM_ELE = bl3am_c * BL3AM_X;
   [_BL3A_LGC]        BL3_X = BL3AM_X + BL3AT_X;
                      @BIN(Bl3AM_X);

! Steam demand & reserve constraint;
   [_MPS_RSV]         MPS_RSV = BL1_RSV;
   [_HPS_RSV]         HPS_RSV = BL2_RSV + BL3_RSV;
   [_MPS_DMND]        BL1_STM > mps_demand;
   [_MPS_RSRV]        MPS_RSV > mps_reserve;
   [_HPS_DMND]        BL2_STM + BL3_STM > hps_demand;
   [_HPS_RSRV]        HPS_RSV > hps_reserve;

END MODEL BLSS-v2
```

Fig. 4.23: Boilers subsystem model: BLSS-v2: in LINGO.

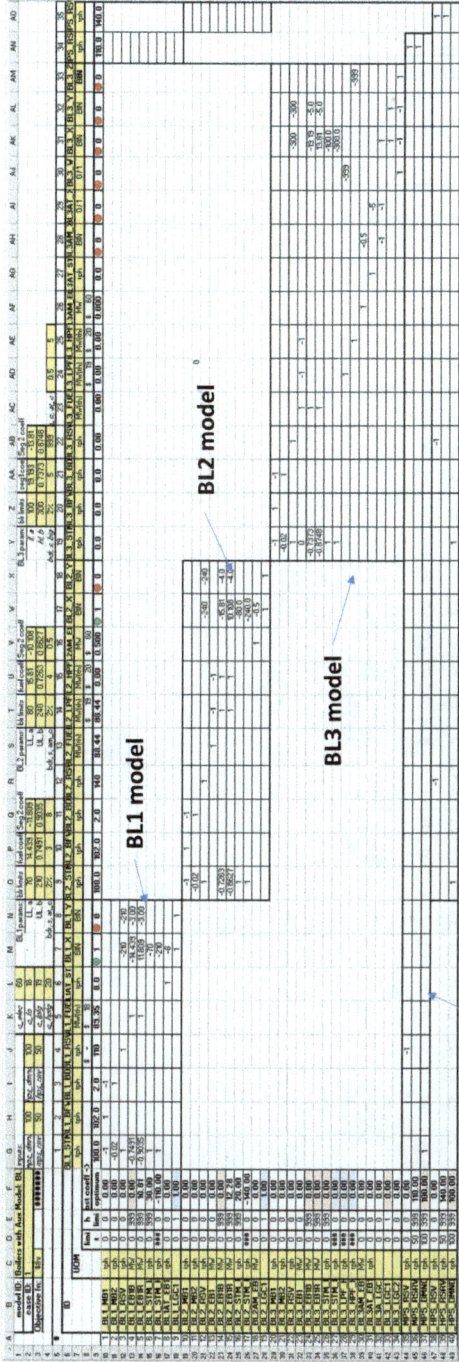

Fig. 4.24: Boiler subsystem model: BLSS–v2: Microsoft Excel model structure.

Tab. 4.3: Boilers with Aux subsystem model: BLSS-v2: Base case solution.

Boilers with Aux Subsystem Model: BLSS-v2						
Global optimal solution found.						
Objective value:					$ 3,318.6	
#	variable	UOM	cost coeff.		value	cost
1	BL1_STM	tph			100.0	$ -
2	BL1_BFW	tph			102.0	$ -
3	BL1_BD	tph			2.0	$ -
4	BL1_RSV	tph	$ -		110	$ -
5	BL1_FUEL	MW(th)	$ 18		89.35	$ 1,608.3
6	BL1AT_STM	tph			8.0	$ -
7	BL1_X	BIN		🟢	1	$ -
8	BL1_Y	BIN		🟠	0	$ -
9	BL2_STM	tph			100.0	$ -
10	BL2_BFW	tph			102.0	$ -
11	BL2_BD	tph			2.0	$ -
12	BL2_RSV	tph			140.0	$ -
13	BL2_FUEL	MW(th)			88.44	$ -
14	BL2_LPFG	MW(th)	$ 19		88.44	$ 1,680.4
15	BL2_HPFG	MW(th)	$ 20		0.00	$ -
16	BL2AM_ELE	MW	$ 60		0.500	$ 30.0
17	BL2_X	BIN		🟢	1	$ -
18	BL2_Y	BIN		🟠	0	$ -
19	BL3_STM	tph			0.0	$ -
20	BL3_BFW	tph			0.0	$ -
21	BL3_BD	tph			0.0	$ -
22	BL3_RSV	tph			0.00	$ -
23	BL3_FUEL	MW(th)			0.00	$ -
24	BL3_LPFG	MW(th)	$ 19		0.00	$ -
25	BL3_HPFG	MW(th)	$ 20		0.00	$ -
26	BL3AM_ELE	MW	$ 60		0.000	$ -
27	BL3AT_STM	tph			0.0	$ -
28	BL3AM_X	BIN		🟠	0	$ -
29	BL3AT_X	0 / 1		🟠	0	$ -
30	BL3_W	0 / 1		🟠	0	$ -
31	BL3_X	BIN		🟠	0	$ -
32	BL3_Y	BIN		🟠	0	$ -
33	BL3_Z	BIN		🟠	0	$ -
34	MPS_RSV	tph			110.0	$ -
35	HPS_RSV	tph			140.0	$ -

In the following section, we will model compressors; these devices are used to transport gases across longer distances and against large pressure drops.

4.9 Modeling compressors

Compressors are similar to pumps, except they transport compressible gases and hence have much faster dynamics. Compressors and pumps develop very high pressures at low flow rates that could lead to backflow. To mediate this problem for pumps, a bypass recirculation is commonly used. Conversely, compressors can become violently unstable at low flow rates, which could severely damage the equipment. At high flow rates, the operations becomes unstable as the flow enters the choked flow regime, which can again seriously damage the equipment.

The minimum stable flow is referred to as the surge point and the maximum is referred to as the stonewall point; within this operating window, performance is similar to a pump: the head drops as flow increases, efficiency goes through to a peak, and similar to pumps, power vs. flow can be adequately represented by a linear relationship.

Modeling a compressor is similar to that of a pump, except that there are strict operability limits; the low limit is usually 10% above the surge point and the high flow limit is usually 10% below the stonewall point. Figures 4.25 and 4.26 show the generic compressor model, C01-v1, in LINGO and Microsoft Excel, respectively.

The model is simple; we will test it with just one test case to confirm no apparent errors. Figure 4.26 shows the solution to the test problem; note that the two optimizers produce identical results.

```
MODEL:    TITLE Generic Compressor Model: C01-v1: Base Case;
! ;
! Nath, "Industrial Process Plants: Global Optimization of Utility Systems", Chap. 4, (2023).
! ;
 DATA:!inputs;     gas_demand = 8;         ! compressor inlet flow, kM/h;

     !params;      c01_sp_a = 0.132; c01_sp_b = 0.029; ! power[MW.sp] v gas flow[kM3/h] coeff;
                   c01_gas_ll = 3;   c01_gas_hl = 9;   ! gas flow limita, kM3/h;
  ENDDATA

  ! Objective;    MIN = C01_SP;

  ! Performance curves;
  [_C01_EB1]      C01_SP  = c01_sp_a * C01_X + c01_sp_b * C01_GAS;

  ! Capacity constraints;
  [_C01_GAS_LL]   C01_GAS > c01_gas_ll * C01_X;
  [_C01_GAS_HL]   C01_GAS < c01_gas_hl * C01_X;

  ! Inlet flow requirement;
  [_GAS_DMND]     C01_GAS > gas_demand;

  ! Integers;     @BIN( C01_X );

END MODEL C01-v1
```

Fig. 4.25: Generic compressor model: C01-v1: in LINGO.

	A	B	C	D	E	F	G	H	I	J
1		model ID:	C01-v1: generic compressor				inputs:	params:	big	999
2		case ID:	1				gas_dmnd	c01_gas_ll, hl	3	10
3		Objective fn:	MW.sp			0.364	8	c01_sp_a, b	0.1320	0.0290
4										
5	#						1	2	3	
6		ID					C01_GAS	C01_SP	C01_X	
7			UOM				kM3/h	MW.sp	BIN	
8				low	high	cost coeff ->		$ 1.0		
9				limit	limit	optimum	8.0	0.364	● 1	
10	1	_C01_EB1	MW.sp	0	0	0.0	-0.029	1	-0.132	
11	2	_C01_GAS_LL	kM3/h	0	999	5.0	1		-3.0	
12	3	_C01_GAS_HL	kM3/h	-999	0	-2.0	1		-10.0	
13	4	_GAS_DMND	kM3/h	8	999	8.0	1			

Fig. 4.26: Generic compressor model: C01-v1: in Microsoft Excel, showing test case #1 solution.

Compressors, like pumps, also require power from connected motors, turbines, or engines. A compressor could be powered by a dedicated motor, a dedicated turbine, a dedicated internal combustion engine, or a dual drive arrangement that allows a selection between the connected drivers. As discussed earlier in this chapter, modeling for each case is very similar to that for the pump. We will only discuss the case with an internal combustion engine drive for brevity. All model files are included in the digital media associated with this book.

A compressor with an internal combustion engine drive model combines the compressor model with the internal combustion engine model, and eliminates duplicate variables. Figures 4.27 and Fig. 4.28 show the LINGO model, and Fig. 4.29 shows the Excel model.

```
! Compressor C01 with ICE drive C01E: model constraints file: C01ED-v1.lng;

  ! C01;
  [_C01_EB1]        C01_SP  = c01_sp_a * C01_X + c01_sp_b * C01_GAS;
  [_C01_GAS_LL]     C01_GAS > c01_gas_ll * C01_X;
  [_C01_GAS_HL]     C01_GAS < c01_gas_hl * C01_X;

  ! C01E;
  [_C01E_EB1]       C01E_FG = c01e_fg_a * C01_X + c01e_fg_b * C01_SP;

  ! Integers;       @BIN( C01_X );
!;
```

Fig. 4.27: Generic compressor with ICE drive model constraints file: C01ED-v1.lng.

In the next chapter, we will discuss the modeling of turbogenerators, rotating equipment used exclusively to generate electrical power.

```
MODEL: TITLE C01ED-v1: Compressor with ICE drive: Base Case;
! ;
! R Nath, "Industrial Process Plants: Global Optimization of Utility Systems", Chapter 4, (2023);
! ;
DATA:!inputs;      gas_demand = 8;                              ! flow demand, kM3/h;
     !params;      c01_sp_a = 0.1320;    c01_sp_b = 0.0290;     ! SP [MW.sp] vs gas flow [kM3/h];
                   c01_gas_ll = 3;       c01_gas_hl = 10;       ! flow limits [kM3/h];
                   c01e_fg_a = 0.1161;   c01e_fg_b = 2.1242;    ! ICE fuel [MW.th] vs SP [MW.sp];

  ENDDATA

  ! Objective;     MIN = C01E_FG;

  ! constraints;   @FILE('C:\BookDG\LingoModels\C01ED-v1.lng');

  ! demand constraint;
  [_GAS_DMND]      C01_GAS > gas_demand;

END MODEL C01ED-v1
```

Fig. 4.28: Generic compressor with ICE drive model: C01ED-v1: in LINGO.

▲	A	B	C	D	E	F	G	H	I	J
1		model ID:	C01ED-v1: compressor w engine driver				inputs:	params:	*big*	999
2		case ID:	1				gas_dmnd	c01_gas_ll, hl	3	10
3		Objective fn:	MW.th			0.889	8	c01_sp_a, b	0.1320	0.0290
4								c01e_fg_a, b	0.1161	2.1242
5	#						1	2	3	4
6		ID					C01_GAS	C01_SP	C01_X	E01_FG
7			UOM				kM3/h	MW.sp	BIN	MW.th
8				low	high	cost coeff ->				$ 1.00
9				limit	limit	optimum	8.0	0.364	● 1	0.889
10	1	_C01_EB1	MW.sp	0	0	0.0	-0.03	1	-0.13	
11	2	_C01_GAS_LL	kM3/h	0	999	5.0	1		-3.0	
12	3	_C01_GAS_HL	kM3/h	-999	0	-2.0	1		-10.0	
13	4	_GAS_DMND	kM3/h	8	999	8.0	1			
14	5	_C01E_EB1	MW.th	0	0	0.0		-2.12	-0.12	1

Fig. 4.29: Generic compressor with ICE drive model: C01ED-v1: in Excel, showing test case #1 solution.

Summary

In this chapter, we discussed modeling equipment used to transport fluids. Pumps transport liquid streams such as boiler feed water and cooling water; fans and blowers transport gases such as boiler combustion air or boiler flue gas across tiny pressure drops, and compressors transport gases such as compressed air and fuel gas across large pressure drops. Such equipment, commonly called rotating equipment, is a consumer of shaft power that is produced by a dedicated electric motor, a dedicated steam turbine, a dedicated internal combustion engine, or a dual drive arrangement that permits choosing between two drives connected to the device.

This chapter presents MILP models for a generic pump powered by an electrical motor, a steam turbine, a dual drive, and a generic compressor powered by an internal combustion engine.

Nomenclature

Symbol	UOM	Description
BFW	–	Boiler Feed Water
bljam_c	MW	Boiler #j: Auxiliary motor: electric power consumption
bljat_c	tph	Boiler #j: Auxiliary turbine: steam usage
c01_sp_a	MW.sp	Compressor C01, shaft power consumption vs. gas flow: intercept
c01_sp_b	MWh.sp/kM3	Compressor C01, shaft power consumption vs. gas flow: slope
c01E_fg_a	MW.th	Compressor C01, engine: fuel gas consumption vs. shaft power: intercept
c01E_fg_b	MW.th/MW.sp	Compressor C01, engine: fuel gas consumption vs. shaft power: slope
c01E_gas_hl	kM3/h	Compressor C01: compressed gas: high limit
c01E_gas_ll	kM3/h	Compressor C01: compressed gas: low limit
e01_fg_a	MW.th	Engine E01, flue consumption vs. shaft power regression: intercept
e01_fg_b	MW.th/MW.sp	Engine E01, flue consumption vs. shaft power regression: slope
e01_sp_hl	MW.sp	Engine E01, shaft power production: high limit
e01_sp_ll	MW.sp	Engine E01, shaft power production: low limit
flo_demand	tph	Pump liquid flow demand
gas_demand	kM3/h	Compressed gas demand
m01_eta	–	Motor M01, efficiency
m01_sp_hl	MW.sp	Motor M01, shaft power production: high limit
m01_sp_ll	MW.sp	Motor M01, shaft power production: low limit
p01_des_ll	tph	Pump P01, inlet liquid flow: design low limit
p01_flo_hl	tph	Pump P01, inlet liquid flow: high limit
p01_sp_a	MW.sp	Pump P01, shaft power consumption vs inlet liquid flow: intercept
p01_sp_b	MW.sp/tph	Pump P01, shaft power consumption vs. inlet liquid flow: slope
p01m_eta	–	Pump P01, motor: efficiency
p01t_stm_a	tph	Pump P01, turbine: steam vs. shaft power: intercept
p01t_stm_b	tph/MW.sp	Pump P01, turbine: steam vs. shaft power: slope
sp_demand	MW.sp	Shaft power demand
t01_stm_a	tph	Turbine T01, inlet steam flow vs. shaft power regression: intercept
t01_stm_b	tph/MW.sp	Turbine T01, inlet steam flow vs. shaft power regression: slope
t01_stm_hl	tph	Turbine T01, inlet steam flow: high limit
t01_stm_ll	tph	Turbine T01, inlet steam flow: low limit
BLj_X	0/1	Boiler #j: operational status: 0 = OFF, 1=ON
BLjAM_ELE	MW	Boiler #j: Auxiliaries motor: electrical power consumption
BLjAT_STM	tph	Boiler #j: Auxiliaries turbine: steam usage
BLjM_X	0/1	Boiler #j: Auxiliaries motor: operational status: 0 = OFF, 1=ON
BLjT_X	0/1	Boiler #j: Auxiliaries turbine: operational status: 0 = OFF, 1=ON
C01_X	0/1	Compressor C01: operational status: 0 = OFF, 1=ON
C01E_FG	MW.th	Compressor C01: engine fuel consumption
E01_FG	MW.th	Engine E01, fuel consumption
E01_SP	MW.sp	Engine E01, shaft power production
E01_X	0/1	Engine E01: operational status: 0 = OFF, 1=ON
M01_ELE	MW	Motor M01, electric power consumption
M01_SP	MW.sp	Motor M01, shaft power production
M01_X	0/1	Motor M01: operational status: 0 = OFF, 1=ON
P01_FLO	tph	Pump P01, inlet liquid flow
P01_SP	MW.sp	Pump P01, shaft power requirement

(continued)

Symbol	UOM	Description
P01_X	0/1	Pump P01: operational status: 0 = OFF, 1=ON
P01M_ELE	MW	Pump P01, motor: electric power usage
P01M_SP	MW.sp	Pump P01, motor: shaft power produced
P01M_X	0/1	Pump P01: motor: operational status: 0 = OFF, 1=ON
P01T_SP	MW.sp	Pump P01, turbine: shaft power produced
P01T_STM	tph	Pump P01, turbine: steam usage
P01T_X	0/1	Pump P01: turbine: operational status: 0 = OFF, 1=ON
T01_SP	MW.sp	Turbine T01, shaft power production
T01_STM	tph	Turbine T01, steam usage
T01_X	0/1	Turbine T01: operational status: 0 = OFF, 1=ON

References

[1]   "Electric Motor Efficiency under Variable Frequencies and Loads", http://www.itrc.org/reports/pdf/r06004.pdf (2006).

[2]   North American Electric Company, web site: https://www.naemotors.com/

[3]   Ekwonu, M. C., S. Perry and E. A. Oyedoh. "Modelling and simulation of gas engines using aspen HYSYS", Journal of Engineering Science and Technology Review, 6(3):1–4, (2013).

[4]   "What is a Centrifugal Pump Minimum Flow Bypass Line?" Blog article, https://blog.craneengineering.net/what-is-a-centrifugal-pump-minimum-flow-bypass-line

5 Modeling turbogenerators

A turbogenerator comprises a turbine coupled to an electrical generator. Steam and gas turbogenerators are two common turbogenerators in process plant utility systems. This chapter will discuss the modeling of turbogenerators and develop MILP models for commonly used turbogenerators.

Steam turbogenerators are more common and come in a variety of configurations. Two commonly used configurations are [1]:
– Non-condensing, back pressure steam turbogenerators (STG), and
– Condensing, extraction steam turbogenerators (CTG)

Gas turbogenerators (GTG) are typically large machines with sizable investments. They come in various sizes, producing tens to hundreds of megawatts of power. Two commonly used configurations are:
– Simple-cycle, and
– Combined-cycle

In a single-cycle operation, the GTG exhaust heat is not recovered. In contrast, in a combined-cycle operation, the GTG exhaust heat is recovered to produce high-pressure steam that is passed through an STG or a CTG. In this book, we will discuss MILP modeling of only the simple-cycle operation, though the techniques presented also apply to modeling of a combined-cycle operation.

The following sections will discuss and develop models for the three types of turbogenerators mentioned above.

5.1 Modeling non-condensing, back pressure, steam turbogenerators

A non-condensing, back pressure, steam turbogenerator, STG, takes high-pressure steam and discharges it to a lower pressure; the expansion of the high-pressure steam spins the turbine blades on a shaft connected to a generator, which produces electrical power.

The classical definition of steam turbine efficiency, called thermal efficiency is defined for a closed thermodynamic cycle and is the conversion efficiency of inlet steam thermal energy to the output mechanical energy of the rotating shaft; it is the ratio of the output shaft work to the inlet steam energy [2]. The thermal efficiency of STGs calculated this way is relatively low, typically in the 30% range, which may make you think that STGs are inefficient.

However, this is an incorrect assessment in the context of a process plant utility system that utilizes the discharged steam and avoids the corresponding generation in

https://doi.org/10.1515/9783111020679-006

a boiler. A more appropriate assessment in such situations is to assess the STG operation incrementally, i.e., compare the system wide optimal operating cost, with and without the STG. Analyzed this way, the incremental efficiency of STGs could be very high. So, in a well-balanced utility system, STGs are highly efficient [3], which is a reason for their popularity.

Modeling the turbine component of an STG is the same as modeling a generic turbine discussed in Chapter 4, and modeling the electrical generation component is straightforward in the form of a performance equation. Thus, a steam turbine produces shaft power and converts it to electrical energy in the generator. Since shaft power is an internal variable in the turbogenerator, it need not be an explicit variable in the optimization model; instead, we can replace it with the variable representing the electrical power generation.

From a modeling perspective, an STG has two continuous and a binary variable. Continuous variables are **STG_STM** and **STG_ELE,** representing the inlet steam flow and the output electrical power generation, respectively, and the binary variable **STG_X** indicates its ON/OFF status. The equipment manufacturer provides the performance curve between the inlet steam flow and power generation; alternatively, it could be obtained by a plant test for existing equipment. This relationship in the normal operating range is reasonably linear and can be expressed by the following equation.

$$\textbf{STG\_STM} = stg\_stm\_a*\textbf{STG\_X} + stg\_stm\_b*\textbf{STG\_ELE} \qquad (5.1)$$

An STG could also be operated in a standby mode, sometimes called "slow roll" mode, in which case a small flow of steam, *stg_stm_s,* is maintained to keep the STG warm and ready to produce power on short notice, but there is no power generation. Adding the standby capability to the STG model would require the introduction of an additional binary variable, **STG_Y**, to indicate its standby operation. The modeling of standby operations is similar to that discussed in section 3.1.5.

Depending on the energy costs, if the credit for power is relatively low, it may not be economical to produce power, in which case, the model would generate the trivial solution of zero steam usage and zero power production.

Figures 5.1 and 5.2 show the complete model for an STG, as configured in LINGO. The model is simple. STG should produce power only when the economics is favorable; at low electric energy cost, it is expected to be turned OFF and turned ON at higher power cost. We will test the model for sensitivity to electric power cost. Table 5.1 shows the 4 cases we will use to test the model. Figure 5.3 shows the STG model in Microsoft Excel Solver and the results of test problem #1.

Note that both optimizers produce identical and expected results. The following section will develop a model for a condensing extraction steam turbogenerator, CTG.

Tab. 5.1: Non-condensing back pressure steam turbogenerator model, STG-v1: test cases.

#	Attribute	Symbolic name	Value	UOM	Expected key results
all	steam cost	c_stm	5	$/ton	
1	power cost	c_elec	40	$/MWh	**STG_ELE** = 0, **STG_X** = 0, obj fn = 0
2	power cost	c_elec	60	$/MWh	**STG_ELE** = 0, **STG_X** = 0, obj fn = 0
3	power cost	c_elec	80	$/MWh	**STG_ELE** = 0, **STG_X** = 0, obj fn = 0
4	power cost	c_elec	100	$/MWh	**STG_ELE** = max, **STG_X** = 1, obj fn < 0

```
! Non-condensing Back Pressure Turbo-generator STG: model constraints file: STG-v1.lng;

 ! energy balance;
 [_STG_EB1]        STG_STM = stg_stm_a *STG_X + stg_stm_b *STG_ELE + stg_stm_s *STG_Y;

 ! capacity limits;
 [_STG_STM_HL]     STG_STM < stg_stm_hl * STG_X + stg_stm_s *STG_Y;
 [_STG_ELE_LL]     STG_ELE > stg_ele_ll * STG_X;
 [_STG_ELE_HL]     STG_ELE < stg_ele_hl * STG_X;

 ! Logic;
 [_STG_LGC1]       STG_X + STG_Y < 1;

 ! Integers;      @BIN(STG_X);      @BIN(STG_Y);
!;
```

Fig. 5.1: Non-condensing back pressure steam turbo-generator model constraints file: STG-v1.lng.

```
MODEL:    TITLE Non-condensing Back Pressure Model: STG-v1 : Base Case;
! :
! Nath, "Industrial Process Plants: Global Optimization of Utility Systems", Chapter 5, (2023);
! ;
DATA:!inputs;     c_stm = 5; c_elec = 40;              ! costs;
     !params;     stg_stm_a = 8; stg_stm_b = 14;        ! inlet steam vs power gen;
                  stg_stm_s = 4;                        ! standby inlet steam , tph;
                  stg_stm_hl = 50; stg_ele_ll = 1; stg_ele_hl = 3;! capacity limits;
ENDDATA

  ! Objective;    MIN = c_stm * STG_STM - c_elec * STG_ELE;

  ! constraints;  @FILE('C:\BookDG\LingoModels\STG-v1.lng');

END MODEL STG-v1
```

Fig. 5.2: Non-condensing back pressure steam turbo-generator model, STG-v1: in LINGO.

5.2 Modeling condensing extraction turbogenerators

A condensing extraction steam turbogenerator (CTG) is similar to an STG, except that it has multiple stages; each stage discharges a part of the incoming steam and passes on the remainder to the next stage. In general, the first stage intakes high-pressure steam and discharges part of it to a medium-pressure header; the next stage expands the incoming medium-pressure stream and discharges a part of it to a low-pressure header, and so on. The last stage discharges to a sub-ambient pressure; this discharge

	A	B	C	D	E	F	G	H	I	J	K	L
1		model ID:	STG-v1: non-condensing back pres. STG				inputs:		params:		stg_stm_hl	50
2		case ID:	1				c_elec	40	stg_stm_a	8.00	stg_ele_ll	1
3		Objective fn:	$/hr			$	c_stm	5	stg_stm_b	14.00	stg_ele_hl	3
4									stg_stm_s	4.00	big	999
5	#						1	2	3	4		
6		ID					STG_STM	STG_ELE	STG_X	STG_Y		
7			UOM				tph	MW	BIN	BIN		
8				low	high	cost coeff ->	$ 5.00	$ (40.0)				
9				limit	limit	optimum	0.00	0.000	● 0	● 0		
10	1	_STG_EB1	tph	0	0	0.00	1	-14.0	-8.0	-4.0		
11	2	_STG_STM_HL	tph	-999	0	0.00	1		-50	-4.0		
12	3	_STG_ELE_LL	MW	0	999	0.00		1	-1.0			
13	4	_STG_ELE_HL	MW	-999	0	0.00		1	-3.0			
14	5	_STG_LGC1	-	0	1	0.00			1	1		

Fig. 5.3: Non-condensing back pressure steam turbo-generator model, STG-v1: in Microsoft Excel.

is only partially condensed, and is fully condensed in an integral condenser unit using the plant cooling water. The expansion of steam in each stage contributes to the spinning of the turbine blades on a common shaft connected to an electrical generator that produces electrical power.

In general, the overall classical thermal efficiency of CTGs is higher than those of back-pressure STGs, typically in the low 40%; however, the incremental efficiency could be lower than those of STGs. However, CTGs provide flexibility in balancing multiple steam headers and have the capability of more significant power generation [3].

From a modeling perspective, for normal operation, a CTG with N stages has N + 3 continuous variables and a binary variable. The continuous variables represent the high-pressure inlet steam to the first stage, the discharge steam flows from each stage, condenser cooling water flow, and the electrical power generation. The binary variable represents its ON/OFF state.

In this section, we will consider a three-stage CTG with continuous variables **CTG_HPS, CTG_MPS, CTG_LPS, CTG_CND, CTG_CW** and **CTG_ELE,** representing the inlet HP steam flow, first-stage MP steam extraction flow, second-stage LP steam extraction flow, third-stage partially condensed steam flow, condenser cooling water flow and the output power, respectively. The third-stage exhaust is fully condensed in a condenser unit, integral to the CTG. The condenser uses cooling water that comes from the cooling water subsystem; the amount of cooling water, **CTG_CW**, is directly proportional to the **CTG_CND** flow, with a proportionality constant, ctg_cw_b. The proportionality constant could be calculated by enthalpy balance around the condenser or by empirical observations.

The equipment manufacturer usually provides the performance curves relating to inlet steam flow, extraction flows, and power generation in a graphical or tabular form. A linear correlation such as the following equation adequately represents the relationship.

$$\mathbf{CTG\_ELE} = ctg\_ele\_a + ctg\_ele\_b * \mathbf{CTG\_HPS} + ctg\_ele\_c * \mathbf{CTG\_MPS} + ctg\_ele\_d * \mathbf{CTG\_LPS}$$

$$(5.2)$$

Rearranging the above, we get the following equation:

$$\mathbf{CTG\_HPS} = ctg\_hps\_a + ctg\_hps\_b * \mathbf{CTG\_ELE} + ctg\_hps\_c * \mathbf{CTG\_MPS} + ctg\_hps\_d * \mathbf{CTG\_LPS}$$

$$(5.3)$$

We have encountered similar relationships in modeling boilers and turbines. Like boilers and STGs, CTGs can also operate in a standby mode, also called "slow roll" mode, in which case a small flow of steam, *ctg_stm_s*, is maintained, but no power is generated; this keeps the CTG warm and ready to produce power on short notice. Adding a standby capability would require an additional binary variable, **CTG_Y**; so, there will be two binary variables **CTG_X** and **CTG_Y** to indicate the on/off status and standby status, respectively.

Figures 5.4 and 5.5 show the complete model for the CTG in LINGO.

To test this model, we will simplify the objective function by ignoring the credits for MPS, LPS, and Condensate flows, as that would not be necessary in the system-wide model. This model is relatively simple; CTG should produce power only when the economics is favorable. We will test the model for the sensitivity to electric power cost. Table 5.2 shows the 4 cases we will use to test the model. Figure 5.6 shows the CTG model in Microsoft Excel Solver and the results for test problem #1.

Tab. 5.2: Condensing steam turbogenerator model, CTG-v1: test cases.

#	Attribute	Symbolic name	Value	UOM	Expected key results
all	steam cost	*c_stm*	10	$/ton	
1	power cost	*c_elec*	40	$/MWh	**CTG_ELE = 0, CTG_X = 0**, obj fn = 0
2	power cost	*c_elec*	60	$/MWh	**CTG_ELE = 0, STG_X = 0**, obj fn = 0
3	power cost	*c_elec*	80	$/MWh	**CTG_ELE > 0, STG_X = 1**, obj fn < case 2
4	power cost	*c_elec*	100	$/MWh	**CTG_ELE >** case 3, **STG_X = 1**, obj fn < case 3

Note that both solvers give identical and expected results.

In the next section, we will develop a model for a gas turbogenerator, GTG.

5.3 Modeling gas turbogenerators

A gas turbogenerator (GTG) comprises three sections: an integral air compressor, a combustion chamber, and a turbine. Combustion air is compressed to high pressure in the integral air compressor, combined with the fuel stream, and then combusted in a high-pressure chamber; the expansion of the resulting high-temperature, high-

```
! Multi-stage Condensing Turbo-generator CTG: model constraints file: CTG-v1.lng;

    ! Mass and energy balances;
    [_CTG_MB1]      CTG_HPS = CTG_MPS + CTG_LPS + CTG_CND;
    [_CTG_EB1]      CTG_HPS = ctg_hps_a * CTG_X + ctg_hps_b * CTG_ELE
                            + ctg_hps_c * CTG_MPS + ctg_hps_d * CTG_LPS
                            + ctg_hps_s * CTG_Y;

    ! Capacity limits;
    [_CTG_HPS_LL]   CTG_HPS > ctg_hps_ll * CTG_X + ctg_hps_s *CTG_Y;
    [_CTG_HPS_HL]   CTG_HPS < ctg_hps_hl * CTG_X + ctg_hps_s *CTG_Y;
    [_CTG_MPS_HL]   CTG_MPS < ctg_mps_hl * CTG_X;
    [_CTG_LPS_HL]   CTG_LPS < ctg_lps_hl * CTG_X;
    [_CTG_CND_LL]   CTG_CND > ctg_cnd_ll * CTG_X + ctg_hps_s *CTG_Y;
    [_CTG_CND_HL]   CTG_CND < ctg_cnd_hl * CTG_X + ctg_hps_s *CTG_Y;
    [_CTG_CW]       CTG_CW  = ctg_cw_b * CTG_CND;
    [_CTG_ELE_LL]   CTG_ELE > ctg_ele_ll * CTG_X;
    [_CTG_ELE_HL]   CTG_ELE < ctg_ele_hl * CTG_X;

    ! Logic;
    [_CTG_LGC1]     CTG_X + CTG_Y < 1;

    ! Integers;    @BIN(CTG_X);    @BIN(CTG_Y);
!;
```

Fig. 5.4: Condensing steam turbogenerator model constraints file: CTG-v1.lng.

```
MODEL:   TITLE Condensing TG Model: CTG-v1;
!
! Nath, "Industrial Process Plants: Global Optimization of Utility Systems", Chapter 5, (2023).;

    DATA:
        !inputs;    c_stm = 10; c_elec = 40;                    ! costs;

        !params;    ctg_hps_a = 16.997;   ctg_hps_b = 5.4330;   ! HP steam coefficients;
                    ctg_hps_c = 0.7859;   ctg_hps_d = 0.5605;
                    ctg_hps_s = 10;       ctg_cw_b = 0.04;      ! stdby steam, cw coeff;
                    ctg_hps_ll = 50;      ctg_hps_hl = 250;     ! capacity limits;
                    ctg_mps_hl = 180;     ctg_lps_hl = 90;
                    ctg_cnd_ll = 10;      ctg_cnd_hl = 60;
                    ctg_ele_ll = 3;       ctg_ele_hl = 13;
    ENDDATA

    ! Objective;    MIN = c_stm * CTG_HPS - c_elec * CTG_ELE;

    ! constraints;  @FILE('C:\BookDG\LingoModels\CTG-v1.lng');

END MODEL CTG-v1
```

Fig. 5.5: Condensing steam turbogenerator model, CTG-v1: in LINGO.

pressure flue gas spins the rotating blades that drive both the integral air compressor and the electrical generator [4].

In general, the thermal efficiency of simple-cycle CTGs is low, typically 30%. However, combined-cycle GTGs that produce very high-pressure steam and let it down through steam turbo-generators can achieve high incremental efficiencies and provide an acceptable payback on a significant investment [5].

A	B	C	D	E	F	G	H	I	J	K	L	M	N	O
1	model ID:	CTG-v1: condensing turbogenerator				inputs:		ctg_hps_a	16.9970	ctg_hps_s	10	ctg_hps ll, hl	50	250
2	case ID:	1				c_elec	40	ctg_hps_b	5.4330	ctg_ele_ll	3.0	ctg_mps_hl		180
3	Objective fn:	$/hr			50.00	c_stm	10	ctg_hps_c	0.7859	ctg_ele_hl	13.0	ctg_lps_hl		90
4						big	999	ctg_hps_d	0.5605	ctg_cw_b	0.04	ctg_cnd ll, hl	10	60
#						1	2	3	4	5	6	8	9	
6	ID					CTG_HPS	CTG_MPS	CTG_LPS	CTG_CND	CTG_ELE	CTG_CW	CTG_X	CTG_Y	
7		UOM	low limit	high limit		tph	tph	tph	tph	MW	ktph	BIN	BIN	
8					cost coeff ->	$ 10.00				$ (40.00)				
9					optimum	0.0	0.0	0.0	0.0	0.000	0.00	0	0	
10	1	CTG_MB1	tph	0	0	0.00	1	-1	-1	-1					
11	2	CTG_EB1	tph	0	0	0.00	1	-0.8	-0.6		-5.4				
12	3	CTG_HPS_LL	tph	0	999	0.00	1						-17.0	-10.0	
13	4	CTG_HPS_HL	tph	-999	0	0.00	1						-50.0	-10.0	
14	5	CTG_MPS_HL	tph	-999	0	0.00		1					-250.0	-10.0	
15	6	CTG_LPS_HL	tph	-999	0	0.00			1				-180.0		
16	7	CTG_CND_LL	tph	0	999	0.00				1			-90.0	-10.0	
17	8	CTG_CND_HL	tph	-999	0	0.00				1			-10.0	-10.0	
18	9	CTG_ELE_LL	MW	0	999	0.00					1		-60.0		
19	10	CTG_ELE_HL	MW	-999	0	0.00					1		-3.0		
20	11	CTG_LGC1	-	0	1	0.00				-0.04			-13.0	1	
21	12	CTG_CW	ktph	0	0	0.00						1	1		

Fig. 5.6: Condensing steam turbogenerator model, CTG-v1: in Microsoft Excel, showing test case #1 solution.

From a modeling perspective, a GTG has two continuous and binary variables. The continuous variables are **GTG_FG** and **GTG_ELE,** representing the fuel gas flow and the output power, respectively, and a binary variable **GTG_X** indicates its on/off status.

The equipment manufacturer provides performance curves at design conditions, relating the fuel flow and power generation as graphs. They contain sufficient information to get a functional relationship between the fuel consumed and the power generated, such as shown by the dots in Fig. 5.7. Note that the performance of GTG is sensitive to ambient temperature; it generates more power at higher efficiencies at lower ambient temperatures. The manufacturer also provides this sensitivity information as separate graphs to correct the performance curve for non-design conditions. We have used such information for the performance curve in Fig. 5.7.

The performance curve shown by dots in Fig. 5.7 is somewhat nonlinear, especially toward the lower half of the curve; so linearization is required to keep the problem linear and thus ensure global optimum. We will do a piecewise linearization of the performance curve. In this case, a three-segment curve seems adequate; it covers the entire range with less than 1.6% error. In essence, the performance curve is split into three regions. Figure 5.7 also shows the regression coefficients of each linear segment.

These segments have increasing slopes, similar to the boiler fuel vs. steam curve. Note that curves in other regions underestimate the fuel consumption in that region, as shown by the dashed extension lines. Since fuel is a cost, three ≥ inequality constraints will represent the performance curve without using binary variables, resulting in an efficient formulation.

Unlike other equipment, GTGs can operate over the entire range – from zero to maximum power generation; so, standby operations need not be explicitly modeled.

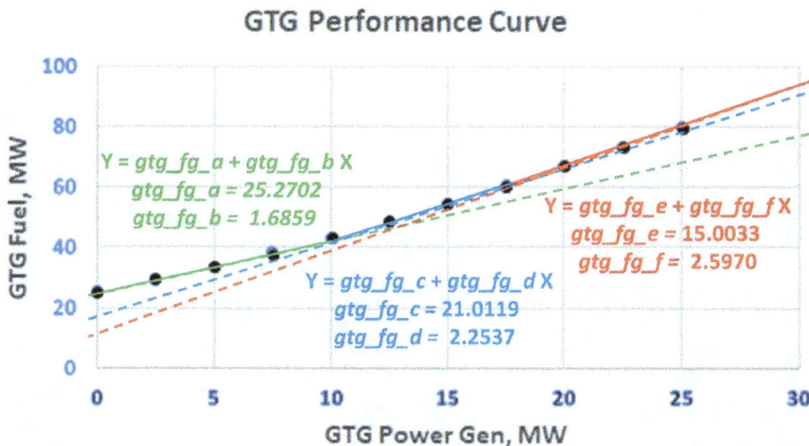

GTG Performance Curve

$Y = gtg\_fg\_a + gtg\_fg\_b\,X$
$gtg\_fg\_a = 25.2702$
$gtg\_fg\_b = 1.6859$

$Y = gtg\_fg\_e + gtg\_fg\_f\,X$
$gtg\_fg\_e = 15.0033$
$gtg\_fg\_f = 2.5970$

$Y = gtg\_fg\_c + gtg\_fg\_d\,X$
$gtg\_fg\_c = 21.0119$
$gtg\_fg\_d = 2.2537$

Fig. 5.7: Gas turbogenerator performance curve.

Gas turbine operations are analogous to other internal combustion engines, such as those in automobiles; standby operation is equivalent to engine idling.

Modeling the GTG with two continuous and one binary variable will be straight-forward. It would be similar to the STG model. Figures 5.8 and 5.9 show the complete model for GTG in LINGO. The model is simple. GTG should produce power when the economics is favorable; at low electric costs, GTG should be OFF, but turned ON for higher power costs. We will test the model for its sensitivity to electric power cost. Table 5.3 shows the 4 test cases along with expected results. Figure 5.10 shows the CTG model in Microsoft Excel Solver and the results for the test problem #1.

Tab. 5.3: Gas turbogenerator, simple-cycle model, GTG-v1: test cases.

#	Attribute	Symbolic name	Value	UOM	Expected key results
all	steam cost	*c_stm*	20	$/ton	
1	power cost	*c_elec*	40	$/MWh	**GTG_ELE** = 0, **GTG_X** = 0, obj fn = 0
2	power cost	*c_elec*	60	$/MWh	**GTG_ELE** = 0, **GTG_X** = 0, obj fn = 0
3	power cost	*c_elec*	80	$/MWh	**GTG_ELE** = max, **GTG_X** = 1, obj fn < 0
4	power cost	*c_elec*	100	$/MWh	**GTG_ELE** = max, **GTG_X** = 1, obj fn < case 3

```
! Gas Turbo-genereator without heat recovry GTG: model constraints file: GTG-v1.lng;

  ! energy balance;
  [_GTG_EB1G]      GTG_FG > gtg_fg_a * GTG_X + gtg_fg_b * GTG_ELE;
  [_GTG_EB1B]      GTG_FG > gtg_fg_c * GTG_X + gtg_fg_d * GTG_ELE;
  [_GTG_EB1R]      GTG_FG > gtg_fg_e * GTG_X + gtg_fg_f * GTG_ELE;

  ! capacity limits;
  [_GTG_ELE_HL]    GTG_ELE < gtg_elec_hl * GTG_X;

  ! Integers:      @BIN(GTG_X);
!;
```

Fig. 5.8: Gas turbogenerator, simple cycle, LINGO model constraint file, GTG-v1.lng.

```
MODEL:    TITLE Gas Turbo Generator Simple Cycle Model: GTG-v1;
! ;
! Nath, "Industrial Process Plants: Global Optimization of Utility Systems", Chapter 5, (2023).;
! ;
  DATA:
      !inputs;     c_fuel = 20; c_elec = 40;          ! costs;

      !para:;      gtg_fg_a = 25.2701; gtg_fg_b = 1.6859;  ! fuel vs power coeff, low range;
                   gtg_fg_c = 21.0119; gtg_fg_d = 2.2537;  ! fuel vs power coeff, mid range;
                   gtg_fg_e = 15.0033; gtg_fg_f = 2.5970;  ! fuel vs power coeff, hi range;
                   gtg_elec_hl = 25;                       ! power genenration, hi limit, MW;
  ENDDATA

  ! Objective;    MIN = c_fuel * GTG_FG - c_elec * GTG_ELE;

  ! constraints;  @FILE('C:\BookDG\LingoModels\GTG-v1.lng');

END MODEL GTG-v1
```

Fig. 5.9: Gas turbogenerator, simple-cycle, model, GTG-v1: in LINGO.

	A	B	C	D	E	F	G	H	I	J	K
1		model ID:	GTG-v1: GTG Simple cycle				inputs:		params:	gtg_ele_hl	25
2		case ID:	1				c_elec	40	gtg_fg_a, b	25.27	1.69
3		Objective fn:	$/hr			$0.00	c_fuel	20	gtg_fg_c, d	21.01	2.25
4							big	999	gtg_fg_e, f	15.00	2.60
5	#						1	2	3		
6		ID					GTG_FG	GTG_ELE	GTG_X		
7			UOM				tph	MW	BIN		
8				low	high	cost coeff ->	$ 20.00	-$40.00			
9				limit	limit	optimum	0.00	0.000	0		
10	1	_GTG_EB1G	tph	0	999	0.00	1	-1.7	-25.3		
11	2	_GTG_EB1B	tph	0	999	0.00	1	-2.3	-21.0		
12	3	_GTG_EB1R	MW	0	999	0.00	1	-2.6	-15.0		
13	4	_GTG_ELE_HL	MW	-999	0	0.00		1	-25.0		

Fig. 5.10: Gas turbogenerator, simple-cycle, model, GTG-v1: in Microsoft Excel, showing test case #1 solution.

Summary

This chapter discussed the modeling of turbogenerators used to generate electrical power. A turbo generator comprises a turbine coupled to an electrical generator. Commonly used turbo generators in process plant utility systems are steam turbogenerators and gas turbogenerators. Steam turbogenerators come in two varieties: non-condensing steam turbogenerators (STG) and condensing turbogenerators (CTG). Gas turbogenerators (GTG) come in two configurations: simple-cycle, where the exhaust heat is not recovered, and combined-cycle, where the exhaust heat generates additional steam and power. In this chapter, we have developed MILP models for generic STG, CTG, and a simple cycle GTG.

Nomenclature

Symbol	UOM	Description
c_elec	$/MWh	cost: electric power
c_stm	$/t	cost: steam
CTG	–	Condensing steam turbo-generator
ctg_ele_a	MW	CTG: power gen vs. steam flow regression: intercept
ctg_ele_b	MW/tph	CTG: power gen vs. HPS steam flow regression: slope
ctg_ele_c	MW/tph	CTG: power gen vs. MPS steam flow regression: slope
ctg_ele_d	MW/tph	CTG: power gen vs. LPS steam flow regression: slope
ctg_hps_a	tph	CTG: hps vs. other variables: intercept
ctg_hps_b	tph/tph	CTG: hps vs. power gen regression: slope
ctg_hps_c	tph/tph	CTG: hps vs. MPS steam flow regression: slope
ctg_hps_d	tph/tph	CTG: hps vs. LPS steam flow regression: slope

(continued)

Symbol	UOM	Description
GTG	–	Gas turbogenerator
STG	–	Non-condensing steam turbogenerator
stg_ele_hl	MW	STG: power generation: high limit
stg_ele_ll	MW	STG: power generation: low limit
stg_stm_a	tph	STG: inlet steam vs. power regression: intercept
stg_stm_b	tph/MW	STG: inlet steam vs. power regression: slope
stg_stm_hl	tph	STG: inlet steam: high limit
stg_stm_s	tph	STG: standby mode: inlet steam flow
CTG_ELE	MW	CTG: electric power generation
CTG_HPS	tph	CTG: inlet HP steam flow
CTG_HPS	tph	CTG: outlet MP steam flow
CTG_HPS	tph	CTG: outlet LP steam flow
CTG_X	0/1	CTG: operational status: 0 = OFF, 1=ON
CTG_Y	0/1	CTG: standby status: 1 = standby, 0 otherwise
GTG_ELE	MW	GTG: electric power generation
GTG_FG	MW.th	GTG: fuel consumption
GTG_X	0/1	GTG: operational status: 0 = OFF, 1=ON
STG_ELE	MW	STG: electric power generation
STG_STM	tph	STG: inlet steam flow
STG_X	0/1	STG: operational status: 0 = OFF, 1=ON
STG_Y	0/1	STG: standby status: 1 = standby, 0 otherwise

References

[1] Steam turbine flow & operation, Processing Magazine, (July 329, 2016), https://www.processingmagazine.com/home/article/15586711/steam-turbine-flow-operation
[2] Thermal efficiency, Wikipedia article, https://en.wikipedia.org/wiki/Thermal_efficiency
[3] Consider Installing High Pressure Boilers with Backpressure Turbine-Generators, Advanced Manufacturing, US Department of Energy, https://www.energy.gov/sites/prod/files/2014/05/f16/steam22_backpressure.pdf
[4] How Turbine Power Plants Work, Office of Fossil Energy and Carbon Management, US Department of Energy, https://www.energy.gov/fecm/how-gas-turbine-power-plants-work.pdf
[5] Kirana, Marten Y and Putra AI, "Investment Analysis of Gas Turbine Combined Cycle to reduce the production cost of PT XYZ using Monte Carlo Simulation", South East Asia Journal of Contemporary Business, Economics and Law, 24(5), 2021.

6 Modeling the steam subsystem

So far in the book, we have developed models for some of the components of a utility system, viz., boilers, pumps, compressors, and turbogenerators. We are ready to model the steam subsystem shown in Fig. 1.3; however, there are a few missing components, as shown below:
– Pressure reducing valves (PRVs),
– Blowdown flash unit,
– Deaerator, and
– Headers.

In the following sections, we will discuss and develop models for each of these components, and in the final section, we will create an MILP model for the entire steam subsystem.

6.1 Modeling pressure reducing valves (PRV)

Pressure reducing valves (PRV), also called Pressure reducing stations or Let down stations, are devices for reducing pressure in a material header. They connect a high-pressure material header to a lower-pressure material header. They come in three varieties:
– Pressure reducing valves without de-superheating,
– Pressure reducing valves with de-superheating, and
– Vents and flares.

We will discuss each of them in the following subsections.

6.1.1 Modeling PRVs without de-superheating

A pressure reducing valve without de-superheating, is an automatic valve connecting a higher-pressure material header to a lower-pressure material header. It is installed as a safety device to ensure that the pressure of the high-pressure header stays within the prescribed high pressure limit; if the pressure is too high, it opens and lets down the high-pressure material to the lower-pressure header, and it closes after the pressure has returned to the desired value.

In modeling terms, it is straightforward; it has only one variable for the flow. Consider PRV2 in the sample utility system; steam flow through it is **PRV2_STM**; it flows from the medium-pressure header to the low–pressure header. A utility system may have more than one such device to accommodate the maximum expected flow; from a modeling perspective, its capacity would not be limited, so there would be no con-

https://doi.org/10.1515/9783111020679-007

straints on **PRV2_STM**. However, an appropriate constraint, such as **PRV2_STM** ≤ *maxflow*, could be applied, if it were limiting. Valves may also leak; so a low flow limit, such as **PRV2_STM** ≥ *minflow*, would be appropriate. We will assume no flow limits in our model, and the model will comprise a single unconstrained variable.

The fuel subsystem in the sample utility system in Fig. 1.4 shows another such device, PRVFG, letting down fuel gas from the high-pressure header, HPFG, to the low-pressure header, LPFG.

6.1.2 Modeling PRV with de-superheating

A pressure reducing valve with de-superheating is similar to the one without de-superheating, except this one also controls the temperature of the outlet steam to a desired lower temperature by mixing the inlet steam with an adequate amount of boiler feed water; incidentally, doing so also generates additional lower pressure steam. PRV1 in the sample utility system is one such valve.

In modeling terms, it has three variables, **PRV1_HPS, PRV1_BFW,** and **PRV1_MPS**, representing the inlet high-pressure steam flow, de-superheating water flow, and the outlet medium-pressure steam flow. Such valves are adequately sized to accommodate all expected flows; so, applying flow constraints is unnecessary. In our modeling, we will assume that such is the case.

The three variables are related to each other by mass and energy balances. The following constraint expresses the conservation of mass:

$$\textbf{PRV1\_HPS} + \textbf{PRV1\_BFW} = \textbf{PRV1\_MPS} \qquad (\_PRV1\_MB1)$$

For conservation of energy, we make an enthalpy balance, such as below:

$$h\_hps * \textbf{PRV1\_HPS} + h\_hpbfw * \textbf{PRV1\_BFW} = h\_mps * \textbf{PRV1\_MPS} \qquad (6.1)$$

Where *h_hps, h_mps,* and *h_hpbfw* are the enthalpies of the high-pressure steam, medium-pressure steam, and high–pressure boiler feed water, respectively.

Dividing eq. (6.1) by h_hps gives the following constraint:

$$\textbf{PRV1\_HPS} + prv\_a * \textbf{PRV1\_BFW} = prv\_b * \textbf{PRV1\_MPS} \qquad (\_PRV1\_EB1)$$

The model comprises the two constraints mentioned above.

For testing this model, we must add a demand for MPS steam for testing to avoid getting the trivial zero-flow solution. With three variables, two equations, and one flow specified, there would be zero degrees of freedom and a unique solution. So, any arbitrary objective function would do, for example, to minimize **PRV1_HPS**. Figure 6.1 is a complete model, as configured in LINGO.

```
MODEL:    TITLE PRV with de-superheating Model: PRV1-v1;
! ;
! Nath, "Industrial Process Plants: Global Optimization of Utility Systems", Chapter 6, (2023);
! ;

 DATA:!inputs;     mps_demand = 10;              ! bfw demand, tph;
      !params;     prv1_a = 0.133320;            ! prv parameter, h_hpbfw / h_hps;
                   prv1_b = 0.929206;            ! prv parameter, h_mps / h_hps;
 ENDDATA

  ! Objective;     MIN = PRV1_HPS;

  ! mass and energy balances;
  [_PRV1_MB1]      PRV1_HPS + PRV1_BFW = PRV1_MPS;
  [_PRV1_EB1]      PRV1_HPS + prv1_a * PRV1_BFW = prv1_b * PRV1_MPS;

  ! MPS demand;
  [_MPS_DMND]      PRV1_MPS = mps_demand;

END MODEL PRV1-v1
```

Fig. 6.1: PRV1 with de-superheating model: PRV1-v1: in LINGO.

Since the model is simple, we will test it with just one test case and cross-check it with independent calculations. Figure 6.2 shows the model with a test case solution in Microsoft Excel. Both solvers give identical and expected solutions.

▲	A	B	C	D	E	F	G	H	I
1		model ID:	PRV1-v1: PRV with de-superhearin				inputs:	params:	
2		case ID:	1				mps_dmnd	prv1_a	0.133320
3		Objective fn:	tph			9.18	10	prv1_b	0.929206
4									
5	#						1	2	3
6		ID					PRV1_HPS	PRV1_BFW	PRV1_MPS
7			UOM				tph	tph	tph
8				low	high	cost coeff ->	$ 1.00		
9				limit	limit	optimum	9.2	0.8	10.0
10	1	_PRV1_MB1	tph	0	0	0.00	1	1	-1
11	2	_PRV1_EB1	tph	0	0	0.00	1	0.1	-0.9
12	3	_MPS_DMND	tph	10	10	10.00			1

Fig. 6.2: PRV with de-superheating model, PRV-v1: in Microsoft Excel, showing test case solution.

6.1.3 Modeling a vent

A vent is an automatic pressure reducing device installed for safety reasons to depressurize the header to within specified high pressure limit. At least one such valve must exist on the lowest-pressure material header in a utility system. It is an automatic valve that opens to the ambient or piped to the utility stack. It ensures that the header pressure does not exceed the prescribed high limit. It opens only when the pressure is too high and lets out material till the header pressure is within safe operating pressure limits.

In modeling terms, it is similar to the model of PRV without de-superheating. It has only one variable, **VENT**. In a utility system, such valves are sized to accommodate all expected flows; so constraints are unnecessary. We will assume no flow limits in our model, and the model will comprise a single unconstrained variable.

6.2 Modeling blowdown flash unit

A blowdown flash unit is a two-phase separation vessel maintained at low-pressure header pressure. Blowdown streams from the boilers are collected and enter the blowdown flash unit, reach equilibrium, and produce low-pressure steam and a liquid water stream high in solids. The steam produced enters the LP steam header, and the liquid condensate usually goes to the water treatment facility for purification and reuse.

In modeling terms, there will be four variables, **BDF_MPC**, **BDF_HPC**, **BDF_LPS**, and **BDF_CND**, representing the medium-pressure blowdown, high-pressure blowdown, generated LP steam, and the bottoms condensate stream, respectively.

These variables are related to each other by mass and energy balances. The following equation expresses the conservation of mass.

$$\textbf{BDF\_MPC} + \textbf{BDF\_HPC} = \textbf{BDF\_LPS} + \textbf{BDF\_CND} \qquad \text{(\_BDF\_MB1)}$$

For conservation of energy, we make an enthalpy balance around the flash unit, as shown below:

$$hl\_mps * \textbf{BDF\_MPC} + hl\_hps * \textbf{BFD\_HPC} = hv\_lps * \textbf{BDF\_LPS} + hl\_lps * \textbf{BDF\_CND} \quad \text{(6.2)}$$

Where hl_mps, hl_hps, hv_lps, and hl_lps are the enthalpies of the MP blowdown, HP blowdown, saturated low-pressure steam, and saturated low-pressure condensate, respectively.

We will simplify eq. (6.2) by eliminating **BDF_CND** using the material balance equation and dividing it by ($hl\_mps - hl\_lps$) to get the following:

$$\textbf{BDF\_MPC} + bdf\_a * \textbf{BDF\_HPC} = bdf\_b * \textbf{BDF\_LPS} \qquad \text{(\_BDF\_EB1)}$$

This model comprises the mass and energy balance equations mentioned above.

For testing purposes, we will add constraints to specify the MP and the HP blowdown flows. Note that in this model, there are four variables, two constraints, and two specified values; so there is no degree of freedom, and there will be a unique solution, irrespective of the objective function. However, we will define it as an optimization problem by defining an arbitrary objective function to minimize the blowdown condensate flow, **BDF_CND**.

Figure 6.3 shows a complete model as configured in LINGO.

Since the model is simple, we will test it with just one test case and cross-check it with independent calculations. Figure 6.4 shows the model with a test case solution in Microsoft Excel. Both LINGO and Excel Solvers give identical solutions.

```
MODEL:    TITLE Blow Down Flash Unit Model: BDF-v1;
! ;
! Nath, "Industrial Process Plants: Global Optimization of Utility Systems", Chapter 6, (2023).;
! ;
  DATA:
      !inputs;    mpc_supply = 4; hpc_supply = 6;        ! blow down supply, tph;

      !params:    bdf_a = 1.884648; bdf_b = 8.309105;

  ENDDATA

  ! Objective;    MIN = BDF_CND;

  ! mass & energy balances;
  [_BDF_MB1]      BDF_MPC + BDF_HPC = BDF_LPS + BDF_CND;
  [_BDF_EB1]      BDF_MPC + bdf_a * BDF_HPC = bdf_b * BDF_LPS;

  ! BDF supply;
  [_MPC_SUPLY]    BDF_MPC = mpc_supply;
  [_HPC_SUPLY]    BDF_HPC = hpc_supply;

END MODEL BDF-v1
```

Fig. 6.3: Blow down flash-unit model, BDF-v1: in LINGO.

▲	A	B	C	D	E	F	G	H	I	J
1		model ID:	BDF-v1: blow down flash unit				inputs:		params:	
2		case ID:	1				mpc_supply	4	bdf_a	1.884648
3		Objective fn:	tph			8.16	hpc_supply	6	bdf_b	8.309105
4										
5	#						1		2	3
6		ID					BDF_MPC	BDF_HPC	BDF_LPS	BDF_CND
7			UOM				tph	tph	tph	tph
8				low	high	cost coeff ->				$1.00
9				limit	limit	optimum	4.00	6.00	1.84	8.16
10	1	_BDF_MB1	tph	0	0	0.00	1	1	-1	-1
11	2	_BDF_EB1	tph	0	0	0.00	1	1.885	-8.31	0.00
12	3	_MPC_SUPLY	tph	4	4	4.00	1			
13	3	_HPC_SUPLY	tph	6	6	6.00		1		

Fig. 6.4: Blow down flash-unit model, BDF-v1: in Excel, showing test case solution.

6.3 Modeling deaerator unit

The purpose of a deaerator unit is to produce boiler feed water by removing dissolved air from the combined process condensate, turbine condensate, and demineralized makeup water streams. A deaerator unit is a two-phase separator vessel maintained at a low pressure. Removal of dissolved gases is achieved by preheating the combined water streams with direct injection of LP steam and raising the mixture's temperature to the bubble point temperature. The low-pressure boiler feed water stream thus produced is pumped to high pressure using the boiler feed water pumps.

In modeling terms, there will be four variables, **DEA_BFW**, **CTG_CND**, **DEA_DMW**, and **DEA_LPS**, representing the boiler feed water, turbine condensate, demineralized makeup water, and deaeration LP steam flows, respectively. Note that the process condensate flow, *pc_supply*, is the steam condensate returned by the process back to the utility system; there usually is no continuous measurement of process condensate, and it is generally estimated using an empirical factor as a fraction of total process steam demand; as far as the optimizer is concerned, it is a known quantity.

The mass and energy balances mentioned below relate the four variables and process condensate. The following equation expresses the conservation of mass:

$$\textbf{DEA\_BFW} = \textbf{CTG\_CND} + \textbf{DEA\_DMW} + \textbf{DEA\_LPS} + pc\_supply \qquad (\_DEA\_MB1)$$

For the conservation of energy, we need to make an enthalpy balance around the deaerator, as shown below:

$$hl\_lps * \textbf{DEA\_BFW} = h\_tc * \textbf{CTG\_CND} + h\_dmw * \textbf{DEA\_DMW} + h\_lps * \textbf{DEA\_LPS}$$
$$+ h\_pc * pc\_supply \qquad (6.3)$$

Where *hl_lps, h_tc, h_dmw, h_lps,* and *h_pc* are the enthalpies of saturated LP water, condensing turbo-generator (CTG) condensate, make up demineralized water, low-pressure deaeration steam, and process condensate return, respectively.

We will simplify the above equation by eliminating *pc_supply* using the material balance equation and dividing the resulting equation by (*hl_lps – hl_pc*) to the following energy balance constraint:

$$\textbf{DEA\_BFW} = dea\_a * \textbf{CTG\_CND} + dea\_b * \textbf{DEA\_DMW} + dea\_c * \textbf{DEA\_LPS} \quad (\_DEA\_EB1)$$

For standalone testing of this model, we will add two constraints to specify the CTG condensate flow and the boiler feed water flow to avoid getting a trivial solution. With these specifications, no remaining degrees of freedom and a unique solution will exist, irrespective of the objective function. However, to set it up as an optimization problem, we will need an objective function; in this case, any arbitrary objective function will do; we will minimize the deaeration steam flow, **DEA_LPS**.

Figure 6.5 is the complete model as configured in LINGO. Since the model is simple, we will test it with just one set of input values and cross-check it with independent calculation. Figure 6.6 shows the model with the test case solution in Microsoft Excel. Both solvers gave identical solutions.

```
MODEL:   TITLE Deaerator  Model: DEA-v1;
! ;
! Nath, "Industrial Process Plants: Global Optimization of Utility Systems", Chapter 6, (2023).;
! ;
  DATA:
      !inputs;    tc_supply = 20; pc_supply = 50; bfw_demand = 100;

      !params;    dea_a = -0.11466; dea_b = -0.37934; dea_c = 6.49394;

ENDDATA

  ! Objective;    MIN = DEA_LPS;

  ! mass and energy balances;
  [_DEA_MB1]      DEA_BFW = CTG_CND + DEA_DMW + DEA_LPS + pc_supply;
  [_DEA_EB1]      DEA_BFW = dea_a * CTG_CND + dea_b * DEA_DMW + dea_c * DEA_LPS;

  ! specs for testing;
  [_TC_SPLY]      CTG_CND = tc_supply;
  [_BFW_DMND]     DEA_BFW = bfw_demand;

END MODEL DEA-v1
```

Fig. 6.5: Deaerator model, DEA-v1: in LINGO.

A	B	C	D	E	F	G	H	I	J
1	model ID:	DEA-v1: Deaerator unit				inputs:		params:	
2	case ID:	1				tc_supply	20	dea_a	-0.114660
3	Objective fn:	tph			16.54	bfw_demand	100	dea_b	-0.379340
4						pc_supply	50	dea_c	6.493940
5 #						1	2	3	4
6	ID					DEA_BFW	CTG_CND	DEA_DMW	DEA_LPS
7		UOM				tph	tph	tph	tph
8			low	high	cost coeff ->				$1.00
9			limit	limit	optimum	100.00	20.00	13.46	16.54
10 1	_DEA_MB1	tph	50	50	50.00	1	-1	-1	-1
11 2	_DEA_EB1	tph	0	0	0.00	1	0.11	0.38	-6.49
12 3	_TC_SPLY	tph	20	20	20.00	1			
13 4	_BFW_DMND	tph	100	100	100.00	1			

Fig. 6.6: Deaerator model, DEA-v1: in Microsoft Excel, showing test case solution.

6.4 Modeling headers

Headers are channels for collecting and distributing a particular utility; there are two types: material headers and electrical energy headers.

A material header is a pipe or a network of pipes of appropriate sizes and with a proper number of nozzles for collecting and distributing the utility material of interest to the various producers and consumers of the utility.

An example of material header is a steam header. It collects steam from the various producers and distributes it to the consumers of the steam utility in the process plant and internal utility system consumers. As mentioned earlier, automation maintains steam headers within narrow pressure limits.

An electrical energy header is a network of cables, wires, and switches that connect the external electrical supply and internal generation to the consumers within the utility system and the process plant.

In terms of modeling, a header is a junction with material or energy coming in and out. Although physical material headers can accumulate small amounts, the net accumulation is negligible and can be ignored. A simple **IN = OUT** model is adequate.

For example, let us consider the high-pressure steam header in the sample steam subsystem in Fig. 1.3. HP steam producers are boilers #2 and #3. There are several consumers of HP steam, both within the steam subsystem and in the process plant. The consumers within the steam subsystem are turbine of boiler #3 auxiliaries, BL3AT; turbines of boiler feed water pumps #2 and #4, BP2T, and BP4T, respectively; pressure reducing valve, PRV1; condensing turbogenerator, CTG and non-condensing steam turbogenerator, STG.

All process unit consumers are lumped as a single entity, represented by the symbol *hps_demand*. The high-pressure header model is:

$$\text{BL2\_STM} + \text{BL3\_STM} = \text{PRV1\_HPS} + \text{STG\_STM} + \text{CTG\_HPS} + \text{BL3AT\_STM} + \text{BP2T\_STM}$$
$$+ \text{BP4T\_STM} + hps\_demand$$

$$(6.9)$$

Similarly, the MP steam header model is:

$$\text{BL1\_STM} + \text{PRV1\_MPS} + \text{STG\_STM} + \text{CTG\_MPS} + \text{BP4T\_STM} = \text{PRV2\_STM}$$
$$+ \text{BL1AT\_STM} + mps\_demand$$

$$(6.10)$$

And the LP steam header model is:

$$\text{PRV2\_STM} + \text{BL1AT\_STM} + \text{BP2T\_STM} + \text{BL3AT\_STM} + \text{BDF\_LPS} + \text{CTG\_LPS} = \text{DEA\_LPS}$$
$$+ \text{VENT} + lps\_demand$$

$$(6.11)$$

The boiler feed water header model is:

$$\text{BL1\_BFW} + \text{BL2\_BFW} + \text{BL3\_BFW} + \text{PRV1\_BFW} = \text{BP1\_FLO} + \text{BP2\_FLO}$$
$$+ \text{BP3\_FLO} + \text{BP4\_FLO}$$

$$(6.12)$$

These header models are simply statements of IN = OUT; testing is optional as long as all input and output streams have been included.

6.5 Modeling the steam subsystem

We now have all the unit operations models needed for the steam subsystem and are ready to assemble them to create a model. Some adjustments will be needed as the constraints specifically required for component testing will no longer be required. Also, we will estimate the process condensate to be 50% of the total process steam demand. It also makes sense to streamline the model by eliminating duplication of parameters, such as "big," and moving the MPS and HPS reserve-related constraints to the bottom rows as they span multiple units. Note that any rearrangement of constraints and variables is for the convenience of the human modeler; it makes no difference to the MILP solver as it considers all constraints simultaneously.

Excess HP reserve also serves as the MP steam reserve, as HP steam can flow into the MP header via PRV1. With this understanding, we will modify the MPS reserve constraint to a combined HPS and MPS reserve constraint. We have made these changes to the steam system model, **StmSS-v1**.

This model has 82 variables, including 18 binaries and 93 constraint rows; it now exceeds the capabilities of the basic Solver that comes standard with Microsoft Excel. Solving large problems such as this requires using an upgraded version of the Solver called the "Premium Solver," available from the makers of the basic Excel Solver, Frontline Systems company [1]. The basic Solver that comes standard with Microsoft Excel is a limited version of the "Premium Solver," so all models developed earlier work in the Premium Solver without any modifications [2].

Assembling the steam subsystem model is straightforward. Figure 6.7 shows the Excel model. It is large and requires significant reduction to fit a printed page, so the details are no longer discernable; only the model structure is. Note that the model, as expected, has a block diagonal structure, except for the models shown for blowdown flash, deaerator, steam reserves, and the steam headers that span multiple units and are on the lower part of the spreadsheet.

The LINGO model is shown in two parts in Figs. 6.8 and 6.9. It is compact as it uses the previously created include files for the three boilers, two turbogenerators, and the 4 BFW pumps. Note that the four BFW pumps are identical except for their drives, so we will streamline the model by defining their parameters once and using the CALC section in the LINGO model to assign parameter values to the individual pumps.

Table 6.1 shows the 4 test cases. Both the Premium Solver and the LINGO Solver give identical optimum solutions. Table 6.2 shows the solution details for test case #1.

The results look good, except there is a minor problem. Can you spot it? Note that the four BFW pumps operate in parallel and are identical in size; the only difference being their drivers. In this situation, one would expect that the flow of BFW in the operating pumps is evenly split, but such is not the case. Why? The optimizer has

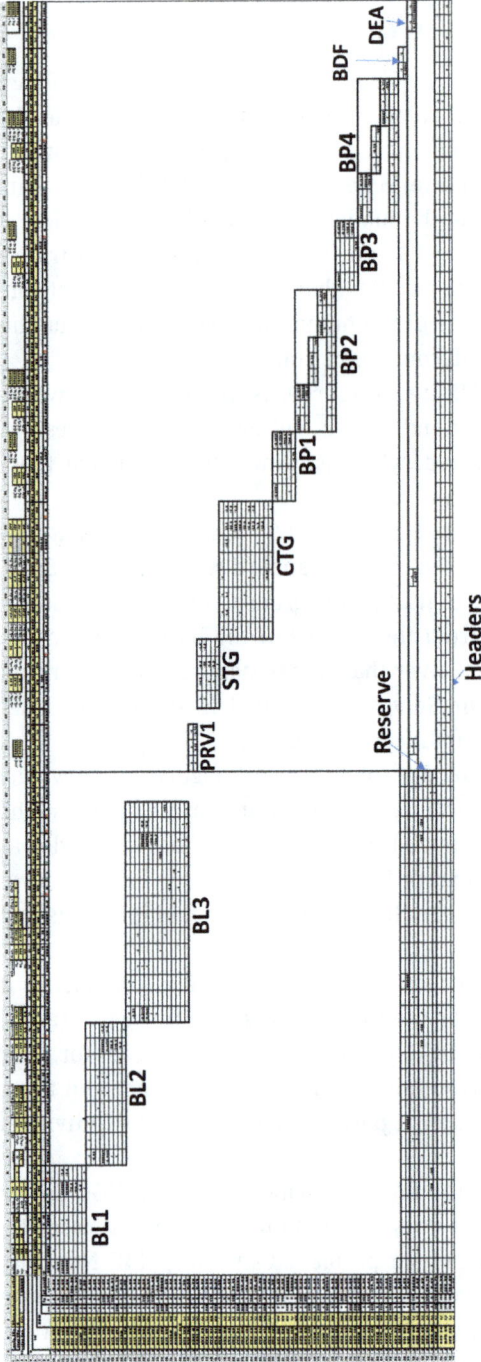

Fig. 6.7: Steam subsystem model: StmSS-v1: model structure.

Tab. 6.1: Steam subsystem model: StmSS-v1: Test cases.

case #	Attribute	Symbolic nam	Value	UOM	Expected key results
	power cost/credit	c_elec	60	$/MWh	
	Fuel oil cost	c_fo	18	$/MWh	
all	LP fuel gas cost	c_lpfg	19	$/MWh	
	HP fuel gas cost	c_hpfg	20	$/MWh	
	demin water cost	d_dmw	3	$/ton	
	HP steam demand	hp_demand	100	tph	produce steam to satisfy process demand, power credit is low
	MP steam demand	mp-demand	100	tph	and export power if profitable
1	LP steam demand	lp_demand	100	tph	
	HP steam reserve	hp_reserve	50	tph	
	MP steam reserve	mp_reserve	50	tph	
	HP steam demand	hp_demand	120	tph	produce steam to satisfy process demand and reserve requiremnets
	MP steam demand	mp-demand	120	tph	and export power if profitable
2	LP steam demand	lp_demand	120	tph	Obj fn > case 1
	PC supply	pc_supply	180	tph	
	HP steam reserve	hp_reserve	50	tph	
	MP steam reserve	mp_reserve	50	tph	
	HP steam demand	hp_demand	80	tph	produce steam to satisfy process demand and reserve requiremnets
	MP steam demand	mp-demand	80	tph	and export power if profitable
3	LP steam demand	lp_demand	80	tph	Obj fn < case 1
	HP steam reserve	hp_reserve	50	tph	
	MP steam reserve	mp_reserve	50	tph	
	HP steam demand	hp_demand	150	tph	produce steam to satisfy process demand and reserve requiremnets
	MP steam demand	mp-demand	150	tph	and export power if profitable
4	LP steam demand	lp_demand	150	tph	Obj fn > case 2
	HP steam reserve	hp_reserve	50	tph	
	MP steam reserve	mp_reserve	50	tph	

even optimized the BFW flows to the pumps. Although the effect of this optimization is minimal, it would be better if the model naturally equalizes the flows to the operating pumps.

Can we equalize these flows? Yes, that would be possible without resorting to NLP and thus maintaining the solution globally optimum. We will, however, have to wait till we discuss a technique to address this problem in Chapter 8.

```
MODEL:  TITLE Steam subsystem Model: StmSS-v1;
! Ref: Nath, "Industrial Process Plants: Global Optimization of Utility Systems", Chapter 6, (2023);

DATA:
 !inputs;
     c_fo =18; c_lpfg =19; c_hpfg =20; c_elec =60; c_dmw =3;                    ! costs;
     hps_demand = 100; mps_demand = 100;  lps_demand = 100;                    ! process demands, tph;
     mps_reserve = 50; hps_reserve = 50;                                       ! steam reserve, tph;

 ! params;
     bl1_fg_c = 15.2371; bl1_fg_d = 0.790615; bl1_fg_e =-12.4525; bl1_fg_f = 0.953495;! BL1;
     bl1_bdr=0.02; bl1_fg_s = 3; bl1_stm_ll = 70; bl1_stm_hl = 210;  bl1at_c = 6;
     bl2_fg_c = 18.1694; bl2_fg_d = 0.834726; bl2_fg_e =-11.6149; bl2_fg_f = 0.991485;  ! BL2;
     bl2_bdr=0.02; bl2_fg_s = 4; bl2_stm_ll = 80; bl2_stm_hl = 240; bl2am_c =0.5;
     bl3_fg_c = 21.1962; bl3_fg_d = 0.814223; bl3_fg_e =-15.2507; bl3_fg_f = 0.966085;  ! BL3;
     bl3_bdr=0.02; bl3_fg_s =5; bl3_stm_ll =100; bl3_stm_hl =300; bl3am_c =0.5; bl3at_c =5;
     prv1_a = 0.133320; prv1_b = 0.929206; bdf_a = 1.884648; bdf_b = 8.309105;    ! PRV, BDF;
     stg_stm_a =8; stg_stm_b =14; stg_stm_s =4; stg_stm_hl =50; stg_ele_ll =1; stg_ele_hl =3;  !STG;
     ctg_hps_a =16.9970; ctg_hps_b =5.4330; ctg_hps_c =0.7859; ctg_hps_d =0.5605; ctg_hps_s=10;!CTG;
     ctg_hps_ll =50; ctg_mps_hl =250; ctg_mps_hl =180; ctg_lps_hl =90; ctg_cnd_ll =10;
     ctg_cnd_hl =60; ctg_ele_ll =3; ctg_ele_hl =13; ctg_cw_b = 0.04;
     bp_sp_a = 0.183910; bp_sp_b =0.000423; bp_des_ll =100; bp_flo_hl =250;     ! common to all BF pumps;
     bp1m_eta =0.95; bp3m_eta =0.95;                                           ! BP1 & BP3;
     bp2m_eta =0.95; bp2t_stm_a =0.27306; bp2t_stm_b =8.48596;                 ! BP2;
     bp4m_eta =0.95; bp4t_stm_a =0.77092; bp4t_stm_b =23.95717;                ! BP4;
     dea_a =-0.114660; dea_b = -0.379340; dea_c =6.493940; big = 999;          !DEA;
 ENDDATA

CALC:
     bp1_sp_a = bp_sp_a;  bp2_sp_a = bp_sp_a; bp3_sp_a = bp_sp_a; bp4_sp_a = bp_sp_a;
     bp1_sp_b = bp_sp_b;  bp2_sp_b = bp_sp_b; bp3_sp_b = bp_sp_b; bp4_sp_b = bp_sp_b;
     bp1_des_ll = bp_des_ll;  bp2_des_ll = bp_des_ll; bp3_des_ll = bp_des_ll; bp4_des_ll = bp_des_ll;
     bp1_flo_hl = bp_flo_hl;  bp2_flo_hl = bp_flo_hl; bp3_flo_hl = bp_flo_hl; bp4_flo_hl = bp_flo_hl;
 ENDCALC

! Objective:     MIN = c_fo*BL1_FUEL + c_lpfg*(BL2_LPFG+BL3_LPFG) + c_hpfg*(BL2_HPFG+BL3_HPFG)
        + c_elec*(BL2AM_ELE+BL3AM_ELE+BP1M_ELE+BP2M_ELE+BP3M_ELE+BP4M_ELE-STG_ELE-CTG_ELE)+c_dmw*DEA_DMW;

 ! BL1;         @FILE('C:\BookDG\LingoModels\BL1-v4.lng');
 [_BL1AT_EB1]   BL1AT_STM - bl1at_c * BL1_X = 0;

 ! BL2;         @FILE('C:\BookDG\LingoModels\BL2-v1.lng');
 [_BL2AM_EB1]   BL2AM_ELE - bl2am_c * BL2_X = 0;
```

Fig. 6.8: Steam subsystem model: StmSS-v1: in LINGO, part 1 of 2.

```
! BL3;              @FILE('C:\BookDG\LingoModels\BL3-v1.lng');
[_BL3AT_EB1]        BL3AT_STM - bl3at_c * BL3AT_X = 0;
[_BL3AM_EB1]        BL3AM_ELE - bl3am_c * BL3AM_X = 0;
[_BL3A_LGC]         BL3AM_X + BL3AT_X - BL3_X = 0;
                    @BIN(B13AM_X);

! PRV1 constraints;
[_PRV1_MB1]         PRV1_HPS + PRV1_BFW = PRV1_MPS;
[_PRV1_EB1]         PRV1_HPS + prv1_a * PRV1_BFW = prv1_b * PRV1_MPS;

! STG;              @FILE('C:\BookDG\LingoModels\STG-v1.lng');

! CTG;              @FILE('C:\BookDG\LingoModels\CTG-v1.lng');

! BFWP1;            @FILE('C:\BookDG\LingoModels\BP1MD-v1.lng');
! BFWP2;            @FILE('C:\BookDG\LingoModels\BP2DD-v1.lng');
! BFWP3;            @FILE('C:\BookDG\LingoModels\BP3MD-v1.lng');
! BFWP4;            @FILE('C:\BookDG\LingoModels\BP4DD-v1.lng');

! Blow down flash unit;
[_BDF_MB1]          BL1_BD + BL2_BD + BL3_BD - BDF_LPS - BDF_CND = 0;
[_BDF_EB1]          BL1_BD + bdf_a * BL2_BD + bdf_a * BL3_BD - bdf_b * BDF_LPS = 0;

! Deaerator unit;
[_DEA_MB1]          BL1_BFW + BL2_BFW + BL3_BFW + PRV1_BFW = CTG_CND +DEA_DMW +DEA_LPS
                        + 0.5*(hps_demand + mps_demand +lps_demand);
[_DEA_EB1]          BL1_BFW + BL2_BFW + BL3_BFW + PRV1_BFW
                        - dea_a * CTG_CND - dea_b * DEA_DMW - dea_c * DEA_LPS = 0;

! Steam reserve constraint;
[_MPS_RSV]          MPS_RSV = BL1_RSV;
[_HPS_RSV]          HPS_RSV = BL2_RSV + BL3_RSV;
[_COMB_RSRV]        MPS_RSV + HPS_RSV > mps_reserve + hps_reserve;
[_HPS_RSRV]         HPS_RSV > hps_reserve;

! Steam header constraints;
[_HPS_HDR]          BL2_STM + BL3_STM
                        = PRV1_HPS + STG_STM + CTG_HPS + BL3AT_STM + BP2T_STM + BP4T_STM + hps_demand;
[_MPS_HDR]          BL1_STM + PRV1_MPS + STG_STM + CTG_MPS + BP4T_STM
                        = PRV2_STM + BL1AT_STM + mps_demand;
[_LPS_HDR]          PRV2_STM + BL1AT_STM + BL3AT_STM + BP2T_STM + CTG_LPS + BDF_LPS
                        = DEA_LPS + VENT + lps_demand;
[_BFW_HDR]          BL1_BFW + BL2_BFW + BL3_BFW + PRV1_BFW = BP1_FLO + BP2_FLO + BP3_FLO + BP4_FLO;
END MODEL StmSS-v1
```

Fig. 6.9: Steam subsystem model: StmSS-v1: in LINGO, part 2 of 2.

Tab. 6.2: Steam subsystem model: StmSS-v1: Test Case #1 solution.

Steam Subsystem Model: SteamSS-v1 : Test case # 1											
Global optimal solution found.											
Opmum objective function value (minimum operating cost, $/hr)										$ 6,321.3	
#	Variable ID	UOM	Cost coeff.	Optimum value	Cost, $/hr	#	Variable ID	UOM	Cost coeff.	Optimum value	Cost, $/hr
1	BL1_STM	tph		0.0	$ -	42	STG_X	BIN		1	$ -
2	BL1_BFW	tph		0.0	$ -	43	STG_Y	BIN		0	$ -
3	BL1_BD	tph		0.0	$ -	44	CTG_HPS	tph		219.0	$ -
4	BL1_RSV	tph		0.0	$ -	45	CTG_MPS	tph		119.0	$ -
5	BL1_FUEL	MW	$ 18.0	0.00	$ -	46	CTG_LPS	tph		90.0	$ -
6	BL1AT_STM	tph		0.0	$ -	47	CTG_CND	tph		10.0	$ -
7	BL1_X	BIN		0	$ -	48	CTG_CW	ktph		0.4	$ -
8	BL1_Y	BIN		0	$ -	49	CTG_ELE	MW	$ (60.0)	10.683	$ (641.0)
9	BL2_STM	tph		136.7	$ -	50	CTG_X	BIN		1	$ -
10	BL2_BFW	tph		139.5	$ -	51	CTG_Y	BIN		0	$ -
11	BL2_BD	tph		2.7	$ -	52	BP1_FLO	tph		134.3	$ -
12	BL2_RSV	tph		103.3	$ -	53	BP1_SP	MW		0.241	$ -
13	BL2_FUEL	MW		132.31	$ -	54	BP1M_ELE	MW	$ 60.0	0.253	$ 15.2
14	BL2_LPFG	MW	$ 19.0	132.31	$2,514.0	55	BP1_X	BIN		1	$ -
15	BL2_HPFG	MW	$ 20.0	0.000	$ -	56	BP2_FLO	tph		250.0	$ -
16	BL2AM_ELE	MW	$ 60.0	0.500	$ 30.0	57	BP2_SP	MW		0.290	$ -
17	BL2_X	BIN		1	$ -	58	BP2_X	BIN		1	$ -
18	BL2_Y	BIN		0	$ -	59	BP2M_SP	MW		0.000	$ -
19	BL3_STM	tph		240.0	$ -	60	BP2M_ELE	MW	$ 60.0	0.000	$ -
20	BL3_BFW	tph		244.80	$ -	61	BP2M_X	BIN		0	$ -
21	BL3_BD	tph		4.8	$ -	62	BP2T_SP	MW		0.290	$ -
22	BL3_RSV	tph		60.0	$ -	63	BP2T_STM	tph		2.7	$ -
23	BL3_FUEL	MW		216.61	$ -	64	BP2T_X	0/1		1	$ -
24	BL3_LPFG	MW	$ 19.0	216.61	$4,115.6	65	BP3_FLO	tph		0.0	$ -
25	BL3_HPFG	MW	$ 20.0	0.00	$ -	66	BP3_SP	MW		0.000	$ -
26	BL3AM_ELE	MW	$ 60.0	0.000	$ -	67	BP3M_ELE	MW	$ 60.0	0.000	$ -
27	BL3AT_STM	tph		5.0	$ -	68	BP3_X	BIN		0	$ -
28	BL3AM_X	BIN		0	$ -	69	BP4_FLO	tph		0.0	$ -
29	BL3AT_X	0/1		1	$ -	70	BP4_SP	MW		0.000	$ -
30	BL3_W	0/1		1	$ -	71	BP4_X	BIN		0	$ -
31	BL3_X	BIN		1	$ -	72	BP4M_SP	MW		0.000	$ -
32	BL3_Y	BIN		0	$ -	73	BP4M_ELE	MW	$ 60.0	0.000	$ -
33	BL3_Z	BIN		0	$ -	74	BP4M_X	BIN		0	$ -
34	MPS_RSV	tph		0.0	$ -	75	BP4T_SP	MW		0.000	$ -
35	HPS_RSV	tph		163.3	$ -	76	BP4T_STM	tph		0.0	$ -
36	PRV1_HPS	tph		0.0	$ -	77	BP4T_X	0/1		0	$ -
37	PRV1_BFW	tph		0.0	$ -	78	BDF_LPS	tph		1.7	$ -
38	PRV1_MPS	tph		0.0	$ -	79	BDF_CND	tph		5.8	$ -
39	PRV2_STM	tph		69.0	$ -	80	VENT	tph		0.0	$ -
40	STG_STM	tph		50.0	$ -	81	DEA_DMW	tph	$ 3.0	155.8	$ 467.5
41	STG_ELE	MW	$ (60.0)	3.000	$ (180.0)	82	DEA_LPS	tph		68.5	$ -

Summary

In this chapter, we developed generic MILP models for the remaining components in the steam subsystem, viz. pressure reducing valve with de-superheating, PRV1; blow-down flash unit, BDF; deaerator, DEA, and steam and boiler feed water headers. We then assembled the component models to create a MILP model for the entire steam subsystem, StmSS-v1.

This model has 82 variables, including 18 binaries and 93 constraint rows. This model exceeds the limits of the basic Solver that comes with Microsoft Excel and requires using a larger version of Solver called the "Premium Solver." The two Solvers are identical, except that the basic Solver limits the number of variables and constraints. Previously developed models require no modifications as they are fully compatible with the Premium Solver.

Nomenclature

Symbol	UOM	Description
BDF_CND	Tph	Blowdown flash: LP condensate discharge
BDF_HPC	Tph	Blowdown flash: inlet HP blowdown
BDF_LPS	Tph	Blowdown flash: LP steam generation
BDF_MPC	Tph	Blowdown flash: inlet MP blowdown
BL3_W	0/1	Boiler #3: LPFG indicator: 1=using LPFG, 0=using HPFG
BL3_Z	0/1	Boiler #3: HPFG indicator: 1=using HPFG, 0=using LPFG
BLj_BFW	Tph	Boiler #j: boiler feed water inflow
BLj_FUEL	MW.th	Boiler #j: total fuel consumption
BLj_HPFG	MW.th	Boiler #j: HP fuel gas consumption
BLj_LPFG	MW.th	Boiler #j: LP fuel gas consumption
BLj_RSV	Tph	Boiler #j: steam reserve
BLj_STM	Tph	Boiler #j: steam generation
BLj_X	0/1	Boiler #j: operational status: 0 = OFF, 1=ON
BLj_Y	0/1	Boiler #j: standby status: 1=in standby mode, 0 otherwise
BLjAM_ELE	MW	Boiler #j: auxiliaries motor drive: electric power usage
BLjAM_X	0/1	Boiler #j: auxiliaries motor drive: 0 = OFF, 1=ON
BLjAT_STM	tph	Boiler #j: Auxiliaries turbine: steam flow
BLjAT_STM	tph	Boiler #j: auxiliaries turbine drive: steam flow
BLjAT_X	0/1	Boiler #j: auxiliaries turbine drive: 0 = OFF, 1=ON
BPj_FLO	tph	BFW Pump #j: boiler feed water inflow
BPj_SP	MW.sp	BFW Pump #j: shaft power required
BPj_X	0/1	BFW Pump #j: operational status: 0 = OFF, 1=ON
BPjM_ELE	MW	BFW Pump #j: motor drive: power consumption
BPjM_SP	MW.sp	BFW Pump #j: motor drive: shaft power produced
BPjM_X	0/1	BFW Pump #j: motor drive: operational status: 0 = OFF, 1=ON
BPjT_SP	MW.sp	BFW Pump #j: turbine drive: shaft power produced
BPjT_STM	tph	BFW Pump #j: turbine: steam flow
BPjT_X	0/1	BFW Pump #j: turbine drive: operational status: 0 = OFF, 1=ON
CTG_CND	tph	CTG: condensate from condenser
CTG_CW	ktph	CTG: condenser: cooling water flow
CTG_ELE	MW	CTG: power generation
CTG_HPS	tph	CTG: inlet HP steam flow
CTG_MPS	tph	CTG: outlet MP steam flow
CTG_X	0/1	CTG: operational status: 0 = OFF, 1=ON
CTG_Y	0/1	CTG: standby status: 1=in standby mode, 0 otherwise

(continued)

Symbol	UOM	Description
DEA_BFW	tph	Deaerator: boiler feed water production
DEA_DMW	tph	Deaerator: fresh demineralized water intake
DEA_LPS	tph	Deaerator: deaeration steam
DEA_LPS	tph	Deaerator: deaeration steam
HPS_RSV	tph	HP Boilers: total steam reserve
MPS_RSV	tph	MP Boiler: steam reserve
PRV1_BFW	tph	PRV: de-superheating water flow
PRV1_HPS	tph	PRV: HP steam flow
PRV1_MPS	tph	PRV: MP steam flow
PRV2_STM	tph	PRV2: steam flow
STG_ELE	MW	STG: power generation
STG_STM	tph	STG: steam flow
STG_X	0/1	STG: operational status: 0 = OFF, 1=ON
STG_Y	0/1	STG: standby status: 1=in standby mode, 0 otherwise
VENT	tph	VENT: steam flow
bdf_a	–	Enthalpy ratio: (hl_hps - hl_lps)/hl_mps
bdf_b	–	Enthalpy ratio: (hv_lps - hl_lps)/hl_mps
bp_flo_hl		BFW Pump: inlet flow: high limit
dea_a	–	Enthalpy ratio: (h_tc - h_pc)/(hl_mps - h_pc)
dea_b	–	Enthalpy ratio: (h_dmw - h_pc)/(hl_mps - h_pc)
dea_c	–	Enthalpy ratio: (h_lps - h_pc)/(hl_mps - h_pc)
h_hpbfw	MW.th/t	Enthalpy: HP boiler feed water
h_hps	MW.th/t	Enthalpy: HP steam
h_mps	MW.th/t	Enthalpy: MP steam
hl_hps	MW.th/t	Enthalpy, staurated liquid: HP blowdown
hl_lps	MW.th/t	Enthalpy, staurated liquid: LP condensate
hl_mps	MW.th/t	Enthalpy, saturated liquid: MP blowdown
hpc_supply		Supply for testing: HP blowdown
hps_demand	tph	Process Demand: HP steam
hv_lps	MW.th/t	Enthalpy, vapor: LP steam
lps_demand	tph	Process Demand: LP steam
mpc_supply		Supply for testing: MP blowdown
mps_demand	tph	Process Demand: MP steam
pc_supply		Supply: process condensate return
prv1_a	–	Enthalpy ratio: h_hpbfw /h_hps
prv1_b	–	Enthalpy ratio: h_mps /h_hps

References

[1]   Frontline Systems Inc https://www.solver.com
[2]   Anonymous, "Frontline Solvers Reference Guide, Version 2023 Q2", Frontline Systems, Inc. (2023).

7 Expanding the model

In the last chapter, we developed a model for the steam subsystem, the biggest subsystem with most equipment, it forms a base to expand the model to include the remaining subsystems. In this chapter, we will develop models for the remaining subsystems: the fuel, the electric power, the cooling water, and the compressed air subsystems, and then integrate them with the steam subsystem model to develop a combined model for the entire sample utility system. However, in this chapter, we will not model the external energy suppliers, i.e., fuel suppliers and the local power grid, as we have yet to discuss the vital topic of modeling contracts, which is the subject of the next chapter.

The fuel subsystem, shown in Fig. 1.4, comprises the following components:
- Fuel oil, LP fuel gas, and HP fuel gas Headers
- A proposed Gas Turbogenerator, GTG
- Fuel Gas Compressor, FC
- Fuel PRV, PRFG,
- Fuel gas to flare, FGFLARE
- Fuel oil to flare, FOFLARE
- Internal fuel consumers:
  - Boilers 1, 2, and 3
  - Air compressor ICE drivers AC1E, AC2E, AC3E and AC4E; and
- External suppliers of Fuel oil, LP, and HP fuel gas.

Some of the internal fuel consumers, viz., the boilers, have already been modeled in the steam subsystem model, and the air compressor drivers will be modeled when we develop a model for the air compression subsystem. So, for the sake of simplicity, we will ignore the internal consumers and account for them properly in the combined utility system model. Note that generic models for all the components in the fuel subsystem have already been developed; so, modeling the fuel subsystem would be a relatively straightforward task, as discussed in the next section.

The electrical subsystem, shown in Fig. 1.6, comprises a single header that connects all electrical power generators and consumers within the utility system to the process consumers and the local power grid. The utility system could export excess power to the grid, depending on the contractual agreement with the local power provider. We will model for such a possibility.

The cooling water subsystem, shown in Fig. 1.7, comprises the following:
- CWP1, pump with electric motor drive
- CWP2, pump with turbine drive
- CWP3, pump with dual (motor and turbine) drives
- Internal CW consumer, CTG Condenser
- Cooling water header

https://doi.org/10.1515/9783111020679-008

To model the cooling water subsystem, we will use the generic pump models developed in Chapter 4 and add a cooling water header; this modeling would be straightforward, as detailed later in this chapter.

The compressed air subsystem, shown in Fig. 1.8, will be modeled in the subsequent section; it comprises:
- Air compressors AC1, 2, 3, and 4; rotary compressors with internal combustion engine drivers,
- Air compressor AC5, a centrifugal compressor with electrical motor drive, and
- Compressed air header.

We will use the generic compressor models developed in Chapter 4 to model the compressed air subsystem and add a compressed air header; this modeling would be relatively straightforward, as detailed later in this chapter.

The gas turbogenerator GTG in the fuel subsystem currently does not exist but is considered a possible investment opportunity. From a modeling perspective, this means that we need to have the capability to either include GTG when optimizing, such as when exploring the impact of GTG on operations, but also be able to exclude GTG when optimizing the existing operations. How can we add the capability of including or excluding the GTG within the same model? We will explore this issue in the next section.

7.1 Available, unavailable, and required specifications

In the fuel subsystem, we will be modeling the GTG – an equipment that currently does not exist but is considered a possible investment opportunity. This functionality can be achieved by applying a suitable upper limit to the GTG binary variable, **GTG_X**. GTG would be excluded from optimization by setting the upper limit on **GTG_X** to 0, and GTG would be included in optimization by restoring the upper limit of **GTG_X** to 1, as summarized in Tab. 7.1.

Tab. 7.1: Summary of GTG inclusion and exclusion strategy.

#	Attribute	Symbolic name	Value	Comment
1	Include in optimization	gtg_x_hl	1	Available for optimization, i.e., ON if profitable, OFF otherwise
2	Exclude from optimization	gtg_x_hl	0	Always OFF, irrespective of economc considerations

For the GTG, the "exclude" option effectively makes **GTG_X** to be 0 and turns the GTG OFF. The "include" capability is achieved by resetting **GTG_X** hi-limit to 1, in which case, the optimizer determines whether or not to turn on the GTG, based on economic considerations.

Such a capability to "include" or "exclude" equipment in optimization would be helpful for all equipment in the utility system. "Include" capability is what we have used so far; it lets the optimizer use economics to determine whether or not to use the equipment. The "exclude" capability would make equipment unavailable for optimization; this capability would be particularly useful when the equipment is undergoing maintenance or repairs.

There is yet another possibility, a "require" option, which would mandate that the equipment be operating in the optimum solution, irrespective of the economics; this would enable the capability to evaluate the penalty associated with forcing the use of a particular equipment.

Such capability to specify the possibility of inclusion, exclusion, or forced inclusion of equipment is worth adding to all major equipment models; we will do so from now on. This option would be an input to optimization; we will name it with a symbolic name, created by adding the suffix "_aru" to the equipment ID. Note that in LINGO, variable values could only be numerical. So, we will also have a similar variable with the equipment name appended with "_123" for use by LINGO.

Note that we also have some equipment with standby operation capability, such as the boilers and the turbogenerators. It would make sense to modify these three possibilities mentioned above and include the standby operation for such cases. Table 7.2 shows these three possibilities and the associated constraint strategy, considering the standby possibility.

Tab. 7.2: Summary of the available, required, and unavailable specification strategy.

#	Attribute	Symbolic name	Value	Sum of _X + _Y lo limit	Sum of _X + _Y hi limit	Comment
1	Available to optimization	_aru, _123	A 1	0	1	Available to optimization, i.e., ON or standby operation if profitable, OFF otherwise
2	Required in optimum solution	_aru, _123	R 2	1	1	ON or standby operation, irrespective of economc considerations
3	Unavaiable to optimization	_aru, _123	U 3	0	0	OFF, irrespective of economc considerations

In Excel, we will implement this strategy by the use of the Excel IF function to modify the low and high limits of the associated constraint; for example, for boiler #1, the constraint eq. (_BL1_LGC1), is the sum **BL1_X** + **BL1_Y**, its low limit is set using the logic: IF($bl1\_aru$ = "R", 1, 0), and its high limit is set using the logic IF($bl1\_aru$ = "U", 0, 1).

In LINGO, we will implement this logic in the CALC block with the following three statements.

```
bl1_lgc_ll = 0; bl1_lgc_hl = 1;         ! default settings as for
Available;
@IFC (bl1_123 #EQ# 2: bl1_lgc_ll = 1;); ! set lo limit to 1 for
Required;
@IFC (bl1_123 #EQ# 3: bl1_lgc_hl = 0;); ! set hi limit to 0 for
Unavailable;
```

7.2 Modeling fuel subsystem

Major pieces of equipment in the fuel subsystem are the proposed GTG and the existing fuel compressor FC. We will add the flexibility to include, exclude, or force inclusion to both. Doing so will enable us to use the same model for analyzing the fuel system operations with and without the proposed GTG. Such a comparison would give a realistic assessment of the benefits associated with the GTG operation, helpful information when evaluating the GTG as an investment.

The fuel subsystem model is created by assembling models for its components: Gas turbogenerator, GTG-v1; motor-driven compressor, C01MD-v1, for fuel compressor FC; fuel PRVFU and fuel vents as single-variable models. The fuel headers will be simple IN = OUT models. The objective function would be to minimize the cost of purchased energy (fuel and electric power).

The complete fuel subsystem model, FuelSS-v1, is shown in Figs. 7.1 and 7.2 for LINGO. Figure 7.3 shows the model in Excel. Table 7.3 shows the test cases; these cases confirm that the operations of GTG and Fuel compressor are sensitive to economic data (costs of LP fuel and electric power). This model is small enough for the basic Solver in Microsoft Excel. Figure 7.3 also shows the solution for the test case #1. As expected, the Excel and LINGO models give identical solutions to all test problems.

7.3 Modeling electrical power subsystem

The electrical subsystem is simple, comprising a single header, connecting all electrical power generators and power consumers in the plant site to the local power supplier. Depending on the power generation capability of the utility system and the contractual agreement with the local power grid, the utility system could export excess power back to the grid. We will model for this possibility.

Tab. 7.3: Fuel subsystem model, FuelSS-v1: Test cases.

Case #	Attribute	Symbolic name	Value	UOM	Expected key results
	HPFG cost	c_hpfg	20	$/MWh.th	
All	LPFG cost	c_lpfg	19	$/MWh.th	
	Fuel oil cost	c_fo	18	$/MWh.th	
1	Elec power cost	c_elec	60	$/MWh	power cheap, **GTG_X** = 0, **FC_X** = 1
	GTG inclusion	gtg_aru	A	A/R/U	
2	Elec power cost	c_elec	80	$/MWh	power expensive, export power,
	GTG inclusion	gtg_aru	A	A/R/U	**GTG_X** = 1, **FC_X** = 0
3	Elec power cost	c_elec	40	$/MWh	power cheap, **GTG_X** = 0, **FC_X** = 1
	GTG inclusion	gtg_aru	A	A/R/U	
4	Elec power cost	c_elec	80	$/MWh	power expensive, but cannot use
	GTG inclusion	gtg_aru	U	A/R/U	GTG, **GTG_X** = 0, **FC_X** = 0

```
! Fuel subsystem model constraints file: FuelSS-v1.lng;

! GTG;
[_GTG_EB1G]      GTG_FG  > gtg_fg_a * GTG_X + gtg_fg_b * GTG_ELE;
[_GTG_EB1B]      GTG_FG  > gtg_fg_c * GTG_X + gtg_fg_d * GTG_ELE;
[_GTG_EB1R]      GTG_FG  > gtg_fg_e * GTG_X + gtg_fg_f * GTG_ELE;
[_GTG_ELE_HL]    GTG_ELE < gtg_ele_hl * GTG_X;
! ARU ;
[_GTG_LGC_LL]    GTG_X   > gtg_lgc_ll;
[_GTG_LGC_HL]    GTG_X   < gtg_lgc_hl;
! Integers;       @BIN(GTG_X);

! Fuel Compressor;
[_FC_EB1]        FC_SP = fc_sp_a * FC_X + fc_sp_b * FC_FG;
[_FC_EB2]        FC_SP = fc_eta * FC_ELE;
[_FC_FG_LL]      FC_FG > fc_fg_ll * FC_X;
[_FC_FG_HL]      FC_FG < fc_fg_hl * FC_X;
[_FC_SP_HL]      FC_SP < (fc_sp_a + fc_sp_b * fc_fg_hl) * FC_X;
! ARU ;
[_FC_LGC_LL]     FC_X  > fc_lgc_ll;
[_FC_LGC_HL]     FC_X  < fc_lgc_hl;
! Integers;       @BIN(FC_X);

!;
```

Fig. 7.1: Fuel subsystem constraints file, FuelSS-v1-.lng.

The utility system could import or export power but not do both simultaneously. Almost always, the credit for power export is less than the cost of power import. If the export credit were greater than the import cost, the optimizer would give the trivial solution – import cheaper power and export the same back at higher credit to the maximum extent possible. We will assume that export power gets less credit than the cost of import power, which is realistic and almost always the case. However, if such were not the case, a modeling technique, yet to be discussed in Chapter 8 to restrict

```
MODEL:  TITLE Fuel subsystem Model: FuelSS-v1: Base Case;
! ;
! Ref: Nath, "Industrial Process Plants: Global Optimization of Utility Systems", Chapter 7, (2023);
! ;

DATA: !inputs;     c_fo  = 18;  c_lpfg = 19;  c_hpfg = 20;  c_elec = 60; ! energy costs;
                   hpfg_demand = 30; lpfg_demand = 30; fo_demand = 30;  ! fuel demand, MW.th;
                   gtg_123 = 1; fc_123 = 1;                  ! 1=avial,2=requre,3=unavail;

! GTG params;      gtg_fg_a = 25.2701; gtg_fg_b = 1.6859; gtg_fg_c = 21.0119; gtg_fg_d = 2.2537;
                   gtg_fg_e = 15.0033; gtg_fg_f = 2.5970; gtg_ele_hl = 25;
! FC params;
                   fc_sp_a = 0.01; fc_sp_b = 0.015; fc_fg_ll = 25; fc_fg_hl = 75; fc_eta = 0.95;
ENDDATA

CALC:
       gtg_lgc_ll = 0; gtg_lgc_hl = 1; fc_lgc_ll = 0; fc_lgc_hl = 1;
       @IFC (gtg_123 #EQ# 3: gtg_lgc_hl = 0;); @IFC (gtg_123 #EQ# 2: gtg_lgc_ll = 1;);
       @IFC (fc_123 #EQ# 3: fc_lgc_hl = 0;); @IFC (fc_123 #EQ# 2: fc_lgc_ll = 1;);
ENDCALC

! Objective;    MIN = c_fo * BUY_FO + c_lpfg * BUY_LPFG + c_hpfg * BUY_HPFG
                    + c_elec * ( FC_ELE - GTG_ELE);

! constraints;    @FILE('C:\BookDG\LingoModels\FuelSS-v1.lng');

! Fuel Headers;
   [_HPFG_HDR]    BUY_HPFG + FC_FG = PRVFG + GTG_FG + hpfg_demand;
   [_LPFG_HDR]    BUY_LPFG + PRVFG = FC_FG + FGFLARE + lpfg_demand;
   [_FO_HDR]      BUY_FO = FOFLARE + fo_demand;

END MODEL FuelSS-v1
```

Fig. 7.2: Fuel subsystem model, FuelSS-v1.lg4, in LINGO.

the model to import or export under all circumstances using a binary variable, could be used.

We will consider the usual case where the import cost exceeds the export credit. Modeling-wise, we will need two power costs, one for power import and another for power export. Also, we will need two separate variables, one representing power import, **BUY_ELEC**, and another representing power export, **SEL_ELEC**.

An overall balance will be:

$$\textbf{BUY\_ELEC} + \text{internal generation} = \textbf{SEL\_ELEC} + \text{internal consumption} + \textit{elec\_demand}$$

(7.1)

Where "internal generation" refers to power generation within the utility system, such as in the STG, the CTG, and the GTG units; "internal consumption" refers to all power consumption within the utility system, such as in the various operating pumps, fans, and compressors with electric motor drives, and "elec_demand" refers to the electrical power demand by the process plant.

This model is trivially simple, with just the above-mentioned single constraint; so model development and testing are unnecessary.

We will add the constraint mentioned above to the overall model.

model ID:	Fuel subsystem model: FuelSS-v1		
case ID:	4		
Objective fm:	$/hr		$1,709.05

inputs:

	cost	demand	c_elec	GTG_aru	A
hpfg	20	30	60	FC_aru	A
lpfg	19	30			
fo	18	30			

params:

tg_fg_o,b	25.2701	1.6859
gtg_fg_c,d	21.0119	2.2537
gtg_fg_e,f	15.0033	2.5970
gtg_ele_hl	25	
fc_sp_o,b	0.0100	0.0150
fc_fg_ll,hl	25	75
fcm_eta	95%	
big	999	

Main model table:

#	ID	UOM	low limit	high limit	cost coeff -> / optimum	BUY_HPFG	BUY_LPFG	BUY_FO	PRVFG	FGFLARE	FOFLARE	GTG_FG	GTG_X	GTG_ELE	FC_FG	FC_SP	FC_X	FC_ELE
						MW.th	MW.th	MW.th	MW.th	MW.th	MW.th	MW.th	BIN	MW	MW.th	MW.sp	BIN	MW
						1	2	3	4	5	6	7	8	9	10	11	12	13
	cost coeff ->					$ 20.00	$ 19.00	$ 18.00						-$60.00				$ 60.0
	optimum					0.000	60.00	30.000	0.000	0.000	0.000	0.00	0	0.000	30.00	0.460	1	0.484
1	GTG_EB1G	MW.th	0	999	0.00							1	-25.3	-1.7				
2	GTG_EB1B	MW.th	0	999	0.00							1	-21.0	-2.3				
3	GTG_EB1R	MW.th	0	999	0.00							1	-15.0	-2.6				
4	GTG_ELE_HL	MW	-999	0	0.00								-25.0	1				
5	GTG_LGC	MW.sp	0.0	1.0	0.00								1					
6	FC_EB1	MW.th	0	0	0.00										-0.0150	1		
7	FC_FLO_LL	MW.th	-999	999	5.00										1		-25.0	
8	FC_FLO_HL	MW.th	-999	0	-45.00										1		-75.0	
9	FC_SP_HL	MW.sp	-999	0	-0.68											1	-1.1	-0.95
10	FC_LGC	MW	0.0	1.0	1.00												1.0	
11	FC_EB2	MW	0	0	0.00											1		1
12	HPFG_HDR	MW.th	30	30	30.00	1			-1			-1						
13	LPFG_HDR	MW.th	30	30	30.00		1		1	-1					-1			
14	FO_HDR	MW.th	30	30	30.00			1			-1							

Fig. 7.3: Fuel subsystem model in Microsoft Excel Solver, FuelSS-v1.xslx.

7.4 Modeling cooling water subsystem

We will ignore the single internal cooling water consumer, the CTG condenser, as it is already modeled in the steam system and will be included when we integrate the models. We will model all other components as shown below:

- CWP1, pump with electric motor drive
- CWP2, pump with HPS to MPS turbine drive
- CWP3, pump with dual (motor and HPS to LPS turbine) drives, and
- Cooling water header

Modeling is straightforward; we will use the generic pump models developed in Chapter 4 for the three cooling water pumps and the **IN = OUT** model for the header. The objective function would be to minimize electric power and steam costs only, ignoring the cooling water cost as it would not be necessary for the combined model. The model is small and can be solved using the basic Solver in Microsoft Excel.

For testing, we will use a cooling water demand of 15 ktph; and test for operations sensitivity to economic data, i.e., steam and electric power costs. Table 7.4 shows the test cases.

Figure 7.4 shows the LINGO model, and Fig. 7.5 shows the model in the basic Microsoft Excel Solver. Both models give identical solutions for all test problems. Figure. 7.5 also shows the solution to test case #1.

Tab. 7.4: Cooling water subsystem, CwSS-v1: Test cases.

Case #	Attribute	Symbolic name	Value	UOM	Expected key results
All	Cooling water demand	cw_demand	15	ktph	
1	elec cost	c_elec	60	$/MWh	Power is cheap, use electric drives
	hps2mps cost	c_hps2mps	5	$/t	
	hps2lps cost	c_hps2lps	10	$/t	
2	elec cost	c_elec	100	$/MWh	Power expensive, use turbine driven pumps
	hps2mps cost	c_hps2mps	5	$/t	
	hps2lps cost	c_hps2lps	10	$/t	
3	elec cost	c_elec	100	$/MWh	hps2mps expensive, use electric & hp->lp driven pumps
	hps2mps cost	c_hps2mps	10	$/t	
	hps2lps cost	c_hps2lps	10	$/t	

7.5 Modeling with general integers

So far in the book, we have dealt with one kind of discrete variable, the binary variable that can take one of two values: 0 or 1. We have used this variable to indicate a binary property of the equipment, such as ON/OFF state and standby/normal operation. In each case, the equipment was unique, i.e., no other equipment with identical characteristics existed. For instance, there are three boilers; they are similar, but each has unique characteristics.

```
MODEL:    TITLE Cooling Water subsystem Model: CWSS-v1: base case;
! Ref: Nath, "Industrial Process Plants: Global Optimization of Utility Systems", Chapter 7, (2023).;

DATA:!inputs;        c_elec =60; c_hps2mps =5; c_hps2lps =10; cw_demand =15;        ! costs, demand, ktph;

      !params;        cwp_sp_a =0.1;        cwp_sp_b =0.13; cwp_des_ll =2; cwp_flo_hl =8;    ! cwp common;
                      cwp1m_eta =0.95;      cwp1_aru = 1;                                     ! cwp1;
                      cwp2t_stm_a = 7.89; cwp2t_stm_b = 11.86; cwp2_aru = 1;                  ! cwp2;
                      cwp3m_eta = 0.95;    cwp3t_stm_a = 3.94; cwp3t_stm_b = 5.94; cwp3_aru = 1;  ! cwp3;
ENDDATA

CALC:
      cwp1_sp_a =cwp_sp_a; cwp1_sp_b =cwp_sp_b; cwp1_des_ll =cwp_des_ll; cwp1_flo_hl =cwp_flo_hl;
      cwp2_sp_a =cwp_sp_a; cwp2_sp_b =cwp_sp_b; cwp2_des_ll =cwp_des_ll; cwp2_flo_hl =cwp_flo_hl;
      cwp3_sp_a =cwp_sp_a; cwp3_sp_b =cwp_sp_b; cwp3_des_ll =cwp_des_ll; cwp3_flo_hl =cwp_flo_hl;
      cwp1_lgc_ll = 0; cwp1_lgc_hl = 1;
      @IFC (cwp1_aru #EQ# 3: cwp1_lgc_hl = 0;); @IFC (cwp1_aru #EQ# 2: cwp1_lgc_ll = 1;);
      cwp2_lgc_ll = 0; cwp2_lgc_hl = 1;
      @IFC (cwp2_aru #EQ# 3: cwp2_lgc_hl = 0;); @IFC (cwp2_aru #EQ# 2: cwp2_lgc_ll = 1;);
      cwp3_lgc_ll = 0; cwp3_lgc_hl = 1;
      @IFC (cwp3_aru #EQ# 3: cwp3_lgc_hl = 0;); @IFC (cwp3_aru #EQ# 2: cwp3_lgc_ll = 1;);
ENDCALC

  ! Objective;        MIN = c_elec*(CWP1M_ELE+CWP3M_ELE)+ c_hps2mps*CWP2T_STM + c_hps2lps*CWP3T_STM;

  ! CWP1;             @FILE('C:\BookDG\LingoModels\CWP1MD-v1.lng');
  [_CWP1_LGC_LL]     CWP1_X > cwp1_lgc_ll;
  [_CWP1_LGC_HL]     CWP1_X < cwp1_lgc_hl;

  ! CWP2;             @FILE('C:\BookDG\LingoModels\CWP2TD-v1.lng');
  [_CWP2_LGC_LL]     CWP2_X > cwp2_lgc_ll;
  [_CWP2_LGC_HL]     CWP2_X < cwp2_lgc_hl;

  ! CWP3;             @FILE('C:\BookDG\LingoModels\CWP3DD-v1.lng');
  [_CWP3_LGC_LL]     CWP3_X > cwp3_lgc_ll;
  [_CWP3_LGC_HL]     CWP3_X < cwp3_lgc_hl;

  ! CW header;
  [_CWP_DMND]        CWP1_FLO + CWP2_FLO + CWP3_FLO = cw_demand;

END MODEL CWSS-v1
```

Fig. 7.4: Cooling water subsystem model in LINGO, CWSS-v1.

We have yet to encounter situations where we have multiple instances of equipment with identical characteristics. In such cases, one could repeat the individual model multiple times, and each equipment model has its own ON/OFF binary variable; however, MILP solvers provide a better option for such situations. MILP solvers offer something called "General Integer" variables that could take not just two but any number of discrete values, such as integer values between 0 and 4 for example, a value of 2 would indicate that two pieces of equipment are operational, without saying which two, as that is a moot issue when multiple equipment are providing identical service. With this capability, we could have a single equipment model and associate a general integer variable that could take any integer value up to the number of available instances, representing the number of operating equipment.

In LINGO, General Integers are specified using the **@GIN()** function, instead of the **@BIN()** function [1]. In the Excel Solver, we specify such variables as **INT** instead of **BIN** [2].

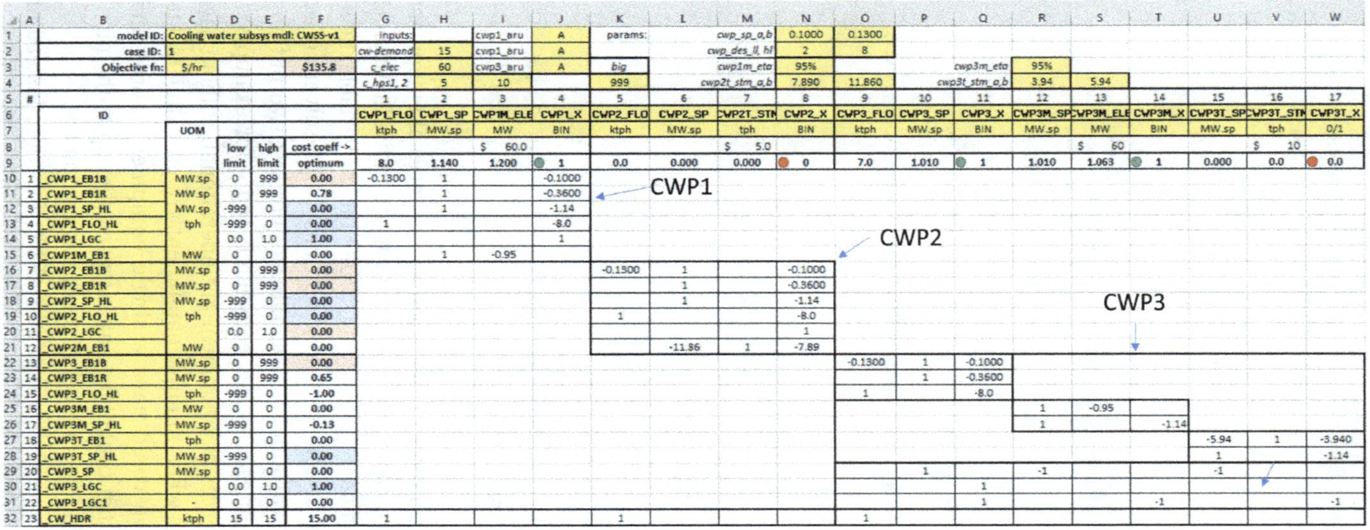

ID			params:	cwp_sp_a,b	0.1000	0.1300				cwp1_aru		
model ID: Cooling water subsys mdl: CWSS-v1			inputs:	cw-demand	C_elec	c_hps1,2			cwp1_aru	cwp2_aru	cwp3_aru	
case ID: 1					15	60						
Objective fnc: $/hr $135.8			Big	999			cwp1m_eta	95%		cwp3m_eta	95%	
							cwp2t_stm_a,b	7.890	11.860	cwp3t_stm_a,b	3.94	5.94

ID	UOM	low limit	high limit	cost coeff -> optimum	CWP1_FLO ktph	CWP1_SP MW.sp	CWP1M_ELE MW	CWP1M_X BIN	CWP2_FLO ktph	CWP2_SP MW.sp	CWP2T_STN tph	CWP2M_X BIN	CWP3_FLO ktph	CWP3_SP MW.sp	CWP3M_ELE BIN	CWP3M_X MW	CWP3_SP BIN	CWP3T_STN MW.sp	CWP3T_SP tph	CWP3T_X 0/1
						$ 60.0	$ 1.200	1		$ 5.0	$ 0.000	0		$ 1.010	$ 1.063	$ 60		$ 10	$ 0.0
					8.0	1.140			0.0	0.000		0	7.0	1.010	1.010		1	0.000		0.0
CWP1_EB1B	MW.sp	0	999	0.00	-0.1300	1	-0.1000			-0.1000			-0.1300	1	-0.1000					
CWP1_EB1R	MW.sp	0	999	0.78		1	-0.3600			-0.3600				1	-0.3600					
CWP1_SP_HL	MW.sp	-999	0	0.00	1	1	-1.14			-1.14			1		-8.0					
CWP1_FLO_HL	tph	-999	0	0.00			-8.0			-8.0					-1					
CWP1_1GC	MW	0.0	1.0	1.00			1			1										
CWP1M_EB1	MW	0	0	0.00		-0.95														
CWP2_EB1B	MW.sp	0	999	0.00				-0.1300		-0.1000										
CWP2_EB1R	MW.sp	0	999	0.00				1		-0.3600										
CWP2_SP_HL	MW.sp	-999	0	0.00				1		-1.14										
CWP2_FLO_HL	tph	-999	0	0.00	1					-8.0										
CWP2_1GC	MW	0.0	1.0	0.00						-1										
CWP2M_EB1	MW	0	0	0.00					-11.86	-7.89										
CWP3_EB1B	MW.sp	0	999	0.65									-0.1300	1	-0.1000					
CWP3_EB1R	MW.sp	0	999	-1.00										1	-0.3600					
CWP3_FLO_HL	tph	0	0	0.00									1		-8.0					
CWP3M_EB1	MW	0	0	-0.13													-0.95			
CWP3M_SP_HL	MW.sp	-999	0	0.00													-1.14	-5.94	-3.940	
CWP3T_EB1	tph	-999	0	0.00											1	1	1	1	-1.14	
CWP3T_SP_HL	MW.sp	-999	0	1.00											1		-1	-1		
CWP3_SP	MW.sp	0.0	1.0	0.00												-1				
CWP3_1GC	-	0.0	0.0	0.00											1	1				
CW_HDR	ktph	15	15	15.00	1				1				1				-1			-1

CWP1 CWP2 CWP3

Fig. 7.5: Cooling water subsystem model, CWSS-v1: in basic Excel Solver.

In the next section, we will encounter such a situation that will serve as an illustration of the use of General Integer variables.

7.6 Compressed air subsystem model

The compressed air subsystem, shown in Fig. 1.8, comprises the following:
- Air compressors AC1, 2, 3, and 4; rotary compressors with internal combustion engine drivers
- Air compressor AC5, centrifugal compressor with electrical motor drive, and
- Compressed air header

We will use the generic compressor models developed in Chapter 4 for these compressors. For the compressed air header, we will use a simple IN = OUT model.

Note that the air compressors 1, 2, 3, and 4 are identical units that operate in parallel, providing the same service; so which ones to use is a moot issue. Moreover, in online optimization implementation, this decision could be left to the judgment of the board operator, who would also consider other relevant factors.

These four compressors have internal combustion engine (ICE) drives and will be modeled as a single unit like the generic compressor with ICE drive model, C01ED-v1. We will, however, associate a general integer variable, **AC1234_NX**, which could take any integer value between 0 and 4 to indicate the number of operating units. Doing so makes the model compact and efficient, reducing the number of variables by a factor of 4. The A/R/U specification will be replaced by two inputs: the number of required compressors, *ac1234_r*, and the total number of available compressors, *ac1234_a*.

Air compressor # 5 is a large unit driven by an electrical motor; it would be modeled just like the generic compressor with motor drive model C01MD-v1.

We will test this model with several values of compressed air demand ranging from 5 to 50 km^3/hr.

Figures 7.6 and 7.7 show the complete LINGO model for the Compressed Air subsystem. Figure 7.8 is the Microsoft Excel implementation, showing the solution to test problem #1. Both Solvers give identical solutions.

7.7 The integrated utility system model

In this chapter, so far, we have discussed and developed MILP models for the remaining four subsystems: fuel, electrical power, cooling water, and compressed air. We are now ready to integrate them with the steam subsystem model developed in the previous chapter to create a model of the entire utility system.

Note that this integration will almost complete the utility system model, but we still need the models for purchased fuel and electrical power contracts. However, we

```
! Compressed Air Subsystem CaSS: model constraints file: CaSS-v1.lng;

    ! AC1,2,3 & 4;
    [_AC1234_EB1]     AC1234_SP  = ac1_sp_a * AC1234_NX + ac1_sp_b * AC1234_CA;
    [_AC1234E_EB1]    AC1234E_FG = e01_fg_a * AC1234_NX + e01_fg_b * AC1234_SP;
    [_AC1234_CA_LL]   AC1234_CA > ac1_ca_ll * AC1234_NX;
    [_AC1234_CA_HL]   AC1234_CA < ac1_ca_hl * AC1234_NX;
    [_AC1234_NX_LL]   AC1234_NX > ac1234_r;
    [_AC1234_NX_HL]   AC1234_NX < ac1234_a;
    !integers;        @GIN( AC1234_NX );

    ! AC5;
    [_AC5_EB1]        AC5_SP = ac5_sp_a * AC5_X + ac5_sp_b * AC5_CA;
    [_AC5M_EB1]       AC5_SP = ac5m_eta * AC5M_ELE;
    [_AC5_CA_LL]      AC5_CA > ac5_ca_ll * AC5_X;
    [_AC5_CA_HL]      AC5_CA < ac5_ca_hl * AC5_X;
    [_AC5_LGC_LL]     AC5_X  > ac5_lgc_ll;
    [_AC5_LGC_HL]     AC5_X  < ac5_lgc_hl;
    !integers;        @BIN( AC5_X );
!;
```

Fig. 7.6: Compressed air subsystem LINGO constraints file: CaSS-v1.lng.

```
MODEL: TITLE CaSS-v1: Compressed Air Sub System model: Base Case;
!
! Nath, "Industrial Process Plants: Global Optimization of Utility Systems", Chapter 7, (2023).;

    DATA: !inputs;    ca_demand = 30; c_lpfg = 19; c_elec = 60; ! demand in kM3/h, costs;
                      ac1234_r = 1; ac1234_a = 4;               ! AC1-4 required, avaialable;
                      ac5_aru = 1;                              !1 =Avaialble,2= required,3= unavail;

          !params;    ac1_sp_a = 0.1320; ac1_sp_b = 0.0290;    ! AC1 CA flow vs sp coefficients;
                      ac1_ca_ll = 3; ac1_ca_hl = 9;            ! AC1 CA flow limits, kM3/h;
                      e01_fg_a = 0.1161; e01_fg_b = 2.1242;    ! ICE fuel vs sp coefficients;
                      ac5_sp_a = 1.044; ac5_sp_b = 0.041;      ! AC5 CA flow vs sp coefficients;
                      ac5_ca_ll = 20; ac5_ca_hl = 55;          ! AC5 CA flow limits, kM3/h;
                      ac5m_eta = 0.95;                         ! AC 5 motor efficiency;
    ENDDATA

    CALC:
        ac5_lgc_ll = 0; ac5_lgc_hl = 1;
        @IFC (ac5_aru #EQ# 3: ac5_lgc_hl = 0;); @IFC (ac5_aru #EQ# 2: ac5_lgc_ll = 1;);
    ENDCALC

    ! Objective;     MIN = c_lpfg * AC1234E_FG + c_elec * AC5M_ELE;

    ! CaSS;          @FILE('C:\BookDG\LingoModels\CaSS-v1.lng');

    ! CA header;
    [_CA_HDR]        AC1234_CA + AC5_CA = VENT_CA + ca_demand;

END MODEL CASS-v1
```

Fig. 7.7: Compressed air subsystem model, CASS-v1: in LINGO.

need to discuss a few new techniques before modeling the contracts, which we will do in the next chapter.

Combining the five subsystems is straightforward and similar to what was done earlier. This integrated system model, **UtilSys-v1**, has 123 variables, including 25 binary, one general integer, and 146 constraint rows. This model size exceeds the capa-

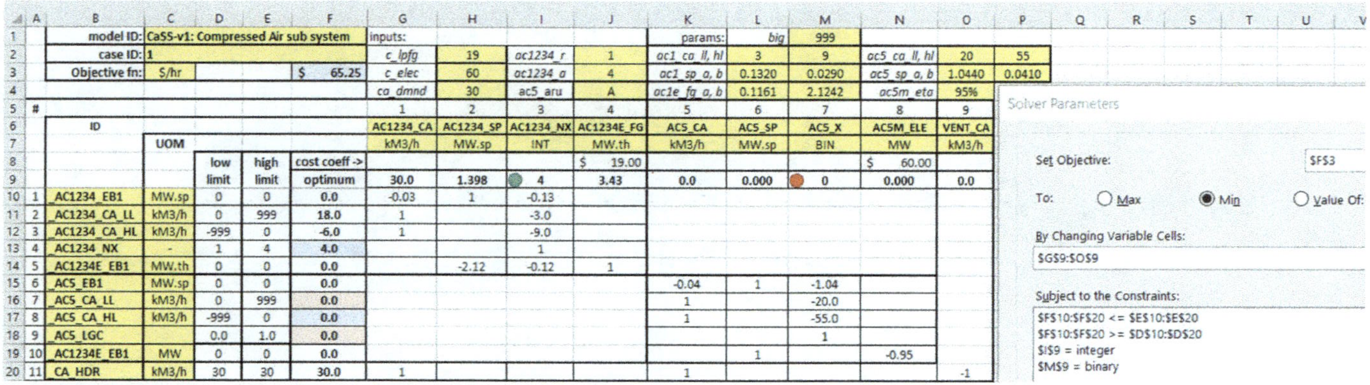

model ID: CaSS-v1: Compressed Air sub system
case ID: 1
Objective fn: $/hr $ 65.25

inputs:

	value
c_lqfg	19
c_elec	60
ca_dmnd	30

Additional header values: oc1234_r = 1, oc1234_a = 4, ac5_aru = A

params:

oc1_ca_ll,hl	3	big	999	oc5_ca_ll,hl	20	55
oc1_sp_o,b	0.1320		0.0290	oc5_sp_o,b	1.0440	0.0410
oc1e_fg_o,b	0.1161		2.1242	oc5m_eta	95%	

Decision variables (row 9, changing cells $G$9:$O$9):

	AC1234_CA	AC1234_SP	AC1234_NX	AC1234E_FG	ACS_CA	ACS_SP	ACS_X	ACSM_ELE	VENT_CA
UOM	kM3/h	MW.sp	INT	MW.th	kM3/h	MW.sp	BIN	MW	kM3/h
price				$ 19.00				$ 60.00	
value	30.0	1.398	4	3.43	0.0	0.000	0	0.000	0.0

Constraints (rows 10–20):

#	ID	UOM	low limit	high limit	optimum	AC1234_CA	AC1234_SP	AC1234_NX	AC1234E_FG	ACS_CA	ACS_SP	ACS_X	ACSM_ELE	VENT_CA
1	AC1234_EB1	MW.sp	0	0	0.0	-0.03	1	-0.13						
2	AC1234_CA_LL	kM3/h	0	999	18.0	1		-3.0						
3	AC1234_CA_HL	kM3/h	-999	0	-6.0	1		-9.0						
4	AC1234_NX	-	1	4	4.0			1						
5	AC1234E_EB1	MW.th	0	0	0.0		-2.12	-0.12	1	-0.04		-1.04		
6	ACS_EB1	MW.sp	0	999	0.0					1	1	-20.0		
7	ACS_CA_LL	kM3/h	0	0	0.0					1		-55.0		
8	ACS_CA_HL	kM3/h	-999	999	0.0						1	1		
9	ACS_LGC	kM3/h	0.0	1.0	0.0									
10	AC1234E_EB1	MW	0	0	0.0								-0.95	
11	CA_HDR	kM3/h	30	30	30.0	1				1				-1

Solver Parameters

Set Objective: $F$3

To: ○ Max   ● Min   ○ Value Of: [$F$3]

By Changing Variable Cells:
$G$9:$O$9

Subject to the Constraints:
$F$10:$F$20 <= $E$10:$E$20
$F$10:$F$20 >= $D$10:$D$20
$I$9 = integer
$M$9 = binary

Fig. 7.8: Compressed Air subsystem model, CaSS-v1: in Excel Standard Solver.

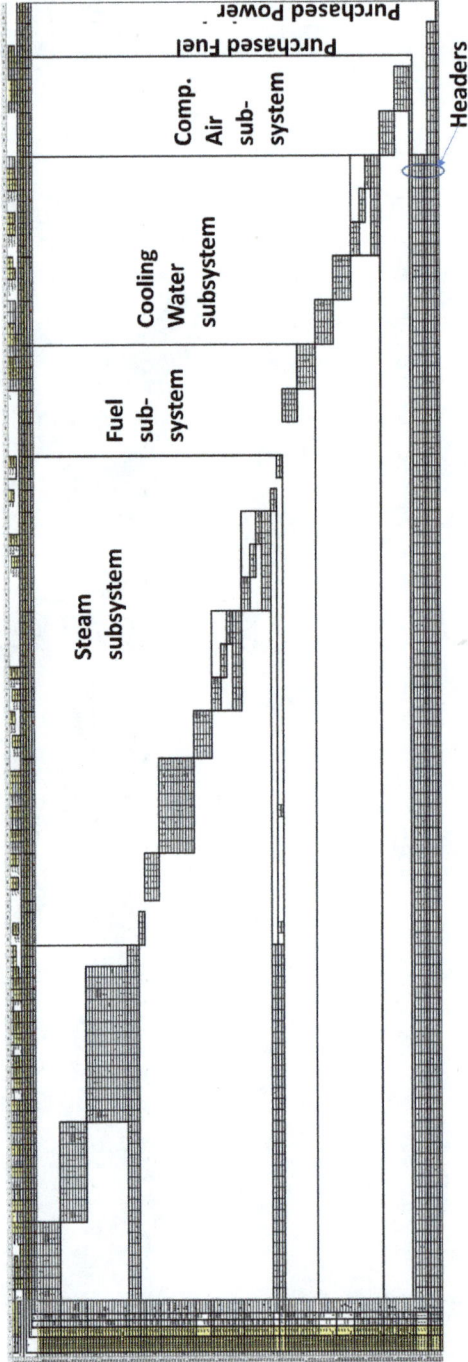

Fig. 7.9: Utility system model: UtilSys-v1: in Microsoft Excel, showing model structure.

bilities of the Basic Solver in Microsoft Excel, so we will use the "Premium Solver" instead [2].

Figure 7.9 shows the Excel model for the integrated system; however, it is so large that we had to shrink it significantly to fit the printed page. Details are no longer discernable; only the model structure is. Note that the model, as expected, has a block diagonal structure except for the header constraints that span multiple units or subsystems and are on the lower part of the spreadsheet.

The LINGO model will be easier to see as it is in algebraic format and also easier to fit on printed pages. Furthermore, LINGO allows the inclusion of external files within a LINGO model [1]. This feature makes the model compact and easier to manage. We will include the following include files in the utility system model;

- **UtilSys-params.lng:** It contains all equipment parameters. Figure 7.10 shows the file.
- **UtilSys-calcs.lng:** It contains all calculations before optimization. Figure 7.11 shows the file.
- **StmSS-v1.lng:** It contains the steam subsystem constraints; these are the steam subsystem constraints, as shown in Figs. 6.8 and 6.9.
- **FuelSS-v1.mdl:** It contains the fuel subsystem constraints, as shown in Fig. 7.1.

```
!Steam ss params;
bl1_fg_c = 15.2371; bl1_fg_d = 0.790615; bl1_fg_e =-12.4525; bl1_fg_f = 0.953495;        ! BL1;
bl1_bdr=0.02; bl1_fg_s = 3; bl1_stm_ll = 70; bl1_stm_hl = 210;  bl1at_c = 6;

bl2_fg_c = 18.1694; bl2_fg_d = 0.834726; bl2_fg_e =-11.6149; bl2_fg_f =0.991485;          ! BL2;
bl2_bdr=0.02; bl2_fg_s = 4; bl2_stm_ll = 80; bl2_stm_hl = 240; bl2am_c =0.5;
bl3_fg_c = 21.1962; bl3_fg_d = 0.814223; bl3_fg_e =-15.2507; bl3_fg_f = 0.966085;         ! BL3;
bl3_bdr=0.02; bl3_fg_s = 5; bl3_stm_ll = 100; bl3_stm_hl = 300; bl3am_c =0.5; bl3at_c =5;

prv1_a = 0.133320; prv1_b = 0.929206; bdf_a = 1.884648; bdf_b = 8.309105;                  ! PRV, BDF;

stg_stm_a =8; stg_stm_b =14; stg_stm_s =4; stg_stm_hl =50; stg_ele_ll =1; stg_ele_hl =3; !STG;

ctg_hps_a =16.997; ctg_hps_b =5.433; ctg_hps_c =0.7859; ctg_hps_d =0.5605; ctg_hps_s=10;!CTG;
ctg_hps_ll =50; ctg_hps_hl =250; ctg_mps_hl =180; ctg_lps_hl =90; ctg_cnd_ll =10;
ctg_cnd_hl =60; ctg_ele_ll =3; ctg_ele_hl =13; ctg_cw_b = 0.04;

bp_sp_a =0.18391; bp_sp_b =0.000423; bp_des_ll =100; bp_flo_hl =250;                       ! BP common;
bp1m_eta =0.95; bp2m_eta =0.95; bp2t_stm_a =0.27306; bp2t_stm_b =8.48596;                  ! BP1, BP2;
bp3m_eta =0.95; bp4m_eta =0.95; bp4t_stm_a =0.77092; bp4t_stm_b =23.95717;                 ! BP3, BP4;
dea_a =-0.114660; dea_b = -0.379340; dea_c =6.493940;                                      ! DEA;

!Fuel ss params;
gtg_fg_a = 25.2701; gtg_fg_b = 1.6859; gtg_fg_c = 21.0119; gtg_fg_d = 2.2537;              ! GTG;
gtg_fg_e =15.0033; gtg_fg_f =2.5970; gtg_ele_hl = 25; big = 999;
fc_sp_a = 0.01; fc_sp_b = 0.015; fc_fg_ll = 25; fc_fg_hl = 75; fc_eta = 0.95;              ! FC;

!CW ss params;
cwp_sp_a = 0.1; cwp_sp_b = 0.13; cwp_des_ll = 2; cwp_flo_hl = 8;                           ! CWP common;
cwp1m_eta = 0.95; cwp2t_stm_a = 7.89; cwp2t_stm_b = 11.86;                                 ! CWP1, CWP2;
cwp3m_eta = 0.95; cwp3t_stm_a = 3.94; cwp3t_stm_b = 5.94;                                  ! CWP3;

!CA params,AC1234 first;
ac1_sp_a =0.1320; ac1_sp_b =0.0290; ac1_ca_ll =3; ac1_ca_hl =9; e01_fg_a =0.1161; e01_fg_b =2.1242;
ac5_sp_a = 1.044; ac5_sp_b = 0.041; ac5_ca_ll = 20; ac5_ca_hl = 55; ac5m_eta = 0.95;       !AC 5;

! ;
```

Fig. 7.10: LINGO Utility system parameters file: UtilSys-params.lng.

```
CALC:
    bp1_sp_a = bp_sp_a; bp1_sp_b = bp_sp_b; bp1_des_11 = bp_des_11; bp1_flo_hl = bp_flo_hl;
    bp2_sp_a = bp_sp_a; bp2_sp_b = bp_sp_b; bp2_des_11 = bp_des_11; bp2_flo_hl = bp_flo_hl;
    bp3_sp_a = bp_sp_a; bp3_sp_b = bp_sp_b; bp3_des_11 = bp_des_11; bp3_flo_hl = bp_flo_hl;
    bp4_sp_a = bp_sp_a; bp4_sp_b = bp_sp_b; bp4_des_11 = bp_des_11; bp4_flo_hl = bp_flo_hl;

    cwp1_sp_a =cwp_sp_a; cwp1_sp_b =cwp_sp_b; cwp1_des_11 =cwp_des_11; cwp1_flo_hl =cwp_flo_hl;
    cwp2_sp_a =cwp_sp_a; cwp2_sp_b =cwp_sp_b; cwp2_des_11 =cwp_des_11; cwp2_flo_hl =cwp_flo_hl;
    cwp3_sp_a =cwp_sp_a; cwp3_sp_b =cwp_sp_b; cwp3_des_11 =cwp_des_11; cwp3_flo_hl =cwp_flo_hl;
    cwp1_lgc_11 = 0; cwp1_lgc_hl = 1;

    b11_lgc_11 = 0; b11_lgc_hl = 1; b12_lgc_11 = 0; b12_lgc_hl = 1; b13_lgc_11 = 0; b13_lgc_hl = 1;
    @IFC (b11_123 #EQ# 3: b11_lgc_hl = 0;); @IFC (b11_123 #EQ# 2: b11_lgc_11 = 1;);
    @IFC (b12_123 #EQ# 3: b12_lgc_hl = 0;); @IFC (b12_123 #EQ# 2: b12_lgc_11 = 1;);
    @IFC (b13_123 #EQ# 3: b13_lgc_hl = 0;); @IFC (b13_123 #EQ# 2: b13_lgc_11 = 1;);
    stg_lgc_11 = 0; stg_lgc_hl = 1; ctg_lgc_11 = 0; ctg_lgc_hl = 1;
    @IFC (stg_123 #EQ# 3: stg_lgc_hl = 0;); @IFC (stg_123 #EQ# 2: stg_lgc_11 = 1;);
    @IFC (ctg_123 #EQ# 3: ctg_lgc_hl = 0;); @IFC (ctg_123 #EQ# 2: ctg_lgc_11 = 1;);
    bp1_lgc_11 = 0; bp1_lgc_hl = 1; bp2_lgc_11 = 0; bp2_lgc_hl = 1;
    bp3_lgc_11 = 0; bp3_lgc_hl = 1; bp4_lgc_11 = 0; bp4_lgc_hl = 1;
    @IFC (bp1_123 #EQ# 3: bp1_lgc_hl = 0;); @IFC (bp1_123 #EQ# 2: bp1_lgc_11 = 1;);
    @IFC (bp2_123 #EQ# 3: bp2_lgc_hl = 0;); @IFC (bp2_123 #EQ# 2: bp2_lgc_11 = 1;);
    @IFC (bp3_123 #EQ# 3: bp3_lgc_hl = 0;); @IFC (bp3_123 #EQ# 2: bp3_lgc_11 = 1;);
    @IFC (bp4_123 #EQ# 3: bp4_lgc_hl = 0;); @IFC (bp4_123 #EQ# 2: bp4_lgc_11 = 1;);
    gtg_lgc_11 = 0; gtg_lgc_hl = 1; fc_lgc_11 = 0; fc_lgc_hl = 1;
    @IFC (gtg_123 #EQ# 3: gtg_lgc_hl = 0;); @IFC (gtg_123 #EQ# 2: gtg_lgc_11 = 1;);
    @IFC (fc_123 #EQ# 3: fc_lgc_hl = 0;); @IFC (fc_123 #EQ# 2: fc_lgc_11 = 1;);
    cwp1_lgc_hl =0; cwp1_lgc_hl =1; cwp2_lgc_11 =0; cwp2_lgc_hl =1; cwp3_lgc_11 =0; cwp3_lgc_hl =1;
    @IFC (cwp1_123 #EQ# 3: cwp1_lgc_hl = 0;); @IFC (cwp1_123 #EQ# 2: cwp1_lgc_11 = 1;);
    @IFC (cwp2_123 #EQ# 3: cwp2_lgc_hl = 0;); @IFC (cwp2_123 #EQ# 2: cwp2_lgc_11 = 1;);
    @IFC (cwp3_123 #EQ# 3: cwp3_lgc_hl = 0;); @IFC (cwp3_123 #EQ# 2: cwp3_lgc_11 = 1;);
    ac5_lgc_11 = 0; ac5_lgc_hl = 1;
    @IFC (ac5_123 #EQ# 3: ac5_lgc_hl = 0;); @IFC (ac5_123 #EQ# 2: ac5_lgc_11 = 1;);
ENDCALC
!;
```

Fig. 7.11: LINGO Utility system CALCS include file: UtilSys-calcs.lng.

- **CwSS-v1.mdl:** It contains the cooling water subsystem constraints, as shown in Fig. 7.4.
- **CaSS-v1.mdl:** It contains the compressed air subsystem constraints. Figure 7.6 shows the file.

The LINGO model using these include files is shown in Fig. 7.12.

Table 7.5 shows the four test cases and the expected results for the test cases. The Premium Solver and the LINGO Solver find globally optimum solutions; the solutions are essentially identical. Table 7.6 shows the Excel Solver solution with BFW pumps #2 and #4 in operation. Note that the two solvers give globally optimum solutions that have identical objective functions but differ in the usage of boiler feed water pumps; the two solutions have swapped boiler feed water pump #1 with motor drive and boiler feed water pump #4 with motor drive that are essentially identical units. This situation of multiple globally optimum solutions is not uncommon, especially for large models and models with identical equipment. Also, the multiplicity of global optimum solutions poses no problem, as we discuss in online implementation strategies in Part II of this book.

In the next chapter, we will discuss the modeling of purchased energy contracts and thus complete the sample utility system model.

```
MODEL:  TITLE Utility System Model: UtilSys-v1;
! ;
! Ref: Nath, "Industrial Process Plants: Global Optimization of Utility Systems", Chapter 8, (2023);

  DATA:
 !inputs;    cw_demand = 15; ca_demand = 30; c_dmw =3;
             c_fo =18; c_lpfg =19; c_hpfg =20; c_buyelec =60; c_selelec=50; ! purchased energy costs;
             hps_demand =100; mps_demand=100; lps_demand=100; mps_reserve=50; hps_reserve=50;
             hpfg_demand = 30; lpfg_demand = 30; fo_demand = 30; elec_demand = 30;

! Equipment ARU specs:                               1=Available, 2=Required, 3=Unavailable;
             bl1_aru =1; bl2_aru =1; bl3_aru =1; stg_aru =1; ctg_aru =1; bp1_aru =1; bp2_aru =1;
             bp3_aru =1; bp4_aru =1; gtg_aru =3; fc_aru =1; cwp1_aru =1; cwp2_aru =1; cwp3_aru =1;
             ac1234_r =1; ac1234_a =4; ac5_aru =1;

! params;           @FILE('C:\BookDG\LingoModels\UtilSys-params.lng')
   ENDDATA

! calcs;            @FILE('C:\BookDG\LingoModels\UtilSys-calcs.lng')

! Objective;       MIN = c_fo * BUY_FO + c_lpfg * BUY_LPFG + c_hpfg * BUY_HPFG
                      + c_buyelec * BUY_ELEC - c_selelec * SEL_ELEC + c_dmw*DEA_DMW;

! stm SS;           @FILE('C:\BookDG\LingoModels\stmSS-v1.lng');
! fuel SS;          @FILE('C:\BookDG\LingoModels\fuelSS-v1.lng');
! CW SS;            @FILE('C:\BookDG\LingoModels\cwSS-v1.lng');
! CA SS;            @FILE('C:\BookDG\LingoModels\caSS-v1.lng');

! Headers;
  [_HPS_HDR]       BL2_STM + BL3_STM = PRV1_HPS + STG_STM + CTG_HPS + BL3AT_STM + BP2T_STM
                      + BP4T_STM + CWP2T_STM + CWP3T_STM + hps_demand;
  [_MPS_HDR]       BL1_STM + PRV1_MPS + STG_STM + CTG_MPS + BP4T_STM + CWP2T_STM
                      = PRV2_STM + BL1AT_STM + mps_demand;
  [_LPS_HDR]       PRV2_STM + BL1AT_STM + BL3AT_STM + BP2T_STM + CTG_LPS + BDF_LPS + CWP3T_STM
                      = DEA_LPS + VENT + lps_demand;
  [_BFW_HDR]       BL1_BFW + BL2_BFW + BL3_BFW + PRV1_BFW = BP1_FLO +BP2_FLO +BP3_FLO +BP4_FLO;
  [_CW_HDR]        CWP1_FLO + CWP2_FLO + CWP3_FLO - CTG_CW = cw_demand;
  [_CA_HDR]        AC1234_CA + AC5_CA = VENT_CA + ca_demand;
  [_HPFG_HDR]      BUY_HPFG + FC_FG = PRVFG + GTG_FG + BL2_HPFG + BL3_HPFG + hpfg_demand;
  [_LPFG_HDR]      BUY_LPFG + PRVFG = FC_FG + FGFLARE +BL2_LPFG +BL3_LPFG +AC1234E_FG +lpfg_demand;
  [_FO_HDR]        BUY_FO = FOFLARE + BL1_FUEL + fo_demand;
  [_ELEC_HDR]      BUY_ELEC -SEL_ELEC +STG_ELE +CTG_ELE +GTG_ELE =FC_ELE +BL2AM_ELE +BL3AM_ELE
             +BP1M_ELE +BP2M_ELE +BP3M_ELE +BP4M_ELE +CWP1M_ELE +CWP3M_ELE +AC5M_ELE +elec_demand;
END MODEL UtilSys-v1
```

Fig. 7.12: LINGO Utility system model: UtilSys-v1: in LINGO.

Tab. 7.5: Utility system model: UtilSys-v1: test cases.

Case #	Attribute	Symbolic name	Value	UOM	Expected key results
1, base case	elec import cost	c_elec	60	$/MWh	
	elec export cost	csel_elec	50	$/MWh	GTG is unavailable, i.e., GTG_X = 0.
	HPFG cost	c_hpfg	20	$/MW(th)h	Elec power import is expensive
	LPFG cost	c_lpfg	19	$/MW(th)h	expect power generation
	FPO cost	c_fo	18	$/MW(th)h	**STG_X = CTG_X = 1, FC_X = 0**
	DMW cost	c_dmw	3	$/ton	at least 2 boilers operating
	HP steam demand	hps_demand	100	tph	
	MP steam demand	mps_demand	100	tph	
	LP steam demand	lps_demand	100	tph	
	HP steam reserve	hs_reserve	50	tph	
	MP steam reserve	mps_reserve	50	tph	
	CW demand	cw_demand	15	ktph	
	CA demand	ca_demand	30	kM3/hr	
	Elec power demand	elec_demand	30	MW	
	HPFG demand	hpfg_demand	30	MW(th)h	
	LPFG demand	mpfg_demand	30	MW(th)h	
	FO demand	fo_demand	30	MW(th)h	
2	HP steam demand	hps_demand	120	tph	
	MP steam demand	mps_demand	120	tph	
	LP steam demand	lps_demand	120	tph	
	CW demand	cw_demand	18	ktph	Similar to base case except higher operating cost
	CA demand	ca_demand	36	kM3/hr	
	Elec power demand	elec_demand	36	MW	
	HPFG demand	hpfg_demand	36	MW(th)h	
	LPFG demand	mpfg_demand	36	MW(th)h	
	FO demand	fo_demand	36	MW(th)h	
3	HP steam demand	hps_demand	80	tph	
	MP steam demand	mps_demand	80	tph	
	LP steam demand	lps_demand	80	tph	
	CW demand	cw_demand	12	ktph	Similar to base case except lower operating cost
	CA demand	ca_demand	24	kM3/hr	
	Elec power demand	elec_demand	24	MW	
	HPFG demand	hpfg_demand	24	MW(th)h	
	LPFG demand	mpfg_demand	24	MW(th)h	
	FO demand	fo_demand	24	MW(th)h	
4	HP steam demand	hps_demand	150	tph	
	MP steam demand	mps_demand	150	tph	
	LP steam demand	lps_demand	150	tph	
	CW demand	cw_demand	15	ktph	Similar to base case except much higher operating cost
	CA demand	ca_demand	30	kM3/hr	
	Elec power demand	elec_demand	30	MW	
	HPFG demand	hpfg_demand	30	MW(th)h	
	LPFG demand	mpfg_demand	30	MW(th)h	
	FO demand	fo_demand	30	MW(th)h	

Tab. 7.6: Utility system model: UtilSys-v1: test case #1 global optimum solution.

Utility system Model: UtilSys-v1 : Test case # 1									
Global optimal solution found.									
Opmum objective function value (minimum operating cost, $/hr)				$ 9,969.34					

#	Variable ID	UOM	Opt. value	#	Variable ID	UOM	cost coeff	Opt. value	Cost, $/hr	#	Variable ID	UOM	cost coeff	Opt. value	Cost, $/hr
1	BL1_STM	tph	0.0	42	STG_X	BIN		1	$ -	83	PRVFG	MW.th		0.000	$ -
2	BL1_BFW	tph	0.0	43	STG_Y	BIN		0	$ -	84	FGFLARE	MW.th		0.000	$ -
3	BL1_BD	tph	0.0	44	CTG_HPS	tph		187.8	$ -	85	FOFLARE	MW.th		0.000	$ -
4	BL1_RSV	tph	0.0	45	CTG_MPS	tph		87.8	$ -	86	GTG_FG	MW.th		0.00	$ -
5	BL1_FUEL	MW.th	0.0	46	CTG_LPS	tph		90.0	$ -	87	GTG_X	BIN		0	$ -
6	BL1AT_STM	tph	0.0	47	CTG_CND	tph		10.0	$ -	88	GTG_ELE	MW		0.000	$ -
7	BL1_X	BIN	0	48	CTG_CW	ktph		0.4	$ -	89	FC_FG	MW.th		30.00	$ -
8	BL1_Y	BIN	0	49	CTG_ELE	MW		9.453	$ -	90	FC_SP	MW.sp		0.460	$ -
9	BL2_STM	tph	136.7	50	CTG_X	BIN		1	$ -	91	FC_X	BIN		1	$ -
10	BL2_BFW	tph	139.5	51	CTG_Y	BIN		0	$ -	92	FC_ELE	MW		0.484	$ -
11	BL2_BD	tph	2.7	52	BP1_FLO	tph		0.0	$ -	93	CWP1_FLO	ktph		0.0	$ -
12	BL2_RSV	tph	103.3	53	BP1_SP	MW.sp		0.0	$ -	94	CWP1_SP	MW.sp		0.000	$ -
13	BL2_FUEL	MW.th	132.31	54	BP1M_ELE	MW		0.000	$ -	95	CWP1M_ELE	MW		0.000	$ -
14	BL2_LPFG	MW.th	132.31	55	BP1_X	BIN		0	$ -	96	CWP1_X	BIN		0	$ -
15	BL2_HPFG	MW.th	0.00	56	BP2_FLO	tph		250.0	$ -	97	CWP2_FLO	ktph		7.4	$ -
16	BL2AM_ELE	MW	0.500	57	BP2_SP	MW.sp		0.290	$ -	98	CWP2_SP	MW.sp		1.1	$ -
17	BL2_X	BIN	1	58	BP2_X	BIN		1	$ -	99	CWP2T_STM	tph		20.485	$ -
18	BL2_Y	BIN	0	59	BP2M_SP	MW.sp		0.000	$ -	100	CWP2_X	BIN		1	$ -
19	BL3_STM	tph	240.0	60	BP2M_ELE	MW		0.000	$ -	101	CWP3_FLO	ktph		8.0	$ -
20	BL3_BFW	tph	244.8	61	BP2M_X	BIN		0	$ -	102	CWP3_SP	MW.sp		1.140	$ -
21	BL3_BD	tph	4.8	62	BP2T_SP	MW.sp		0.290	$ -	103	CWP3_X	BIN		1	$ -
22	BL3_RSV	tph	60.0	63	BP2T_STM	tph		2.7	$ -	104	CWP3M_SP	MW.sp		0.000	$ -
23	BL3_FUEL	MW.th	216.61	64	BP2T_X	0/1		1	$ -	105	CWP3M_ELE	MW		0.000	$ -
24	BL3_LPFG	MW.th	216.61	65	BP3_FLO	tph		0.0	$ -	106	CWP3M_X	BIN		0	$ -
25	BL3_HPFG	MW.th	0.00	66	BP3_SP	MW.sp		0.0	$ -	107	CWP3T_SP	MW.sp		1.140	$ -
26	BL3AM_ELE	MW	0.000	67	BP3M_ELE	MW		0.000	$ -	108	CWP3T_STM	tph	$ -	10.7	$ -
27	BL3AT_STM	tph	5.0	68	BP3_X	BIN		0	$ -	109	CWP3T_X	0/1		1	$ -
28	BL3AM_X	BIN	0	69	BP4_FLO	tph		134.3	$ -	110	AC1234_CA	kM3/h		30.0	$ -
29	BL3AT_X	0/1	1	70	BP4_SP	MW.sp		0.241	$ -	111	AC1234_SP	MW.sp		1.398	$ -
30	BL3_W	0/1	1	71	BP4_X	BIN		1	$ -	112	AC1234_NX	INT		4	$ -
31	BL3_X	BIN	1	72	BP4M_SP	MW.sp		0.241	$ -	113	AC1234E_FG	MW.th		3.43	$ -
32	BL3_Y	BIN	0	73	BP4M_ELE	MW		0.253	$ -	114	AC5_CA	kM3/h		0.0	$ -
33	BL3_Z	BIN	0	74	BP4M_X	BIN		1	$ -	115	AC5_SP	MW.sp		0.000	$ -
34	MPS_RSV	tph	0.0	75	BP4T_SP	MW.sp		0.000	$ -	116	AC5_X	BIN		0	$ -
35	HPS_RSV	tph	163.3	76	BP4T_STM	tph		0.0	$ -	117	AC5M_ELE	MW		0.000	$ -
36	PRV1_HPS	tph	0.0	77	BP4T_X	0/1		0	$ -	118	VENT_CA	kM3/h		0.0	$ -
37	PRV1_BFW	tph	0.0	78	BDF_LPS	tph		1.7	$ -	119	BUY_HPFG	MW.th	$ 20.00	0.00	$ -
38	PRV1_MPS	tph	0.0	79	BDF_CND	tph		5.8	$ -	120	BUY_LPFG	MW.th	$ 19.00	412.36	$ 7,834.80
39	PRV2_STM	tph	58.3	80	VENT	tph		0.0	$ -	121	BUY_FO	MW.th	$ 18.0	30.00	$ 540.00
40	STG_STM	tph	50.0	81	DEA_DMW	tph	$ 3	155.8	$ 467.48	122	BUY_ELEC	MW	$ 60.00	18.784	$ 1,127.06
41	STG_ELE	MW	3.000	82	DEA_LPS	tph		68.5	$ -	123	SEL_ELEC	MW	$ (50.00)	0.000	$ -

Summary

In this chapter, we developed the MILP models for the fuel, electrical power, cooling water, and compressed air subsystems. We added the capability to "include," "exclude," or "require" in optimization for all significant pieces of equipment. We integrated these four subsystem models with the steam subsystem model developed in the previous chapter to create an integrated model for the entire utility system. This model is tested with four test cases using the Premium Solver and the LINGO Solver.

The two solutions had identical objective function values for each case but differed slightly in BFW pumping operations. The phenomena of multiple global optimum solutions are not uncommon and can be expected, especially in models with multiple identical units. Multiple global optimum solutions may appear problematic, but they are not, as will be discussed in Chapter 10 when discussing strategies for online implementation of the optimum solutions.

Nomenclature

Symbol	UOM	Description
AC1234_CA	kM3/h	Air compressor #1234: compressed air flow
AC1234_NX	0 .. 4	Air compressor #1234: number of operating pumps
AC1234_SP	MW.sp	Air compressor #1234: shaft power required
AC1234M_FG	MW.th	Air compressor #1234: ICE drive: fuel consumption
AC5_CA	kM3/h	Air compressor #5: compressed air flow
AC5_SP	MW.sp	Air compressor #5: shaft power required
AC5_X	0/1	Air compressor #5: operational status: 0 = OFF, 1 = ON
AC5M_ELE	MW	Air compressor #5: motor drive: power consumption
BAVG	tph	BFW pump flow equalization: average BFW flow
BDF_CND	tph	Blowdown flash: LP condensate discharge
BDF_HPC	tph	Blowdown flash: inlet HP blowdown
BDF_LPS	tph	Blowdown flash: LP steam generation
BDF_MPC	tph	Blowdown flash: inlet MP blowdown
BL3_W	0/1	Boiler #3: LPFG indicator: 1 = using LPFG, 0 = using HPFG
BL3_Z	0/1	Boiler #3: HPFG indicator: 1 = using HPFG, 0 = using LPFG
BLj_BD	tph	Boiler #j: blowdown flow
BLj_BFW	tph	Boiler #j: boiler feed water inflow
BLj_FUEL	MW.th	Boiler #j: total fuel consumption
BLj_HPFG	MW.th	Boiler #j: HP fuel gas consumption
BLj_LPFG	MW.th	Boiler #j: LP fuel gas consumption
BLj_RSV	tph	Boiler #j: steam reserve
BLj_STM	tph	Boiler #j: steam generation
BLj_X	0/1	Boiler #j: operational status: 0 = OFF, 1 = ON
BLj_Y	0/1	Boiler #j: standby status: 1 = in standby mode, 0 otherwise
BLjAM_ELE	MW	Boiler #j: auxiliaries motor drive: electric power usage
BLjAM_X	0/1	Boiler #j: auxiliaries motor drive: 0 = OFF, 1 = ON
BLjAT_STM	tph	Boiler #j: Auxiliaries turbine: steam flow
BLjAT_X	0/1	Boiler #j: auxiliaries turbine drive: 0 = OFF, 1 = ON
BPj_FLO	tph	BFW Pump #j: boiler feed water inflow
BPj_SP	MW.sp	BFW Pump #j: shaft power required
BPj_X	0/1	BFW Pump #j: operational status: 0 = OFF, 1 = ON
BPjM_ELE	MW	BFW Pump #j: motor drive: power consumption
BPjM_SP	MW.sp	BFW Pump #j: motor drive: shaft power produced
BPjM_X	0/1	BFW Pump #j: motor drive: operational status: 0 = OFF, 1 = ON
BPjT_SP	MW.sp	BFW Pump #j: turbine drive: shaft power produced
BPjT_STM	tph	BFW Pump #j: turbine: steam flow
BPjT_X	0/1	BFW Pump #j: turbine drive: operational status: 0 = OFF, 1 = ON
BUY_ELEC	MW	Purchased energy: buy electric power
BUY_FO	MW.th	Purchased energy: buy fuel oil
BUY_HPFG	MW.th	Purchased energy: buy HP fuel gas
BUY_LPFG	MW.th	Purchased energy: buy LP fuel gas
CAVG	tph	CW pump flow equalization: average CW flow
CTG_CND	tph	CTG: condensate from condenser
CTG_CW	ktph	CTG: condenser: cooling water flow
CTG_ELE	MW	CTG: power generation

(continued)

Symbol	UOM	Description
CTG_HPS	tph	CTG: inlet HP steam flow
CTG_LPS	tph	CTG: outlet LP steam flow
CTG_MPS	tph	CTG: outlet MP steam flow
CTG_X	0/1	CTG: operational status: 0 = OFF, 1 = ON
CTG_Y	0/1	CTG: standby status: 1 = in standby mode, 0 otherwise
CWPj_FLO	tph	CW Pump #j: boiler feed water inflow
CWPj_SP	MW.sp	CW Pump #j: shaft power required
CWPj_X	0/1	CW Pump #j: operational status: 0 = OFF, 1 = ON
CWPjM_ELE	MW	CW Pump #j: motor drive: power consumption
CWPjM_SP	MW.sp	CW Pump #j: motor drive: shaft power produced
CWPjM_X	0/1	CW Pump #j: motor drive: operational status: 0 = OFF, 1 = ON
CWPjT_SP	MW.sp	CW Pump #j: turbine drive: shaft power produced
CWPjT_STM	tph	CW Pump #j: turbine: steam flow
CWPjT_X	0/1	CW Pump #j: turbine drive: operational status: 0 = OFF, 1 = ON
DEA_BFW	tph	Deaerator: boiler feed water production
DEA_DMW	tph	Deaerator: fresh demineralized water intake
DEA_LPS	tph	Deaerator: deaeration steam
DEA_LPS	tph	Deaerator: deaeration steam
FC_ELE	MW	Fuel compressor: power consumption
FC_FG	MW.th	Fuel compressor: fuel compressed
FC_X	0/1	Fuel compressor: operational status: 0 = OFF, 1 = ON
FGFLARE	MW.th	VENT: LPFG
FOFLARE	MW.th	VENT: fuel oil
GTG_ELE	MW	GTG: power generated
GTG_FG	MW.th	GTG: fuel consumed
GTG_X	0/1	GTG: operational status: 0 = OFF, 1 = ON
HPS_RSV	tph	HP Boilers: total steam reserve
MPS_RSV	tph	MP Boiler: steam reserve
PRV1_BFW	tph	PRV: de-superheating water flow
PRV1_HPS	tph	PRV: HP steam flow
PRV1_MPS	tph	PRV: MP steam flow
PRV2_STM	tph	PRV2: steam flow
PRVFG	MW.th	PRVFG: HPFG to LPFG
SEL_ELEC	MW	Purchased energy: sell electric power
STG_ELE	MW	STG: power generation
STG_STM	tph	STG: steam flow
STG_X	0/1	STG: operational status: 0 = OFF, 1 = ON
STG_Y	0/1	STG: standby status: 1 = in standby mode, 0 otherwise
VENT	tph	VENT: steam flow
VENT_CA	kM3/h	VENT: compressed air
ac1234_a	0 .. 4	Input: Air compressors #1 ..4: number of available compressors
ac1234_r	0 .. 4	Input: Air compressors #1 ..4: number of required compressors
ac5_aru	ARU or 123	Input: Air compressor #5: "available'/"required"/"unavailable" spec
bdf_a	–	Parameter: Blowdown flash: Enthalpy ratio: (hl_hps - hl_lps)/hl_mps
bdf_b	–	Parameter: Blowdown flash: Enthalpy ratio: (hv_lps - hl_lps)/hl_mps

(continued)

Symbol	UOM	Description
blj_aru	ARU or 123	Boiler #j: "available'/"required"/"unavailable" spec
blj_bdr	fraction	Parameter: Boiler #j: blowdown ratio
blj_fg_c	MW.th	Parameter: Boiler #j : fuel vs. steam, low range, regression: intercept
blj_fg_d	MWh.th/t	Parameter: Boiler #j : fuel vs. steam, low range, regression: slope
blj_fg_e	MW.th	Parameter: Boiler #j : fuel vs. steam, high range, regression: intercept
blj_fg_f	MWh.th/t	Parameter: Boiler #j : fuel vs. steam, high range, regression: slope
blj_fg_s	MW.th	Parameter: Boiler #j : standby mode: fuel consumption
blj_lgc_hl	0/1	Calculated: BL #j: _LGC constraint: high limit
blj_lgc_ll	0/1	Calculated: BL #j: _LGC constraint: low limit
blj_stm_hl	tph	Parameter: Boiler #j : steam generation : high limit
blj_stm_ll	tph	Parameter: Boiler #j : steam generation : low limit
bljam_c	MW	Parameter: Boiler #j auxiliaries motor: power
bljat_c	tph	Parameter: Boiler #j auxiliaries turbine: steam flow
bpj_des_ll	MW.sp/t	Parameter: BWP Pump #j: inlet flow: design low limit
bpj_flo_hl	MW.sp/t	Parameter: BWP Pump #j: inlet flow: high limit
bpj_lgc_hl	0/1	Calculated: BFW pump #j: _LGC constraint: high limit
bpj_lgc_ll	0/1	Calculated: BFW pump #j: _LGC constraint: low limit
bpj_sp_a	MW.sp	Parameter: BWP Pump #j: SP vs. flow regression: intercept
bpj_sp_b	MW.sp/t	Parameter: BWP Pump #j: SP vs. flow regression: slope
bpjm_eta	tph/tph	Parameter: BWP Pump #j: motor drive: motor efficiency
bpjt_stm_a	tph	Parameter: BWP Pump #j: turbine drive: steam vs. flow: intercept
bpjt_stm_b	tph/tph	Parameter: BWP Pump #j: turbine drive: steam vs. flow: slope
c_buyelec	$/MWh	Input cost: buy electric power
c_dmw	$/ton	Input cost: buy demineralized water
c_fo	$/MWh.th	Input cost: buy fuel oil
c_hpfg	$/MWh.th	Input cost: buy LPFG
c_lpfg	$/MWh.th	Input cost: buy LPFG
c_selelec	$/MWh	Input cost: sell electric power
ca_demand	kM3/h	Process Demand: compressed air
ctg_cnd_hl	tph	Parameter: CTG: condensate: high limit
ctg_cnd_ll	tph	Parameter: CTG: condensate: low limit
ctg_cw_b	ktph/tph	Parameter: CTG: cooling water vs. condensate: slope
ctg_ele_hl	tph	Parameter: CTG: power gen: high limit
ctg_ele_ll	tph	Parameter: CTG: power gen: low limit
ctg_hps_a	tph	Parameter: CTG: hps vs. other variables: intercept
ctg_hps_b	tph/tph	Parameter: CTG: hps vs. power gen regression: slope
ctg_hps_c	tph/tph	Parameter: CTG: hps vs. MPS steam flow regression: slope
ctg_hps_d	tph/tph	Parameter: CTG: hps vs. LPS steam flow regression: slope
ctg_hps_hl	tph	Parameter: CTG: HP steam: high limit
ctg_hps_ll	tph	Parameter: CTG: HP steam: low limit
ctg_hps_s	tph	Parameter: CTG: standby mode: inlet steam flow
ctg_lgc_hl	0/1	Calculated: CTG: _LGC constraint: high limit
ctg_lgc_ll	0/1	Calculated: CTG: _LGC constraint: low limit
ctg_lps_hl	tph	Parameter: CTG: LP steam: high limit
ctg_mps_hl	tph	Parameter: CTG: MP steam: high limit

(continued)

Symbol	UOM	Description
ctg_x_hl	0/1	Calculated: CTG_X: high limit
ctg_x_ll	0/1	Calculated: CTG_X: low limit
cw_demand	ktph	Process Demand: cooling water
cwpj_aru	ARU or 123	Input: CW pump #j: "avaialble'/"required"/"unavailable" spec
dea_a	–	Parameter: Deaerator: Enthalpy ratio: $(h\_tc - h\_pc)/(hl\_mps - h\_pc)$
dea_b	–	Parameter: Deaerator: Enthalpy ratio: $(h\_dmw - h\_pc)/(hl\_mps - h\_pc)$
dea_c	–	Parameter: Deaerator: Enthalpy ratio: $(h\_lps - h\_pc)/(hl\_mps - h\_pc)$
elec_demand	MW	Process Demand: electrical power
fc_lgc_hl	0/1	Calculated: Fuel compressor: _LGC constraint: high limit
fc_lgc_ll	0/1	Calculated: Fuel compressor: _LGC constraint: low limit
fo_demand	MW.th	Process Demand: fuel oil
gtg_fg_a	MW.th	Parameter: GTG : fuel vs. power gen, low range, regression: intercept
gtg_fg_b	MW.th/MW	Parameter: GTG : fuel vs. power gen, low range, regression: slope
gtg_fg_c	MW.th	Parameter: GTG : fuel vs. power gen, mid range, regression: intercept
gtg_fg_d	MW.th/MW	Parameter: GTG : fuel vs. power gen, mid range, regression: slope
gtg_fg_e	MW.th	Parameter: GTG : fuel vs. power gen, high range, regression: intercept
gtg_fg_f	MW.th/MW	Parameter: GTG : fuel vs. power gen, high range, regression: slope
gtg_x_hl	0/1	Calculated: GTG_X: high limit
gtg_x_ll	0/1	Calculated: GTG_X: low limit
hpfg_demand	MW.th	Process Demand: HP fuel gas
hps_demand	tph	Process Demand: HP steam
hps_reserve	tph	Process Demand: HP steam reserve
lpfg_demand	MW.th	Process Demand: LP fuel gas
lps_demand	tph	Process Demand: LP steam
mps_demand	tph	Process Demand: MP steam
mps_reserve	tph	Process Demand: MP steam reserve
pc_supply	tph	Process Supply: process condensate return
prv1_a	–	Parameter: Enthalpy ratio: h_hpbfw /h_hps
prv1_b	–	Parameter: Enthalpy ratio: h_mps /h_hps
stg_ele_hl	MW	Parameter: STG: power generation: high limit
stg_ele_ll	MW	Parameter: STG: power generation: low limit
stg_lgc_hl	0/1	Calculated: STG: _LGC constraint: high limit
stg_lgc_ll	0/1	Calculated: STG: _LGC constraint: low limit
stg_stm_a	tph	Parameter: STG: inlet steam vs. power regression: intercept
stg_stm_b	tph/MW	Parameter: STG: inlet steam vs. power regression: slope
stg_stm_hl	tph	Parameter: STG: inlet steam: high limit
stg_stm_s	tph	Parameter: STG: standby mode: inlet steam flow

Refrences

[1]   Anonymous, "LINGO: The Modeling Language and Optimizer", p 446, LINDO Systems Inc. (2020).
[2]   Anonymous, "Frontline Solvers Reference Guide, version 2023 Q2", Frontline Systems, Inc. (2023).

8 Modeling purchased energy contracts

In the last chapter, we developed an integrated model for the steam, fuel, electrical power, cooling water, and compressed air subsystems, which is almost the complete utility system except for the external energy purchase. This chapter will introduce a powerful new modeling technique called the "Disjunctive Method," which is essential in developing models of certain contracts and flow equalization in identical parallel units. We will develop models for all external energy supply contracts and flow equalization in the BFW and the cooling water pumps. We will then integrate them with the previous model to complete the utility system optimization model. In the following two chapters, we will discuss implementation strategies.

Contracts with external fuel suppliers typically fall into one of the following five categories:
– Spot market contract
– Take or pay contract
– Excess penalty contract
– Volume discount contract
– Multitier pricing contract

A contract with the local power supplier typically has the following two features:
– Energy and demand charges
– Time of use rates

We will discuss each of these in the following sections. The chapter will conclude with implementing purchased energy contracts in the utility system model and thus completing the model.

8.1 Fuel pricing

We will discuss each of the five categories of fuel contracts mentioned in the introduction in the following five subsections.

8.1.1 Spot market contract

This simplest pricing model comprises a constant unit cost of purchased fuel, c_fuel. Market supply and demand conditions set this price, which could fluctuate significantly. Henry Hub price [1] is a commonly used price index for natural gas.

Although the spot price could change daily, it will be a known quantity before the execution of the optimizer. Usually, there are no constraints on the amount of purchased fuel.

https://doi.org/10.1515/9783111020679-009

We have, so far, used a constant unit energy cost similar to the Spot market pricing model. Fuel purchased energy cost, **FUEL_CST**, is computed by multiplying the unit price by the amount of purchased energy, as shown below:

$$\text{FUEL\_CST} = c\_fuel * \text{BUY\_FUEL} \qquad (8.1)$$

where

c_fuel is the unit fuel cost in \$/MWh.th; it is usually based on the higher heating value of the fuel,

BUY_FUEL is the amount of fuel purchased that is determined by the optimizer in MW.th, and

FUEL_CST is the fuel cost contribution to the cost objective function in \$/hr.

8.1.2 Take or pay contract

This is a simple pricing model. This contract agreement is for the supplier to supply a fixed amount of fuel for a fixed charge. There is no flexibility in the fuel purchase or payment amount. The buyer agrees to pay the same amount even if the fuel consumed is less than the contract amount. The model comprises the following constraints:

$$\text{BUY\_FUEL} \leq buy\_fuel\_max \qquad (8.2)$$

$$\text{FUEL\_CST} = fixed\_fuel\ cost \qquad (8.3)$$

where

BUY_FUEL is the amount of fuel purchased, determined by the optimizer in MW.th,

buy_fuel_max is the contract amount of fuel in MW.th,

FUEL_CST is the fuel cost contribution to the cost objective function in \$/h, and

fixed_fuel_cost is the contracted payment amount, in \$/h, usually paid at the end of the month.

Since the amount paid is fixed, including it in the objective function just adds a constant value to the objective function; it does not affect the optimization solution; only the absolute value of the objective function is affected. For this reason, one may choose to exclude this cost from the objective function.

The most cost-effective solution with this contract would be to consume the contracted amount entirely.

8.1.3 Excess penalty contract

The excess penalty contract is a two-tier pricing scheme in which the second tier is purchased at a higher unit price, representing a penalty for exceeding the tier 1 limit. Figure 8.1 is a graphical representation of this contract. The solid red line is the governing cost curve for the first tier with a unit fuel cost of $c1$, and the solid blue line represents the second tier cost curve with unit fuel cost; in this case, $c2 > c1$. The purple dot represents the "breakpoint" where the cost curve transitions from tier 1 pricing to tier 2 pricing.

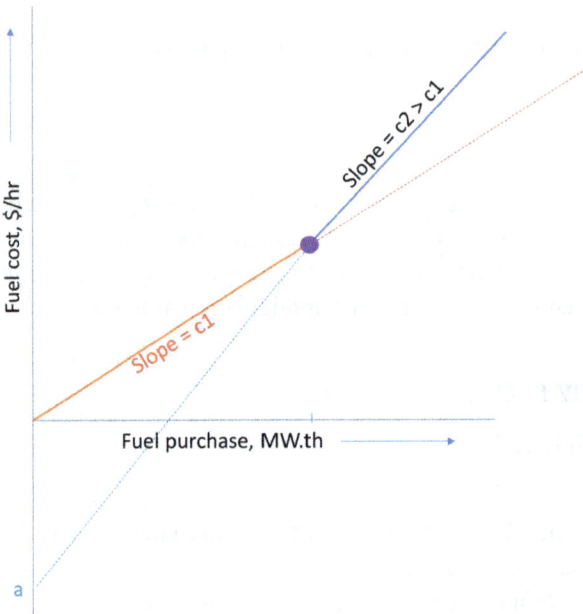

Fig. 8.1: Fuel cost versus fuel purchase: excess penalty contract.

Note the red dashed line is the extension of the tier 1 cost curve above the "breakpoint," and the dashed blue line is the extension of the tier 2 pricing line below the "breakpoint."

How do we model this contract? You may notice a similarity to the boiler fuel vs. steam curve discussed in Section 3.1.6. Since fuel purchase is a cost and the purchase line in a tier segment underestimates the cost in the other segment (the dashed line is below the solid line), we could use two inequalities to model this, as was done for the boiler performance curve.

Note that tier 1 cost line passes through the origin and its equation would be **FUEL_CST** = $c1$ * **BUY_FUEL**; and tier 2 cost line equation would be **FUEL_CST** = $c2$ * **BUY_FUEL** + a, where a is the intercept of the dashed blue line on the cost axis. The following two constraints adequately represent this contract in a cost optimization model.

$$\textbf{FUEL\_CST} \geq c1^*\textbf{BUY\_FUEL} \tag{8.4}$$

$$\textbf{FUEL\_CST} \geq c2^*\textbf{BUY\_FUEL} + a \tag{8.5}$$

8.1.4 Volume discount contract

The volume discount contract is also a two-tier pricing scheme; however, in this case, the unit cost of the second tier is lower than the unit cost of the first tier, representing a discount for exceeding the tier 1 threshold. Figure 8.2 is a graphical representation of this contract. The piecewise linear solid line is the governing contract line; the red segment represents the first tier with a unit fuel cost of $c1$, and the blue segment represents the second tier with a unit fuel cost of $c2 < c1$. The purple dot represents the "breakpoint" where the cost transitions from tier 1 pricing to tier 2 pricing.

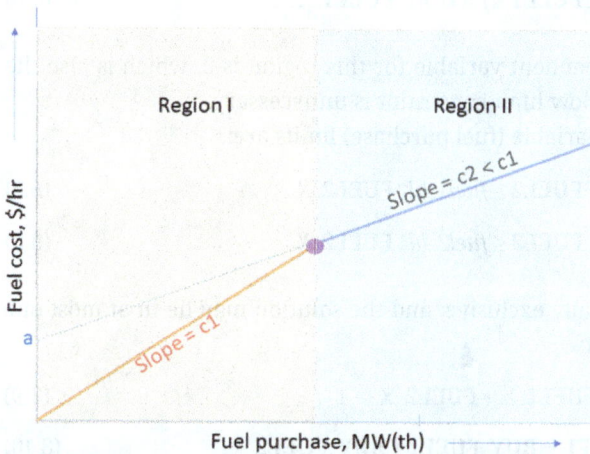

Fig. 8.2: Fuel cost versus fuel purchase: volume discount contract.

We must follow a new technique that uses binary variables. This technique is called the "Disjunctive method" [2]. The method comprises defining multiple regions, one for each linear segment; each region has a continuous and a binary variable associated with it. The MILP model defines the upper and lower bounds on the independent (X-axis) variable for each region using the integer variable and a constraint representing the segment of the region's performance curve (dependent variable or Y-axis variable). An exclusivity constraint restricts the sum of binary variables to unity. The solution has only one nonzero binary variable corresponding to the region in which the optimal value of the independent variable is.

For the two-tier fuel cost curve shown in Fig. 8.2, there will be two regions: one to the left of the "breakpoint" in the pink area and the other to the right, in the blue area. In Fig. 8.2, the pink area is **Region I**, representing fuel purchase in the Tier 1 range with the governing fuel cost curve segment shown in a solid red line. The blue area to the right is **Region II,** representing fuel cost versus fuel purchase in Tier 2 range, with the governing curve shown in a solid blue line. We will designate the "breakpoint" value, which is the upper limit of independent variable in **Region I** by a symbol *fuel1_hl,* and the upper limit of independent variable in **Region II** by a symbol *fuel2_hl.*

Note that the optimum solution would be in **Region I** or **Region II**. We will associate one continuous variable for fuel purchase and a binary variable indicating purchase in each region. **BUY_FUEL1** and **FUEL1_X** are associated with **Region I** and **BUY_FUEL2** and **FUEL2_X** associated with **Region II**. Fuel cost, **FUEL_CST**, will be calculated using these four variables.

In **Region I**, the independent variable (fuel purchase) high limit is:

$$\textbf{BUY\_FUEL1} \le fuel1\_hl^* \textbf{FUEL1\_X} \tag{8.6}$$

Note that low limit of the independent variable for this region is 0, which is also the default in MILP, so specifying a low limit constraint is unnecessary.

In **Region II**, independent variable (fuel purchase) limits are:

$$\textbf{BUY\_FUEL2} \ge fuel1\_hl^* \textbf{FUEL2\_X} \tag{8.7}$$

$$\textbf{BUY\_FUEL2} \le fuel2\_hl^* \textbf{FUEL2\_X} \tag{8.8}$$

Since the two regions are mutually exclusive, and the solution may lie in at most one of the two regions we can assert:

$$\textbf{FUEL1\_X} + \textbf{FUEL2\_X} \le 1 \tag{8.9}$$

$$\textbf{BUY\_FUEL} = \textbf{BUY\_FUEL1} + \textbf{BUY\_FUEL2} \tag{8.10}$$

$$\textbf{FUEL\_CST} = c1^* \textbf{BUY\_FUEL1} + c2^* \textbf{FUEL2\_CST} + a^* \textbf{FUEL2\_X} \tag{8.11}$$

where *a* is the intercept of tier 2 fuel cost line (the blue line) with the fuel cost axis.

The relationships mentioned above will model the volume discount fuel contract. Complete models as implemented in LINGO and Microsoft Excel are shown in Figs. 8.3 and 8.4, respectively.

```
MODEL:    TITLE Fuel cost 2 Tier Model: FUEL_CST_2T-v1: volume discount case;
! Nath, "Industrial Process Plants: Global Optimization of Utility Systems", Chapter 8, (2023);

   DATA:
       c_fuel1 = 25; c_fuel2 = 20;
       fuel1_hl = 20; fuel2_hl = 100;
       fuel_demand = 50;                    ! fuel demand, for testing oinly;
   ENDDATA

   CALC:
       a = fuel1_hl * ( c_fuel1 - c_fuel2);
   ENDCALC

   ! Objective;      MIN = c_fuel1 * BUY_FUEL1 + c_fuel2 * BUY_FUEL2 + a * FUEL2_X;

   ! Region I;
    [_FUEL1_HL]      BUY_FUEL1 < fuel1_hl * FUEL1_X;

   ! Region II;
   [_FUEL2_LL]       BUY_FUEL2 > fuel1_hl * FUEL2_X;
   [_FUEL2_HL]       BUY_FUEL2 < fuel2_hl * FUEL2_X;
                     @BIN( FUEL1_X );  @BIN( FUEL2_X );

   ! Overall;
   [_BUY_FUEL]       BUY_FUEL = BUY_FUEL1 + BUY_FUEL2;
   [_FUEL_LGC]       FUEL1_X + FUEL2_X < 1;

   ! Testing;
   [_FUEL_DMND]      BUY_FUEL = fuel_demand;

END MODEL FUEL_CST_2T-v1
```

Fig. 8.3: Volume discount contract model: Fuel_CST_2 T-v1: in LINGO.

▲	A	B	C	D	E	F	G	H	I	J	K
1		model ID:	FUEL COST 2 TIER model: FUEL_CST_2T-v1				inputs:	unit cost	hl	intercept	params
2		case ID:	1					tier1	25	20	0	Big
3		Objective fn:	$/hr			$ 1,100.00	tier2	20	100	100	999
4							demand	50			
5	#						1	2	3	4	5
6		ID					BUY_FUEL1	BUY_FUEL2	BUY_FUEL	X1	X2
7			UOM				MW.th	MW.th	MW.th	BIN	BIN
8					high	cost coeff ->	$ 25.00	$ 20.0			$ 100.0
9				low limit	limit	optimum	0.0	50.000	50.000	● 0	● 1
10	1	_FUEL1_HL	MW.th	-999	0	0.0	1			-20.00	
11	3	_FUEL2_LL	MW.th	0	999	30.0		1			-20.0
12	2	_FUEL2_HL	MW.th	-999	0	-50.0		1			-100.00
13	4	_BUY_FUEL	MW.th	0	0	0.0	-1	-1	1		
14	5	_LGC			0	1	1.0			1	1
15	6	_DEMAND	MW.th	50.0	50.0	50.0		1			

Fig. 8.4: Volume discount contract model: Fuel_CST_2 T-v1: in Microsoft Excel, showing test case solution.

8.1.5 Multitier pricing contract

In the previous two subsections, we discussed two-tier pricing models; however, fuel contracts can be more complicated with multiple tiers without any particular relationships among the tier unit costs. The "Disjunctive method" in the previous subsection applies to multitier pricing contracts.

As noted earlier, the "Disjunctive method" relies on binary variables, which could be inefficient, especially for a contract with many tiers. Although our fuel cost curve is continuous, i.e., it has no breaks, the disjunctive method applies equally well to discontinuous functions; the technique is versatile. Later in this chapter, we will use it to equalize the flows in boiler feed and cooling water pumps where the performance curve is discontinuous.

There is another method, called the "Lambda Method,", which is also suitable for multitier pricing contracts, using a particular type of constraints called "Special Ordered Sets 2," or SOS2 [3]. It is more efficient than the "Disjunctive method," especially for problems with many regions; its drawback is that it applies only to continuous performance curves. Unfortunately, not all solvers support SOS2-type constraints. For example, while LINGO supports SOS2 constraints [4], Solver in Excel does not support it [5].

We will develop the multitier pricing contract model using both the "Disjunctive method" and the "Lambda method"; however, as we advance, we will only use the "Disjunctive method" as it is more universally applicable.

Let us consider a three-tier contract for multitier pricing, as shown in Fig. 8.5. The piecewise linear solid line is the governing contract line; the red segment represents the first tier with a unit fuel cost of c_1, the blue segment represents the second tier with a unit fuel cost of c_2, and the green segment represents the third tier with unit cost c_3. The purple dots represent the "breakpoints," where the cost transitions from one tier to the next.

The modeling of this contract is similar to that of the Volume discount contract in the previous subsection, except that this one has three regions.

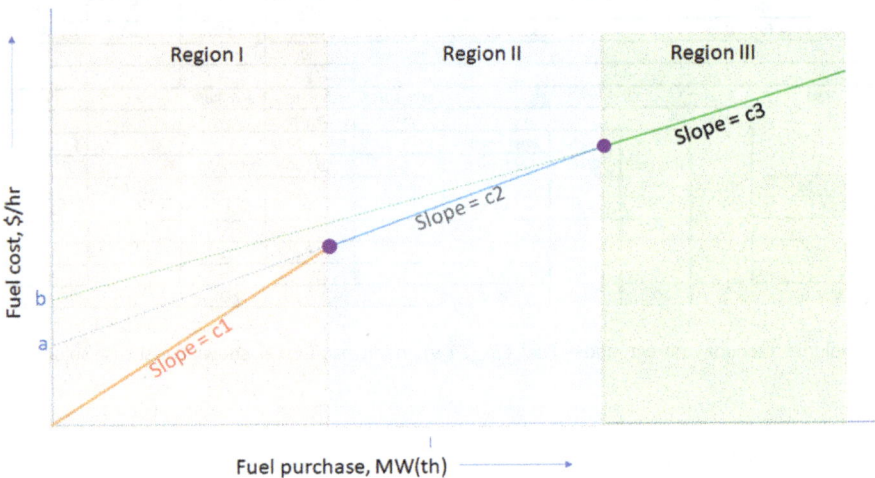

Fig. 8.5: Fuel cost versus fuel purchase: Multitier pricing contract.

The complete models, as implemented in LINGO and Microsoft Excel, are shown in Figs. 8.6 and 8.7, respectively.

```
MODEL:    TITLE Fuel cost 3 Tier Model: FUEL_CST_3T-v1: multi-tier case;
! Nath, "Industrial Process Plants: Global Optimization of Utility Systems", Chapter 8, (2023);

   DATA:
       c_fuel1 = 22; c_fuel2 = 20; c_fuel3 = 18;
       fuel1_hl = 20; fuel2_hl = 70; fuel3_hl = 100;
       fuel_demand = 70;                  ! fuel demand, for testing oinly;
   ENDDATA

   CALC:
       a =    fuel1_hl * ( c_fuel1 - c_fuel2);
       b = a + fuel2_hl * ( c_fuel2 - c_fuel3);
   ENDCALC

   ! Objective;
   [_FG2T_OBJ]     MIN = c_fuel1 * BUY_FUEL1 + c_fuel2 * BUY_FUEL2 + a * FUEL2_X
                         + c_fuel3 * BUY_FUEL3 + b * FUEL3_X;

   ! Region I;
   [_FUEL1_HL]     BUY_FUEL1 < fuel1_hl * FUEL1_X;

   ! Region II;
   [_FUEL2_LL]     BUY_FUEL2 > fuel1_hl * FUEL2_X;
   [_FUEL2_HL]     BUY_FUEL2 < fuel2_hl * FUEL2_X;

   ! Region III;
   [_FUEL3_LL]     BUY_FUEL3 > fuel2_hl * FUEL3_X;
   [_FUEL3_HL]     BUY_FUEL3 < fuel3_hl * FUEL3_X;

   ! Overall;
   [_BUY_FUEL]     BUY_FUEL = BUY_FUEL1 + BUY_FUEL2 + BUY_FUEL3;
   [_FUEL_LGC]     FUEL1_X + FUEL2_X + FUEL3_X < 1;

   ! Intehgers;    @BIN( FUEL1_X );  @BIN( FUEL2_X );  @BIN( FUEL3_X );

   ! Testing;
   [_FUEL_DMND]    BUY_FUEL = fuel_demand;

END MODEL FUEL_CST_3T-v1
```

Fig. 8.6: Three-tier pricing contract model: Fuel_CST_3 T-v1: in LINGO.

⊿	A	B	C	D	E	F	G	H	I	J	K	L	M
1		model ID:	FUEL COST 3 TIER model: FUEL_CST_3T-v1				inputs:	unit cost	hl	intercept	demand	params	Big
2		case ID:	1				tier1	22	20	0	70		999
3		Objective fn:	$/hr			$ 1,440.00	tier2	20	70	40			
4							tier3	18	100	180			
5	#						1	2	3	4	5	6	7
6		ID					BUY_FUEL1	BUY_FUEL2	BUY_FUEL3	BUY_FUEL	X1	X2	X3
7			UOM				MW.th	MW.th	MW.th	MW.th	BIN	BIN	BIN
8						cost coeff ->	$ 22.00	$ 20.0	$ 18.0			$ 40.0	$ 180.0
9				low limit	high limit	optimum	0.0	0.000	70.000	70.000	● 0	● 0	● 1
10	1	FUEL1_HL	MW.th	-999	0	0.0	1				-20.00		
11	3	FUEL2_LL	MW.th	0	999	0.0		1				-20.0	
12	2	FUEL2_HL	MW.th	-999	0	0.0		1				-70.00	
13	5	FUEL3_LL	MW.th	0	999	0.0			1				-70.0
14	4	FUEL3_HL	MW.th	-999	0	-30.0			1				-100.00
15	6	BUY_FUEL	MW.th	0	0	0.0	-1	-1	-1	1			
16	7	LGC		0	1	1.0					1	1	1
17	8	DEMAND	MW.th	70.0	70.0	70.0				1			

Fig. 8.7: Three-tier pricing contract model: Fuel_CST_3 T-v1: in Microsoft Excel, showing test case solution.

Next, we will develop a model for this case using the Lambda method in LINGO. The Lambda method uses SOS2 constraints. The SOS2 constraints represent an ordered set, and the Lambda method ensures that at most two consecutive SOS2 variables are nonzero.

We will add a constraint that the sum of variables in the SOS2 set is one, making this model an interpolation algorithm. The following are the details;

In this case, there are three tiers, so there will be four endpoints of the piecewise linear curve; we need the coordinates of all endpoints of the piecewise linear curve in Fig. 8.5. Call these points by symbolic names (xi, yi). The method comprises four variables, **Wi**, corresponding to each endpoint. These variables will be part of an **SOS2 set**. Figure 8.8 shows the complete model, Fuel_CST_Lambda-v1. Table 8.1 shows the solution for a fuel demand of 80 MW.th.

```
MODEL:    TITLE Fuel cost 3 Tier Model: FUEL_CST_Lambda-v1: base case;
! Nath, "Industrial Process Plants: Global Optimization of Utility Systems", Chapter 8, (2023).;
  DATA:
      c_fuel1 = 22;      c_fuel2 = 20;          c_fuel3 = 18;
      fuel_demand = 70;
      fuel1_hl = 20;     fuel2_hl = 70;     fuel3_hl = 100;

  ENDDATA

  CALC:      ! calculate end points;
      x0 = 0;            y0 = 0;
      x1 = fuel1_hl;     y1 = y0 + (x1-x0) * c_fuel1;
      x2 = fuel2_hl;     y2 = y1 + (x2-x1) * c_fuel2;
      x3 = fuel3_hl;     y3 = y2 + (x3-x2) * c_fuel3;
  ENDCALC

  ! Objective;
  [_FUEL_COST]    MIN = y0*W0 + y1*W1 + y2*W2 + y3*W3;

  ! constraints;
  [_FUEL_DMND]    x0*W0 + x1*W1 + x2*W2 + x3*W3 = fuel_demand;
  [_WT_SUM]       W0 + W1 + W2 + W3 = 1;

  ! SOS2 constraints, order is important;
                  @SOS2('FG_SOS2', W0); @SOS2('FG_SOS2', W1);
                  @SOS2('FG_SOS2', W2); @SOS2('FG_SOS2', W3);

END MODEL FUEL_CST_Lambda-v1
```

Fig. 8.8: Multitier pricing contract model: Fuel_CST_Lambda-v1: in LINGO.

Tab. 8.1: Three-tier pricing contract model using Lambda Method test case: LINGO solution.

Fuel Cost Model: Fuel_CST_Lambda-v1					
Global optimal solution found !					
		Fuel =	80.000	MW.th	
		Cost =	$1,620.0	$/hr	
i	Fuel coeff, xi	Cost coeff, yi	Opt. value, Wi	Fuel = wi*xi MW.th	Cost = wi*yi $/hr
1	0	$ -	0.0000	0.000	$ -
2	20	$ 440.00	0.0000	0.000	$ -
3	70	$ 1,440.0	0.6667	46.667	$ 960.00
4	100	$ 1,980.00	0.3333	33.333	$ 660.00

8.2 Electrical power pricing

This section will discuss the two distinguishing features of electrical power contracts.

8.2.1 Energy and demand charges

So far, for energy purchases, we have encountered only energy costs, i.e., payments based on the amount of fuel energy supplied in MWh. Electrical power purchase contracts also charge for the electrical energy provided in MWh; however, for industrial customers, there is an additional charge called the "Demand charge." Demand in this context refers to the peak in electrical power in MW established over a specified period, usually 12 months.

The demand charge is irrespective of the energy charge, so for example, in a hypothetical situation, even if the utility system were not to purchase any electricity in a particular billing period, there would be no charge for electrical energy; however, there would still be a charge for the demand peak that was established in the prior 11 months. If, on the other hand, a new demand peak in purchased power were to be established in the current billing period, the demand charge corresponding to the newly established peak would apply not only for the current period but also for the subsequent 11 months. In summary, in addition to the electrical energy charge, the utility system is also charged for a rolling 12-month demand peak in electrical power usage.

Establishing a new power peak, especially with a significant increase, is expensive as the increased cost applies to the current month and the subsequent 11 months. Utility systems, in general, closely watch their electrical demand and attempt to take proactive action to prevent or minimize any new demand peak. Some utility systems go as far as to maintain and act on a "Cut list," a prioritized list of electrical users that could be deprived of power for a short period to prevent or minimize a new demand peak.

Think of it this way:
- There is little that we can do now for the existing peak.
- It is a historical fact that one has to live with it.
- At this point, it is a necessary expense.

However, we can be proactive in preventing a new demand peak. Let us see how to make the optimizer aware of this reality.

In the utility optimizer, we have a variable for the electrical power import, **BUY_ELEC**, which is the basis for electrical energy consumption charge. In the objective function, we will also add the demand charge corresponding to the prevailing peak so far, *cur_pkdmnd*. It is a fixed expense and could be ignored without affecting the optimization results; however, we will keep it to get an accurate operating cost value. Also, we need to watch for the possibility of creating a new demand peak, and toward this, we will add a new variable; we will call it the incremental demand peak,

IDEMAND. A new demand peak would be created if the current electrical power import, **BUY_ELEC**, exceeds the prevailing peak, *cur_pkdmnd* that is:

if **BUY_ELEC** > *cur_pkdmnd*
then

$$\text{IDEMAND} = \text{BUY\_ELEC} - cur\_pkdmnd, \tag{8.12}$$

else

$$\text{IDEMAND} = 0 \tag{8.13}$$

How do we implement this logic? Using eq. (8.12) is problematic; however, changing the equality to ≥ inequality would do the trick. You may ask why; it is so because all variables are nonnegative, and using the inequality gives the optimizer the choice to take on zero value (or any other positive value). You may ask why the optimizer would choose 0, not a greater value, because **IDEMAND** is a cost, and the optimizer's objective is to minimize the cost.

8.2.2 Time of use (TOU) rates

Another characteristic of industrial electrical power supply contracts is that the cost of electrical energy usually varies with the time of the day and the season. For electrical energy rate purposes, a day is classified into several rate periods, usually designated as: "on-peak," "mid-peak," and "off peak" hours. Electrical energy is most expensive during the "on-peak" hours, least expensive during the "off-peak" hours, and somewhere in between during the "mid-peak" hours. Similarly, there are usually seasonal rate variations.

What season has high rates? That depends on the regional weather pattern; for instance, rates are high in summer for summer-peaking regions and lower in other months. Also, depending on the electrical purchase contract, new demand peaks may not count during the "off-peak" periods and could only be established during "on-peak" or "mid-peak" periods. Table 8.2 shows an example of the TOU rate structure we will be using.

From an optimization execution perspective, all this means is that the electrical rates depend on the use time. However, the rates are well defined in the governing contract, i.e., they are calculable in advance and are known quantities. We must ensure that the optimization system uses the correct rates for each optimizer execution.

8.3 Flow equalization in equipment operating in parallel

As mentioned in Chapter 6, the four boiler feed water pumps are identical in size and differ only in their drivers. These pumps operate in a parallel configuration and will

Tab. 8.2: A sample electrical power supply TOU rate structure.

Month of Year	TOU Rate ID	Time of Day	Energy Charge, $/MWh	Demand Charge, $/MWpk	Incr. Denand Penalty, $/MWpk	Export Credit, $/MWh
Apr - Sep	Summer_Peak	6 am - 10 am & 4 pm - 8 pm	50	10	100	40
	Summer_Mid	10 am - 4 pm & 8 pm - 10 pm	40	10	100	32
	Summer_Off	10 pm - 6 am	30	-	-	24
Oct - Mar	Winter_High	8 am - 8 pm	40	10	100	32
	Winter_Low	8 pm - 8 am	30	-	-	24

have roughly equal flow of boiler feed water in the operating pumps. However, as we saw earlier in Chapter 6, the optimizer, in pursuit of minimizing the cost, even optimizes these flows to unequal values, which is unrealistic.

How can we implement flow equalization among the operating pumps? A strategy to equalize the flows could be to set the flow in all operating pumps equal to the average flow and ensure that only the expected number of pumps are operating. A very useful piece of information is the average flow in the operating pumps. Typically, one would calculate the average by dividing the total boiler feed water flow by the number of operating pumps. Unfortunately, that would be dividing two variables to calculate a third variable; doing so would render the problem an NLP and jeopardize the guarantee of global optimization.

A very relevant question at this point is: can we calculate the average flow in an MILP formulation? As we shall see next, the "Disjunctive method" can tackle this problem quite elegantly.

8.3.1 Boiler feed water pump flow equalization

The boiler feed water system has four pumps of equal size operating in parallel. Variables associated with these four pumps are **BP1_FLO, BP2_FLO, BP3_FLO**, and **BP4_FLO** for boiler feed water flow and **BP1_X, BP2_X, BP3_X** and **BP4_X** for ON/OFF status. Each pump has a flow capacity of 250 tph, represented by the symbol *bp_flo_hl*.

The "Disjunctive method" will compute the average flow in the operating pumps. In this formulation, we will define four regions corresponding to the number of operating pumps. Continuous variables in the four regions are **BF1**, **BF2**, **BF3**, and **BF4**, representing the region's boiler feed water flow; the binary variables are **BX1**, **BX2**, **BX3**, and **BX4**.

Since only one pump operates in **Region I**, the average flow is also the total flow. In **Region II**, with two operating pumps, the average flow will be total flow divided by 2, and so on, as summarized in Tab. 8.3.

Tab. 8.3: Disjunctive method for BFW pump flow equalization: regions definition.

Region #	Low limit	High limit	Continuous variable	Binary variable	# of pumps	Average flow, BAVG
I	0	bp_flo_hl	BF1	BX1	1	= BF1
II	bp_flo_hl	2 * bp_flo_hl	BF2	BX2	2	= BF2 / 2
III	2 * bp_flo_hl	3 * bp_flo_hl	BF3	BX3	3	= BF3 / 3
IV	3 * bp_flo_hl	4 * bp_flo_hl	BF4	BX4	4	= BF4 / 4

Fig. 8.9 is a graphical view of the average flow versus total flow curves in the four regions; note that this average flow curve is discontinuous (broken), but the Disjunctive method applies nonetheless.

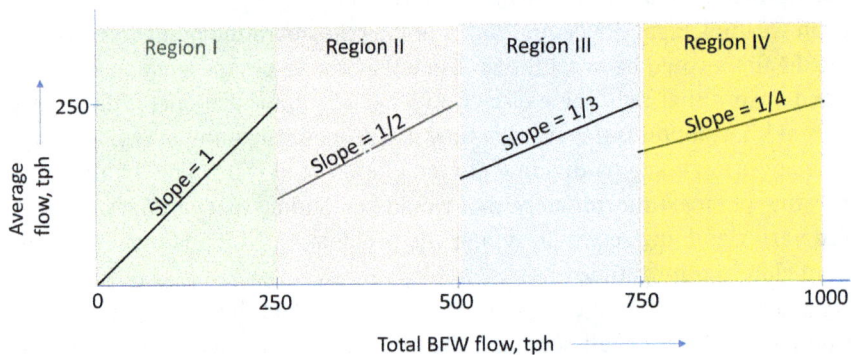

Fig. 8.9: BFW pump flow equalization: regions and average flow vs. total flow plot.

The following is the "Disjunctive method" model for boiler feed water pump flow equalization:

The flow limits on each of the regions are:

$$0 \le \mathbf{BF1} \le bp\_flo\_hl * \mathbf{BX1} \tag{8.14}$$

$$bp\_flo\_hl^* \mathbf{BX2} \leq \mathbf{BF2} \leq 2 * bp\_flo\_hl^* \mathbf{BX2} \tag{8.15}$$

$$2 * bp\_flo\_hl^* \mathbf{BX3} \leq \mathbf{BF3} \leq 3 * bp\_flo\_hl^* \mathbf{BX3} \tag{8.16}$$

$$3 * bp\_flo\_hl^* \mathbf{BX4} \leq \mathbf{BF4} \leq 4 * bp\_flo\_hl^* \mathbf{BX4} \tag{8.17}$$

Since the regions are mutually exclusive, the following constraints apply:

$$\mathbf{BP1\_FLO} + \mathbf{BP2\_FLO} + \mathbf{BP3\_FLO} + \mathbf{BP4\_FLO} = \mathbf{BF1} + \mathbf{BF2} + \mathbf{BF3} + \mathbf{BF4} \tag{8.18}$$

$$\mathbf{BX1} + \mathbf{BX2} + \mathbf{BX3} + \mathbf{BX4} \leq 1 \tag{8.19}$$

The average, **BAVG**, can be calculated as:

$$\mathbf{BAVG} = \mathbf{BF1} + \mathbf{BF2}/\ 2 + \mathbf{BF3}/\ 3 + \mathbf{BF4}/\ 4 \tag{8.20}$$

The following five constraints will ensure the operating pumps operate at flow = **BAVG**.

$$\mathbf{BP1\_\ FLO} \leq \mathbf{BAVG} \tag{8.21}$$

$$\mathbf{BP2\_FLO} \leq \mathbf{BAVG} \tag{8.22}$$

$$\mathbf{BP3\_FLO} \leq \mathbf{BAVG} \tag{8.23}$$

$$\mathbf{BP4\_FLO} \leq \mathbf{BAVG} \tag{8.24}$$

And finally, the constraint to ensure that no more than necessary boiler feed water pumps are operating is:

$$\mathbf{BP1\_X} + \mathbf{BP2\_X} + \mathbf{BP3\_X} + \mathbf{BP4\_X} = \mathbf{BX1} + 2^* \mathbf{BX2} + 3 * \mathbf{BX3} + 4 * \mathbf{BX4} \tag{8.25}$$

Flow equalization constraints, eq. (8.14) through eq. (8.25), will be added to the utility system model, **UtilSys-v2**. We will also add them to the steam subsystem constraints file, **StmSS-v2.lng**, for LINGO.

8.3.2 Cooling water pump flow equalization

The flow equalization model for the cooling water pumps is similar to the boiler feed water pumps case. The cooling water system has three pumps of equal size operating in parallel. Variables associated with these pumps are **CWP1_FLO**, **CWP2_FLO**, and **CWP3_FLO** for flow and **CWP1_X**, **CWP2_X**, and **CWP3_X** for ON/OFF status. Each pump has a flow capacity of 8 ktph, represented by the symbol *cwp_flo_hl*.

The "Disjunctive method" will compute the average cooling water flow in the operating pumps. In the formulation, three regions will correspond to the number of operating pumps. The strategy to equalize the flows is to set the pump flow equal to the average flow and ensure that the expected number of pumps are operating. Table 8.4 summarizes the characteristics of each region.

Tab. 8.4: Disjunctive method for cooling water pump flow equalization: regions definition.

Region #	Low limit	High limit	Continuous variable	Binary variable	# of pumps	Average flow, CAVG
I	0	cwp_flo_hl	CF1	CX1	1	= CF1
II	cwp_flo_hl	2 * cwp_flo_hl	CF2	CX2	2	= CF2 / 2
III	2 * cwp_flo_hl	3 * cwp_flo_hl	CF3	CX3	3	= CF3 / 3

Figure 8.10 is a graphical view of the average flow versus total flow curve in the three regions; note that the curve is discontinuous, but the method applies nonetheless.

The flow limits on each of the ranges are:

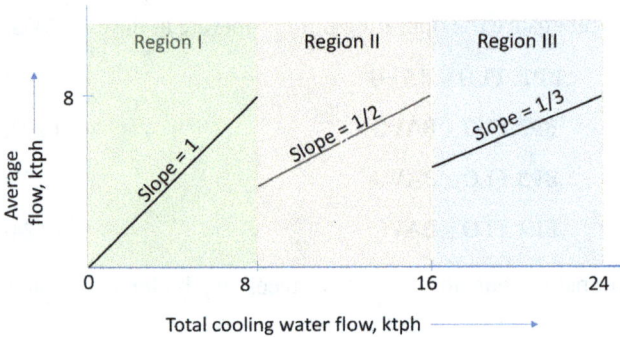

Fig. 8.10: Cooling water pump flow equalization: regions and average flow vs. total flow plot.

$$0 \leq \textbf{CF1} \leq cwp\_flo\_hl^* \textbf{CX1} \tag{8.26}$$

$$cwp\_flo\_hl^* \textbf{CX2} \leq \textbf{CF2} \leq 2 * cwp\_flo\_hl^* CX2 \tag{8.27}$$

$$2 * cwp\_flo\_hl^* \textbf{CX3} \leq \textbf{CF3} \leq 3 * cwp\_flo\_hl^* \textbf{CX3} \tag{8.28}$$

Since the regions are mutually exclusive, the following constraints apply:

$$\textbf{CWP1\_FLO} + \textbf{CWP2\_FLO} + \textbf{CWP3\_FLO} = \textbf{CF1} + \textbf{CF2} + \textbf{CF3} \tag{8.29}$$

$$\textbf{CX1} + \textbf{CX2} + \textbf{CX3} \leq 1 \tag{8.30}$$

The average, **CAVG**, can be calculated as:

$$\textbf{CAVG} = \textbf{CF1} + \textbf{CF2}/\,2 + \textbf{CF3}/\,3 \tag{8.31}$$

The following four constraints will ensure the operating pumps operate at flow = **CAVG**.

$$\text{CWP1\_FLO} \le \text{CAVG} \tag{8.32}$$

$$\text{CWP2\_FLO} \le \text{CAVG} \tag{8.33}$$

$$\text{CWP3\_FLO} \le \text{CAVG} \tag{8.34}$$

And finally, the constraint to ensure that no more than necessary cooling water pumps are operating is:

$$\text{CWP1\_X} + \text{CWP2\_X} + \text{CWP3\_X} = \text{CX1} + 2*\text{CX2} + 3*\text{CX3} \tag{8.35}$$

Flow equalization constraints, eq. (8.26) through eq. (8.35), will be added to the utility system model, **UtilSys-v2**. We will also add them to the cooling water subsystem constraints file, **CwSS-v2.lng**, for LINGO.

8.4 Completing the utility system model

We can now complete the utility system model by adding the various purchased energy contract models. We will use the following contracts for purchased energy in the sample utility system:

Fuel oil supply: There will be a single supplier with an Excess Penalty contract. The base price will be $18 per MWh.th for the first 20 MW.th and $19 for the subsequent purchase up to 200 MW.th.

LPFG supply: There will be two LPFG suppliers: LPF1 and LPF2: LPF1 with Take or Pay Contract for 100 MWh.th at $19 per MWh.th; LPF2 with a Volume Discount contract – the base price will be $19 per MWh.th for the first 100 MW.th and discounted price of $18 thereafter for up to 500 MW.th.

HPFG supply: There will be two HPFG suppliers: HPF1 and HPF2. HPF1 with Spot Purchase Contract, at the current market price that would be determined daily; there is no minimum or maximum limit on the purchase; we will use the current price of $19 per MWh.th. HPF2 has a three-tier pricing contract; the three price tiers will be $20, $19, and $18 per MWh.th with maximum purchases of 20, 60, and 120 MW.th, respectively.

Electrical energy supply: There is a single local supplier, which is usually the case. Electrical energy charges are seasonal and TOU-based as discussed in Section 8.2.2. The electrical demand charge is $7,200 per peak MW per month, which equals $10 per peak MW hourly. The peak created in the previous 11 months is a historical fact and a given for the model, essentially a fixed cost in the current period. Since the demand peak is over a 12-month running period, establishing a new demand peak in the current period will apply to the current month and the subsequent 11 months. Hence, setting a new demand peak carries an incremental cost penalty for 12 months, a hefty penalty totaling $7,200 * 12 = $86,400 per MW. There is a strong case for levying this penalty on the current period alone since it is responsible for making the incremental demand peak, and we will do so to discourage the optimizer from creating an incremental demand peak in the current period. Since the optimizer objective function is $ per hour, we will assign a hefty penalty, such as $100 per hour per incremental MW peak. Note that this penalty value is very high and perhaps somewhat arbitrary and you may wish to choose a different figure.

We will use the purchased energy contract models developed earlier in this chapter and add them to the existing utility system model, UtilSys-v1, and develop an updated utility system model, **UtilSys-v2**. In terms of modeling, only the variables and constraints relating to fuel and power costing and the objective function will be affected by the use of contracts for purchased energy; the rest of the model remains unaffected.

We will use the **@FILE()** function in LINGO that enables the inclusion of subsystem models and other portions of the model from external files, as was done previously; doing so keeps the LINGO model compact and more manageable.

Note that the electric power TOU rate constants will be stored in an external file named "UtilSys-summerPeak.lng" for LINGO and within the model file for Excel. Figure 8.11 shows the full LINGO model. Figure 8.12 is a highly reduced view of the Excel model; although model details are not discernable, we can visualize the overall problem structure.

We will test this model with the same four cases as those used for the testing UtilSys-v1, shown in Tab. 7.5, but now with fuel purchase contracts; in addition, we now have three distinct TOU cases for the electric power pricing, as shown in Tab. 8.2; so there are 12 test cases.

Both the Premium Solver and the LINGO Solver find globally optimum solutions that have identical objective functions but differ in the usage of boiler feed water pumps; the two solutions have swapped boiler feed water pump #1 with motor drive and boiler feed water pump #4 with motor drive that are functionally identical. This situation of multiple globally optimum solutions is not uncommon, especially for large models and models with functionally identical equipment. Also, the multiplicity of global optimum solutions poses no problem, as we discuss in online implementation strategies in Part II of this book.

Table 8.5 is the solution to the test case #1 with "summer_peak" electric power rates, presented as a list of variables.

Although the solution in Tab. 8.5 shows all the relevant information about the optimum solution, it fails to show the relationships between equipment and headers. In this regard, a graphical display of the optimum solution may be more informative. Figure 8.13 shows one such display; it superimposes the solution details on a simplified flow diagram of the utility system. We created this graphic in Excel as an example of the graphical display format; more sophisticated and appealing graphics would be possible. The visual display shows each major piece of equipment and header as a

```
MODEL: TITLE Utility System Model with purchased energy contracts: UtilSys-v2;
! Ref: Nath, "Industrial Process Plants: Global Optimization of Utility Systems", Chapter 8, (2023);

DATA: !inputs;    cw_demand = 15; ca_demand = 30; c_dmw =3;
      c_fo1 =18; fo1_hl =20; c_fo2 =19; fo2_hl =200;                        !FO cost;
      c_lpf1 =19; lpf1_hl =100; c_lpf21 =19; lpf21_hl =100; c_lpf22 =18; lpf22_hl =500; !LPFG cost;
      c_hpf1 =19; c_hpf21 =20; hpf21_hl =20; c_hpf22 =19; hpf22_hl =60; c_hpf23 =18; hpf23_hl =120;!HPFG;
      hps_demand =100; mps_demand=100; lps_demand=100; mps_reserve=50; hps_reserve=50;
      hpfg_demand = 30; lpfg_demand = 30; fo_demand = 30; elec_demand = 30;
      ! Equipment ARU specs:                     1=Available, 2=Required, 3=Unavailable;
      bl1_aru =1; bl2_aru =1; bl3_aru =1; stg_aru =1; ctg_aru =1; bp1_aru =1; bp2_aru =1;
      bp3_aru =1; bp4_aru =1; gtg_aru =3; fc_aru =1; cwp1_aru =1; cwp2_aru =1; cwp3_aru =1;
      ac1234_r =1; ac1234_a =4; ac5_aru =1;
      ! TOU;      @FILE('C:\BookDG\LingoModels\UtilSys-summerPeak.lng')
      ! params;   @FILE('C:\BookDG\LingoModels\UtilSys-params.lng')
      ENDDATA
      ! calcs;    @FILE('C:\BookDG\LingoModels\UtilSys-calcs-v2.lng')

! objective;      MIN = FO_CST + LPF1_CST + c_lpf21 *BUY_LPF21 + c_lpf22 *BUY_LPF22
                  + lpf22_a *LPF22_X + c_hpf1 *BUY_HPF1 + c_hpf21 *BUY_HPF21 + c_hpf22 *BUY_HPF22
                  + hpf22_a *HPF22_X + c_hpf23 *BUY_HPF23 + hpf23_a *HPF23_X + c_dmw *DEA_DMW
                  + cbuy_ener *BUY_ELEC + DMND_CST + cbuy_idem *IDEMAND - csel_elec *SEL_ELEC;

! include files;  @FILE('C:\BookDG\LingoModels\stmSS-v2.lng')      @FILE('C:\BookDG\LingoModels\fuelSS-v1.lng')
                  @FILE('C:\BookDG\LingoModels\cwSS-v2.lng')       @FILE('C:\BookDG\LingoModels\caSS-v1.lng')
                  @FILE('C:\BookDG\LingoModels\Contracts-v1.lng')
! Headers;
  [_HPS_HDR]      BL2_STM + BL3_STM = PRV1_HPS + STG_STM + CTG_HPS + BL3AT_STM + BP2T_STM + BP4T_STM
                       + CWP2T_STM + CWP3T_STM + hps_demand;
  [_MPS_HDR]      BL1_STM + PRV1_MPS + STG_STM + CTG_MPS + BP4T_STM + CWP2T_STM
                       = PRV2_STM + BL1AT_STM + mps_demand;
  [_LPS_HDR]      PRV2_STM + BL1AT_STM + BL3AT_STM + BP2T_STM + CTG_LPS + BDF_LPS + CWP3T_STM
                       = DEA_LPS + VENT + lps_demand;
  [_BFW_HDR]      BL1_BFW + BL2_BFW + BL3_BFW + PRV1_BFW = BP1_FLO + BP2_FLO + BP3_FLO + BP4_FLO;
  [_CW_HDR]       CWP1_FLO + CWP2_FLO + CWP3_FLO = CTG_CW + cw_demand;
  [_CA_HDR]       AC1234_CA + AC5_CA = VENT_CA + ca_demand;
  [_HPFG_HDR]     BUY_HPF1 + BUY_HPF2 + FC_FG = PRVFG + GTG_FG + BL2_HPFG + BL3_HPFG + hpfg_demand;
  [_LPFG_HDR]     BUY_LPF1 + BUY_LPF2 + PRVFG
                       = FC_FG + FGFLARE + BL2_LPFG + BL3_LPFG + AC1234E_FG + lpfg_demand;
  [_FO_HDR]       BUY_FO = FOFLARE + BL1_FUEL + fo_demand;
  [_ELEC_HDR]     BUY_ELEC - SEL_ELEC + STG_ELE + CTG_ELE + GTG_ELE = FC_ELE + BL2AM_ELE + BL3AM_ELE
                  + BP1M_ELE + BP2M_ELE + BP3M_ELE + BP4M_ELE + CWP1M_ELE + CWP3M_ELE + AC5M_ELE
                  + elec_demand;
END MODEL UtilSys-v2
```

Fig. 8.11: Utility System model UtilSys-v2: in LINGO.

simple geometric shape; connected streams are solid lines for material flows and dashed lines for electrical energy flow; numbers along the lines show the material or energy flow. The objective function details are summarized on the top left of the display.

Another way to display the optimum solution is in a tabular form, as shown in Fig. 8.14. Such a table is also easily created in Excel. A text-based table can also be created in LINGO using the @TEXT() function [4], as shown in Fig. 8.15. However, the table in Excel is richer in details and easier to create, so from now on, we will not bother creating it in LINGO. These tables show the details of the objective function on the top half of the table and details of flows associated with each equipment as line items on the lower half.

Each piece of equipment is a row item on the table. In the tables, we use a numerical convention popular among some practicing utility system engineers to show incoming flows to a piece of equipment as positive and outgoing flows as negative values. An advantage of this convention is that summing up similar flows (such as steam and water) along a row provides a quick material balance check: a value of zero indicates a material balance closure, but a nonzero summation is an indication

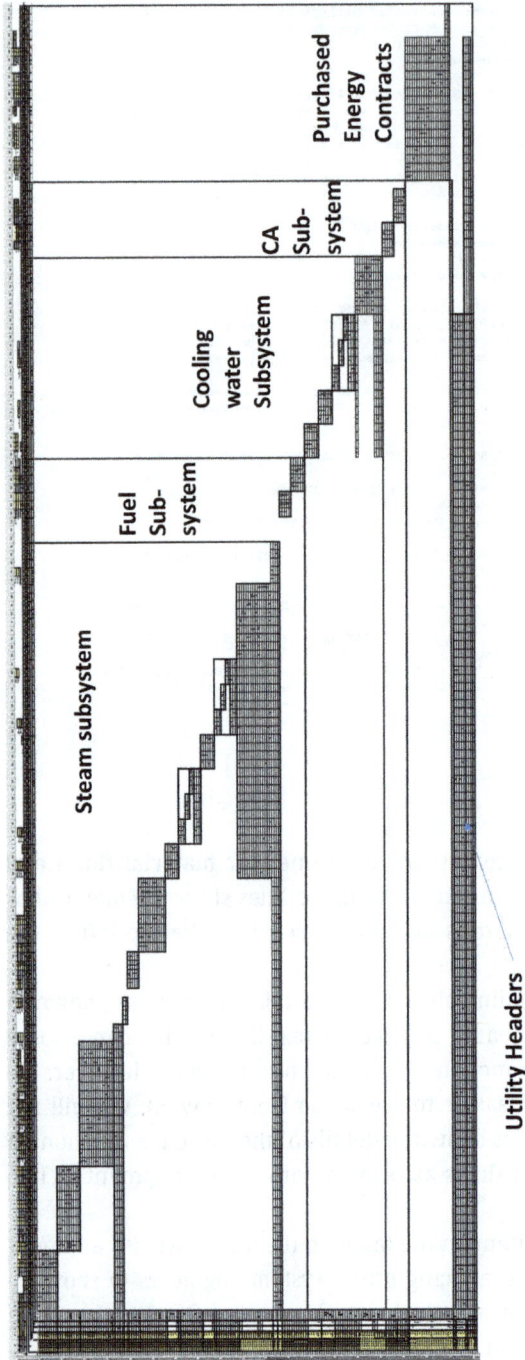

Fig. 8.12: Utility System model: UtilSys-v2: in Excel showing the model structure.

Tab. 8.5: Utility system model UtilSys-v2: test Case #1, summer-peak TOU rates, Excel solution.

Utility system Model: **UtilSys-v2**: Test case # 1
Global optimal solution found.
Opmum objective function value (minimum operating cost, $/hr) $ 9,780.47

#	Variable ID	UOM	Opt. value
1	BL1_STM	tph	0.0
2	BL1_BFW	tph	0.0
3	BL1_BD	tph	0.0
4	BL1_RSV	tph	0.0
5	BL1_FUEL	MW	0.00
6	BL1AT_STM	tph	0.0
7	BL1_X	BIN	0
8	BL1_Y	BIN	0
9	BL2_STM	tph	136.7
10	BL2_BFW	tph	139.5
11	BL2_BD	tph	2.7
12	BL2_RSV	tph	103.3
13	BL2_FUEL	MW	132.31
14	BL2_LPFG	MW	132.31
15	BL2_HPFG	MW	0.00
16	BL2AM_ELE	MW	0.500
17	BL2_X	BIN	1
18	BL2_Y	BIN	0
19	BL3_STM	tph	240.0
20	BL3_BFW	tph	244.8
21	BL3_BD	tph	4.8
22	BL3_RSV	tph	60.0
23	BL3_FUEL	MW	216.61
24	BL3_LPFG	MW	216.61
25	BL3_HPFG	MW	0.00
26	BL3AM_ELE	MW	0.000
27	BL3AT_STM	MW	5.0
28	BL3AM_X	BIN	0
29	BL3AT_X	0/1	1
30	BL3_W	0/1	1
31	BL3_X	BIN	1
32	BL3_Y	BIN	0
33	BL3_Z	BIN	0
34	MPS_RSV	tph	0.0
35	HPS_RSV	tph	163.3
36	PRV1_HPS	tph	0.0
37	PRV1_BFW	tph	0.0
38	PRV1_MPS	tph	0.0
39	PRV2_STM	tph	58.7
40	STG_STM	tph	50.0
41	STG_ELE	MW	3.000
42	STG_X	BIN	1
43	STG_Y	BIN	0
44	CTG_HPS	tph	187.8
45	CTG_MPS	tph	87.8
46	CTG_LPS	tph	90.0
47	CTG_CND	tph	10.0
48	CTG_CW	ktph	0.4
49	CTG_ELE	MW	9.452
50	CTG_X	BIN	1
51	CTG_Y	BIN	0
52	BP1_FLO	tph	192.1

#	Variable ID	UOM	cost coeff	Opt. value	Cost, $/hr
53	BP1_SP	MW		0.265	$ -
54	BP1M_ELE	MW		0.279	$ -
55	BP1_X	BIN		1	$ -
56	BP2_FLO	tph		192.1	$ -
57	BP2_SP	MW		0.265	$ -
58	BP2_X	BIN		1	$ -
59	BP2M_SP	MW		0.000	$ -
60	BP2M_ELE	MW		0.000	$ -
61	BP2M_X	BIN		0	$ -
62	BP2T_SP	MW		0.265	$ -
63	BP2T_STM	tph		2.5	$ -
64	BP2T_X	0/1		1	$ -
65	BP3_FLO	tph		0.0	$ -
66	BP3_SP	MW		0.0	$ -
67	BP3M_ELE	MW		0.000	$ -
68	BP3_X	BIN		0	$ -
69	BP4_FLO	tph		0.0	$ -
70	BP4_SP	MW		0.000	$ -
71	BP4_X	BIN		0	$ -
72	BP4M_SP	MW		0.000	$ -
73	BP4M_ELE	MW		0.000	$ -
74	BP4M_X	BIN		0	$ -
75	BP4T_SP	MW		0.000	$ -
76	BP4T_STM	tph		0.0	$ -
77	BP4T_X	0/1		0	$ -
78	BF1	tph		0.0	$ -
79	BF2	tph		384.3	$ -
80	BF3	tph		0.0	$ -
81	BF4	tph		0.0	$ -
82	BX1	BIN		0	$ -
83	BX2	BIN		1	$ -
84	BX3	BIN		0	$ -
85	BX4	BIN		0	$ -
86	BAVG	tph		192.1	$ -
87	BDF_LPS	tph		1.7	$ -
88	BDF_CND	tph		5.8	$ -
89	VENT	tph		0.0	$ -
90	DEA_DMW	tph	$3.0	155.8	$ 467.48
91	DEA_LPS	tph		68.5	$ -
92	PRVFG	MW(th)		0.00	$ -
93	FGFLARE	MW(th)		0.00	$ -
94	FOFLARE	MW(th)		0.00	$ -
95	GTG_FG	MW(th)		0.00	$ -
96	GTG_X	BIN		0	$ -
97	GTG_ELE	MW		0.000	$ -
98	FC_FG	MW(th)		30.00	$ -
99	FC_SP	MW		0.460	$ -
100	FC_X	BIN		1	$ -
101	FC_ELE	MW		0.484	$ -
102	CWP1_FLO	ktph		0.0	$ -
103	CWP1_SP	MW		0.000	$ -
104	CWP1M_EL	MW		0.000	$ -

#	Variable ID	UOM	cost coeff	Opt. value	Cost, $/hr
105	CWP1_X	BIN		0	$ -
106	CWP2_FLO	ktph		7.7	$ -
107	CWP2_SP	MW		1.101	$ -
108	CWP2T_STM	tph		20.9	$ -
109	CWP2_X	BIN		1	$ -
110	CWP3_FLO	ktph		7.7	$ -
111	CWP3_SP	MW		1.101	$ -
112	CWP3_X	BIN		1	$ -
113	CWP3M_SP	MW		0.000	$ -
114	CWP3M_ELE	MW		0.000	$ -
115	CWP3M_X	BIN		0	$ -
116	CWP3T_SP	MW		1.101	$ -
117	CWP3T_STM	tph	$ -	10.5	$ -
118	CWP3T_X	0/1		1.0	$ -
119	CF1	ktph		0.0	$ -
120	CF2	ktph		15.4	$ -
121	CF3	ktph		0.0	$ -
122	CX1	BIN		0	$ -
123	CX2	BIN		1	$ -
124	CX3	BIN		0	$ -
125	CAVG	ktph		7.7	$ -
126	AC1234_CA	kM3/h		30.0	$ -
127	AC1234_SP	MW		1.398	$ -
128	AC1234_NX	INT		4	$ -
129	AC1234E_FG	MW(th)		3.43	$ -
130	ACS_CA	kM3/h		0.0	$ -
131	ACS_SP	MW		0.000	$ -
132	ACS_X	BIN		0	$ -
133	ACSM_ELE	MW		0.000	$ -
134	VENT_CA	kM3/h		0.0	$ -
135	BUY_FO	MW(th)		30.00	$ -
136	FO_CST	$/hr	1.0	550.0	$ 550.00
137	BUY_LPF1	MW(th)		100.00	$ -
138	LPF1_CST	$/hr	1.0	1900.0	$ 1,900.00
139	BUY_LPF21	MW(th)	19.0	0.000	$ -
140	BUY_LPF22	MW(th)	18.0	312.36	$ 5,622.44
141	BUY_LPF2	MW(th)		312.36	$ -
142	LPF21_X	BIN		0	$ -
143	LPF22_X	0/1	$ 100.0	1	$ 100.00
144	BUY_HPF1	MW(th)	19.0	0.00	$ -
145	BUY_HPF21	MW(th)	20.0	0.00	$ -
146	BUY_HPF22	MW(th)	19.0	0.00	$ -
147	BUY_HPF23	MW(th)	18.0	0.00	$ -
148	BUY_HPF2	MW(th)		0.00	$ -
149	HPF21_X	BIN		0	$ -
150	HPF22_X	BIN	$ 20.0	0	$ -
151	HPF23_X	0/1	$ 80.0	0	$ -
152	BUY_ELEC	MW	$ 50.0	18.811	$ 940.55
153	DMND_CST	MW peak	1.00	200.0	$ 200.00
154	IDEMAND	MW peak	$100.0	0.000	$ -
155	SEL_ELEC	MW	$ (40.0)	0.000	$ -

of material imbalance that should be investigated and rationalized. In this tabular display, the columns represent the headers and can also be summed up for a quick check. A zero summation indicates a material balance closure, while a nonzero summation indicates a net import or export.

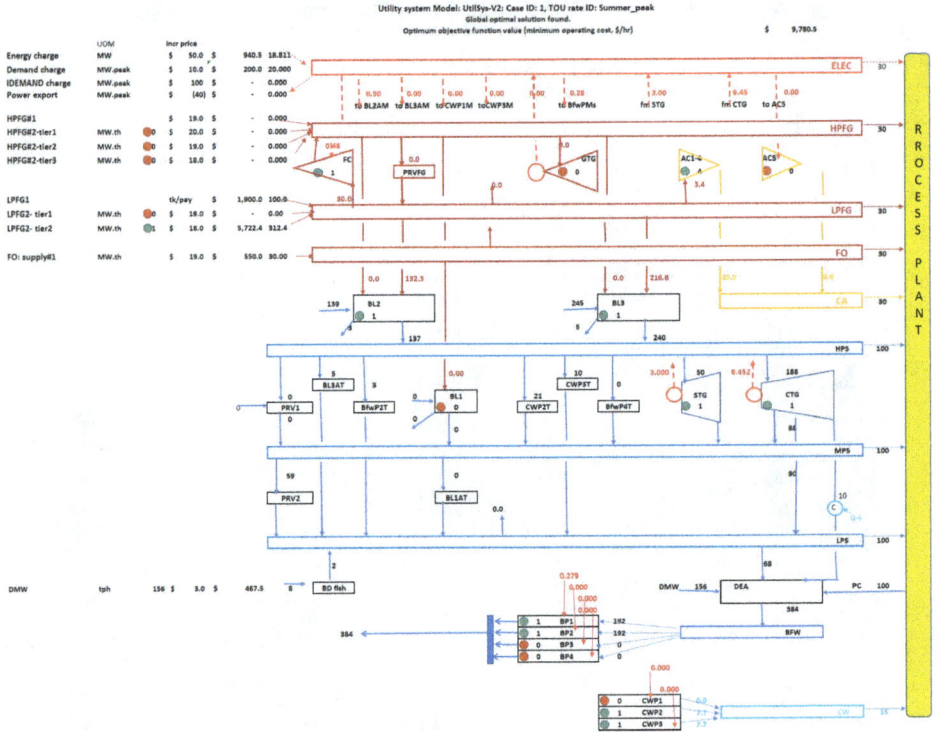

Fig. 8.13: Utility system model, UtilSys-v2: test case #1 solution: Excel generated graphical report.

Utility system Model: UtilSys-V2: Case ID: 1, TOU rate ID: Summer_peak
Global optimal solution found.
Optimum objective function value (minimum operating cost, $/hr) **$ 9,780.5**

Purchased energy:

	contract	min	max	amount		COST, $./hr		incr unit cst
BUY DMW				155.8	tph	$ 467.5		$ 3.00
BUY FO TIER 1&2	XSS PEN		200	30.000	MW.th	$ 550.0		$ 19.00
BUY LPFG - #1	TAKE/PAY		100	100.000	MW.th	$ 1,900.0		take or pay
BUY LPFG #2-tier1	●0		100	0.00	MW.th	$ -		$ 19.00
BUY LPFG #2-tier2	●1		500	312.36	MW.th	$ 5,722.4		$ 18.00
BUY HPFG #1	SPOT			0.000	MW.th	$ -		$ 19.00
BUY HPFG #2-tier1	●0		20	0.000	MW.th	$ -		$ 20.00
BUY HPFG #2-tier2	●0		60	0.000	MW.th	$ -		$ 19.00
BUY HPFG #2-tier3	●0		120	0.000	MW.th	$ -		$ 18.00
BUY ELEC - energy charge				18.811	MW	$ 940.5		$ 50.00
BUY ELEC - demand charge				20.000	MW-peak	$ 200.0		$ 10.00
IDEMAND penalty				0.000	MW	$ -		$ 100.00
Sell ELEC				0.000	MW	$ -		$ (40.00)
TOTAL OP COST						**$ 9,780.5**		

Process DEMAND				100	100	100		-100				15	30	30	30	30	30
(unit inputs +ve, outputs -ve)				**HP**	**MP**	**LP**	**BD**	**PC/TC**	**DMW**	**HP-bfw**	**LP-bfw**	**CW**	**CA**	**FO**	**LPFG**	**HPFG**	**ELEC**
Unit	A/R/U	min	max	tph	tph	tph	tph	tph	tph	tph		ktph	kM3/h	MW.th	MW.th	MW.th	MW
●0 BL1	A	70	210	0		0	0			0				0.0			
	-aux				0	0											
●1 BL2	A	80	240	-137			-3			139					132.31	0.00	
	-aux																0.500
●1 BL3	A	100	300	-240			-5			245					216.61	0.00	
	-aux					-5											0.000
●1 STG	A	1.00	3.00	50	-50												-3.000
●1 CTG	A	3.00	13.00	188	-88	-90		-10				0.4					-9.452
●1 Bfw P1	A		250							-192	192						0.279
●1 Bfw P2	A		250	3		-3				-192	192						0.000
●0 Bfw P3	A		250							0	0						0.000
●0 Bfw P4	A		250	0	0					0	0						0.000
PRV1				0	0						0						
PRV2					59	-59											
BD Flash						-2	8										
Deaerator						68			110	156	-384						
steam VENT						0											
●0 CW P1	A		8									0.0					0.000
●1 CW P2	A		8	21	-21							-7.7					
●1 CW P3	A		8	10		-10						-7.7					0.000
●4 AC1-4	4	12	36										-30.0		3.43		
●0 AC5	A	20	55										0.0				0.000
Air Vent													0.0				
●0 GTG	U		25												0.00	0.00	0.000
●1 FC	A	25	75												30.00	0.00	0.484
PRVFG															0.00	0.00	
Fuel FLARE														0.00	0.00		
sum column (net import)				0.0	0.0	0.0	0.0	0.0	155.8	0.0	0.0	0.0	0.0	30.00	412.36	30.00	18.811

Fig. 8.14: Utility system model, UtilSys-v2: test case #1 solution: Excel generated tabular report.

```
Utility System Model with purchased energy contracts: UtilSys-v2
================================================================: ==========
  Global optimal solution found!
  Objective function:
  ========= ========
  BUY DMW                     156 tph          467.5 $/hr
  --- --------------        ------ -------  ------------  ------
  BUY FO tier1&2             30.0 MW(th)       550.0 $/hr
  --- --------------        ------ -------  ------------  ------
  BUY LPFG 1                100.0 MW(th)      1900.0 $/hr
  BUY LPFG 2 tier1           0.0 MW(th)          0.0 $/hr
  BUY LPFG 2 tier2          312.4 MW(th)      5722.4 $/hr
  --- --------------        ------ -------  ------------  ------
  BUY HPFG 1                  0.0 MW(th)          0.0 $/hr
  BUY HPFG 2 tier1           0.0 MW(th)          0.0 $/hr
  BUY HPFG 2 tier2           0.0 MW(th)          0.0 $/hr
  BUY HPFG 2 tier3           0.0 MW(th)          0.0 $/hr
  --- --------------        ------ -------  ------------  ------
  BUY ELEC energy           18.81 MW           940.5 $/hr
  BUY ELEC demand           20.00 MWpeak       200.0 $/hr
  BUY ELEC Inc demand        0.00 MWpeak         0.0 $/hr
  SEL ELEC energy            0.00 MW             0.0 $/hr
  --- --------------        ------ -------  ------------  ------
  TOT OPER cost                                9780.5 $/hr

  Solution Details: (Positive value indicates Inflow, negative value indicates Outflow)
  ========= ========
  PROCESS      100    100    100        0                  15     30  30.0  30.0  30.0 30.00
  --- -----  -----  -----  -----  -----  -----  -----  -----  -----  -----  -----  -----  -----
  Sta UNIT    -HPS-  -MPS-  -LPS-  - BD-  -CND-  -DMW-  HpBfw  LpBfw  --CW-  --CA-  --FO- -LPFG -HPFG -ELEC
  tus ID       tph    tph    tph    tph    tph    tph MW.th  MW.th  ktph kM3/h  MW.th MW.th MW.th  MW
  --- -----  -----  -----  -----  -----  -----  -----  -----  -----  -----  -----  -----  -----  -----  -----
  0   BL1            0             0                  0                    0.0
           aux       0      0
  --- -----  -----  -----  -----  -----  -----  -----  -----  -----  -----  -----  -----  -----  -----  -----
  1   BL2    -137                  -3                139                        132.3  0.0
           aux                                                                              0.50
  --- -----  -----  -----  -----  -----  -----  -----  -----  -----  -----  -----  -----  -----  -----  -----
  1   BL3    -240                  -5                245                        216.6  0.0
           aux              5       -5
  --- -----  -----  -----  -----  -----  -----  -----  -----  -----  -----  -----  -----  -----  -----  -----
  1
  1   STG      50    -50                                                                          -3.00
  1   CTG     188    -88    -90          -10                       0.4                             -9.45
  --- -----  -----  -----  -----  -----  -----  -----  -----  -----  -----  -----  -----  -----  -----  -----
  0   BFW Pump1,3                               0      0                                            0.00
  1   BFW Pump2    3          -3                  -192    192                                       0.00
  1   BFW Pump4    0     0                        -192    192                                       0.28
  --- -----  -----  -----  -----  -----  -----  -----  -----  -----  -----  -----  -----  -----  -----  -----
      PRV1    0     -0                           0
      PRV2         59    -59
  --- -----  -----  -----  -----  -----  -----  -----  -----  -----  -----  -----  -----  -----  -----  -----
      BD Flash          -2      8
      Deaerator         68          10    156    -384
      Steam Vent         0
  --- -----  -----  -----  -----  -----  -----  -----  -----  -----  -----  -----  -----  -----  -----  -----
  0   CW Pump1                                                      0.0                             0.00
  1   CW Pump2   21    -21                                         -7.7
  1   CW Pump4   10          -10                                   -7.7                             0.00
  --- -----  -----  -----  -----  -----  -----  -----  -----  -----  -----  -----  -----  -----  -----  -----
  4   Air Cmpr1-4                                                        -30.0         3.4
  0   Air Cmpr5                                                           0.0                       0.00
      PRV FG                                                                          0.0   0.0
      Fuel Vent                                                                 0.0   0.0
  --- -----  -----  -----  -----  -----  -----  -----  -----  -----  -----  -----  -----  -----  -----  -----
```

Fig. 8.15: Utility system model, UtilSys-v2: test case #1 solution: LINGO generated report.

Summary

Well-defined and legally binding contracts govern the purchase of fuel and electrical power from external suppliers. Some of these contracts, especially those with local power suppliers, can be complex, incentivizing and disincentivizing power usage during specific periods for an overall more efficient and reliable operation of the local power grid. This chapter introduced a new modeling technique called the "Disjunctive method" and discussed modeling the essential features of all commonly encountered purchased energy contracts. We also improved the BFW and cooling water pump models by using the "Disjunctive method" to equalize flow in the operating pumps.

These contract models are added to the previous version of the utility system model to complete the utility system model for the sample utility system, which was the objective of Part I of this book. The updated model, **UtilitySys-v2**, is a realistic model of a simple process plant utility system and can be used to study various trade-offs in operating a simple industrial utility system. Our testing was limited to 12 test cases using the Premium Solver in Microsoft Excel and the LINGO Solver. For each case, the two solutions gave identical objective function values but sometimes differed in the operations of BFW pumps. Such cases of multiple globally optimum solutions are not uncommon, especially for large models such as this one with multiple functionally identical pieces of equipment, which may appear problematic. However, it is not, as will be discussed in Part II of this book when discussing the online deployment strategies of the optimizer.

Nomenclature

Symbol	UOM	Description
AC1234_CA	kM3/h	Air compressor #1234: compressed air flow
AC1234_NX	0 .. 4	Air compressor #1234: number of operating pumps
AC1234_SP	MW.sp	Air compressor #1234: shaft power required
AC1234M_FG	MW.th	Air compressor #1234: ICE drive: fuel consumption
AC5_CA	kM3/h	Air compressor #5: compressed air flow
AC5_SP	MW.sp	Air compressor #5: shaft power required
AC5_X	0/1	Air compressor #5: operational status: 0 = OFF, 1 = ON
AC5M_ELE	MW	Air compressor #5: motor drive: power consumption
BAVG	tph	BFW pump flow equalization: average BFW flow
BDF_CND	tph	Blowdown flash: LP condensate discharge
BDF_HPC	tph	Blowdown flash: inlet HP blowdown
BDF_LPS	tph	Blowdown flash: LP steam generation
BDF_MPC	tph	Blowdown flash: inlet MP blowdown
BFj	tph	BFW pump flow equalization: flow corresponding to j operating pumps
BL3_W	0/1	Boiler #3: LPFG indicator: 1 = using LPFG, 0 = using HPFG
BL3_Z	0/1	Boiler #3: HPFG indicator: 1 = using HPFG, 0 = using LPFG
BLj_BD	tph	Boiler #j: blowdown flow

(continued)

Symbol	UOM	Description
BLj_BFW	tph	Boiler #j: boiler feed water inflow
BLj_FUEL	MW.th	Boiler #j: total fuel consumption
BLj_HPFG	MW.th	Boiler #j: HP fuel gas consumption
BLj_LPFG	MW.th	Boiler #j: LP fuel gas consumption
BLj_RSV	tph	Boiler #j: steam reserve
BLj_STM	tph	Boiler #j: steam generation
BLj_X	0/1	Boiler #j: operational status: 0 = OFF, 1 = ON
BLj_Y	0/1	Boiler #j: standby status: 1 = in standby mode, 0 otherwise
BLjAM_ELE	MW	Boiler #j: auxiliaries motor drive: electric power usage
BLjAM_X	0/1	Boiler #j: auxiliaries motor drive: 0 = OFF, 1 = ON
BLjAT_STM	tph	Boiler #j: auxiliaries turbine: steam flow
BLjAT_X	0/1	Boiler #j: auxiliaries turbine drive: 0 = OFF, 1 = ON
BPj_FLO	tph	BFW Pump #j: boiler feed water inflow
BPj_SP	MW.sp	BFW Pump #j: shaft power required
BPj_X	0/1	BFW Pump #j: operational status: 0 = OFF, 1 = ON
BPjM_ELE	MW	BFW Pump #j: motor drive: power consumption
BPjM_SP	MW.sp	BFW Pump #j: motor drive: shaft power produced
BPjM_X	0/1	BFW Pump #j: motor drive: operational status: 0 = OFF, 1 = ON
BPjT_SP	MW.sp	BFW Pump #j: turbine drive: shaft power produced
BPjT_STM	tph	BFW Pump #j: turbine: steam flow
BPjT_X	0/1	BFW Pump #j: turbine drive: operational status: 0 = OFF, 1 = ON
BUY_ELEC	MW	Purchased energy: buy electric power
BUY_FO	MW.th	Purchased energy: buy fuel oil
BUY_FUEL	MW.th	amount of fuel purchased
BUY_FUELj	MW.th	amount of fuel purchased in tier #j region
BUY_HPFG	MW.th	Purchased energy: buy HP fuel gas
BUY_LPFG	MW.th	Purchased energy: buy LP fuel gas
BXj	0/1	BFW pump flow equalization: Indicates that j pumps are operating.
CAVG	tph	CW pump flow equalization: average CW flow
CFj	tph	CW pump flow equalization: flow corresponding to j operating pumps
CTG_CND	tph	CTG: condensate from condenser
CTG_CW	ktph	CTG: condenser: cooling water flow
CTG_ELE	MW	CTG: power generation
CTG_HPS	tph	CTG: inlet HP steam flow
CTG_LPS	tph	CTG: outlet LP steam flow
CTG_MPS	tph	CTG: outlet MP steam flow
CTG_X	0/1	CTG: operational status: 0 = OFF, 1 = ON
CTG_Y	0/1	CTG: standby status: 1 = in standby mode, 0 otherwise
CWPj_FLO	tph	CW Pump #j: boiler feed water inflow
CWPj_SP	MW.sp	CW Pump #j: shaft power required
CWPj_X	0/1	CW Pump #j: operational status: 0 = OFF, 1 = ON
CWPjM_ELE	MW	CW Pump #j: motor drive: power consumption
CWPjM_SP	MW.sp	CW Pump #j:: motor drive: shaft power produced
CWPjM_X	0/1	CW Pump #j:: motor drive: operational status: 0 = OFF, 1 = ON
CWPjT_SP	MW.sp	CW Pump #j:: turbine drive: shaft power produced
CWPjT_STM	tph	CW Pump #j:: turbine: steam flow

(continued)

Symbol	UOM	Description
CWPjT_X	0/1	CW Pump #j:: turbine drive: operational status: 0 = OFF, 1 = ON
CXj	0/1	CW pump flow equalization: Indicates that j pumps are operating.
DEA_BFW	tph	Deaerator: boiler feed water production
DEA_DMW	tph	Deaerator: fresh demineralized water intake
DEA_LPS	tph	Deaerator: deaeration steam
DEA_LPS	tph	Deaerator: deaeration steam
DMND_CST	$/h	Electrical power demand charge
FC_ELE	MW	Fuel compressor: power consumption
FC_FG	MW.th	Fuel compressor: fuel compressed
FC_X	0/1	Fuel compressor: operational status: 0 = OFF, 1 = ON
FGFLARE	MW.th	VENT: LPFG
FOFLARE	MW.th	VENT: fuel oil
FUEL_CST	$/h	VENT: fuel oil
FUELj_X	0/1	indicator for fuel purchase in tier #j region
GTG_ELE	MW	GTG: power generated
GTG_FG	MW.th	GTG: fuel consumed
GTG_X	0/1	GTG: operational status: 0 = OFF, 1 = ON
HPS_RSV	tph	HP Boilers: total steam reserve
IDEMAND	MW.pk	incremental demand peak
MPS_RSV	tph	MP Boiler: steam reserve
PRV1_BFW	tph	PRV: de-superheating water flow
PRV1_HPS	tph	PRV: HP steam flow
PRV1_MPS	tph	PRV: MP steam flow
PRV2_STM	tph	PRV2: steam flow
PRVFG	MW.th	PRVFG: HPFG to LPFG
SEL_ELEC	MW	Purchased energy: sell electric power
STG_ELE	MW	STG: power generation
STG_STM	tph	STG: steam flow
STG_X	0/1	STG: operational status: 0 = OFF, 1 = ON
STG_Y	0/1	STG: standby status: 1 = in standby mode, 0 otherwise
VENT	tph	VENT: steam flow
VENT_CA	kM3/h	VENT: compressed air
Wi	–	variable in Lambda model
a, b, c, ci	–	generic constants
ac1234_a	0 … 4	Input: Air compressors #1 … 4: number of available compressors
ac1234_r	0 … 4	Input: Air compressors #1 … 4: number of required compressors
ac5_aru	ARU or 123	Input: Air compressor #5: "available'/"required"/"unavailable" spec
bdf_a	–	Parameter: Blowdown flash: Enthalpy ratio: (hl_hps - hl_lps)/hl_mps
bdf_b	–	Parameter: Blowdown flash: Enthalpy ratio: (hv_lps - hl_lps)/hl_mps
blj_aru	ARU or 123	Input: BL #j: "available'/"required"/"unavailable" spec
blj_bdr	fraction	Parameter: Boiler #j : blowdown ratio
blj_fg_c	MW.th	Parameter: Boiler #j : fuel vs steam, low range, regression: intercept
blj_fg_d	MWh.th/t	Parameter: Boiler #j : fuel vs steam, low range, regression: slope
blj_fg_e	MW.th	Parameter: Boiler #j : fuel vs steam, high range, regression: intercept
blj_fg_f	MWh.th/t	Parameter: Boiler #j : fuel vs steam, high range, regression: slope
blj_fg_s	MW.th	Parameter: Boiler #j : standby mode: fuel consumption

(continued)

Symbol	UOM	Description
blj_lgc_hl	0/1	Calculated: BL #j: _LGC constraint: high limit
blj_lgc_ll	0/1	Calculated: BL #j: _LGC constraint: low limit
blj_stm_hl	tph	Parameter: Boiler #j : steam generation: high limit
blj_stm_ll	tph	Parameter: Boiler #j : steam generation: low limit
bljam_c	MW	Parameter: Boiler #j auxiliaries motor: power
bljat_c	tph	Parameter: Boiler #j auxiliaries turbine: steam flow
bpj_des_ll	tph	Parameter: BWP Pump #j: inlet flow: design low limit
bpj_flo_hl	tph	Parameter: BWP Pump #j: inlet flow: high limit
bpj_lgc_hl	0/1	Calculated: BFW pump #j: _LGC constraint: high limit
bpj_lgc_ll	0/1	Calculated: BFW pump #j: _LGC constraint: low limit
bpj_sp_a	MW.sp	Parameter: BWP Pump #j: SP vs flow regression: intercept
bpj_sp_b	MW.sp/t	Parameter: BWP Pump #j: SP vs flow regression: slope
bpjm_eta	tph/tph	Parameter: BWP Pump #j: motor drive: motor efficiency
bpjt_stm_a	tph	Parameter: BWP Pump #j: turbine drive: steam vs flow: intercept
bpjt_stm_b	tph/tph	Parameter: BWP Pump #j: turbine drive: steam vs flow: slope
buy_fuel_max	MW.th	buy fuel: upper limit
c_buyelec	$/MWh	Input cost: buy electric power
c_dmw	$/ton	Input cost: buy demineralized water
c_fo	$/MWh.th	Input cost: buy fuel oil
c_fuelj	$/MWh.th	fuel cost
c_hpfg	$/MWh.th	Input cost: buy LPFG
c_lpfg	$/MWh.th	Input cost: buy LPFG
c_selelec	$/MWh	Input cost: sell electric power
ca_demand	kM3/h	Process Demand: compressed air
ctg_cnd_hl	tph	Parameter: CTG: condensate: high limit
ctg_cnd_ll	tph	Parameter: CTG: condensate: low limit
ctg_cw_b	ktph/tph	Parameter: CTG: cooling water vs condensate: slope
ctg_ele_hl	tph	Parameter: CTG: power gen: high limit
ctg_ele_ll	tph	Parameter: CTG: power gen: low limit
ctg_hps_a	tph	Parameter: CTG: hps vs. other variables: intercept
ctg_hps_b	tph/tph	Parameter: CTG: hps vs. power gen regression: slope
ctg_hps_c	tph/tph	Parameter: CTG: hps vs. MPS steam flow regression: slope
ctg_hps_d	tph/tph	Parameter: CTG: hps vs. LPS steam flow regression: slope
ctg_hps_hl	tph	Parameter: CTG: HP steam: high limit
ctg_hps_ll	tph	Parameter: CTG: HP steam: low limit
ctg_hps_s	tph	Parameter: CTG: standby mode: inlet steam flow
ctg_lgc_hl	0/1	Calculated: CTG: _LGC constraint: high limit
ctg_lgc_ll	0/1	Calculated: CTG: _LGC constraint: low limit
ctg_lps_hl	tph	Parameter: CTG: LP steam: high limit
ctg_mps_hl	tph	Parameter: CTG: MP steam: high limit
ctg_x_hl	0/1	Calculated: CTG_X: high limit
ctg_x_ll	0/1	Calculated: CTG_X: low limit
cur_pkdemand	MW.pk	Input cost: buy electric power: current peak demand
cw_demand	ktph	Process demand: cooling water
cwpj_aru	ARU or 123	Input: CW pump #j: "available'/"required"/"unavailable" spec
dea_a	–	Parameter: Deaerator: Enthalpy ratio: (h_tc - h_pc)/(hl_mps - h_pc)

(continued)

Symbol	UOM	Description
dea_b	–	Parameter: Deaerator: Enthalpy ratio: (h_dmw - h_pc)/(hl_mps - h_pc)
dea_c	–	Parameter: Deaerator: Enthalpy ratio: (h_lps - h_pc)/(hl_mps - h_pc)
elec_demand	MW	Process demand: electrical power
fc_lgc_hl	0/1	Calculated: Fuel compressor: _LGC constraint: high limit
fc_lgc_ll	0/1	Calculated: Fuel compressor: _LGC constraint: low limit
fixed_fuel_cost	$/hr	buy fuel: agreed upon cost
fo_demand	MW.th	Process demand: fuel oil
fuel_demand	MW.th	fuel demand
fuelj_hl	MW.th	fuel tier #j region high limit
fuelj_ll	MW.th	fuel tier #j region low limit
gtg_fg_a	MW.th	Parameter: GTG : fuel vs. power gen, low range, regression: intercept
gtg_fg_b	MW.th/MW	Parameter: GTG : fuel vs. power gen, low range, regression: slope
gtg_fg_c	MW.th	Parameter: GTG : fuel vs. power gen, mid range, regression: intercept
gtg_fg_d	MW.th/MW	Parameter: GTG : fuel vs. power gen, mid range, regression: slope
gtg_fg_e	MW.th	Parameter: GTG : fuel vs. power gen, high range, regression: intercept
gtg_fg_f	MW.th/MW	Parameter: GTG : fuel vs. power gen, high range, regression: slope
gtg_x_hl	0/1	Calculated: GTG_X: high limit
gtg_x_ll	0/1	Calculated: GTG_X: low limit
hpfg_demand	MW.th	Process demand: HP fuel gas
hps_demand	tph	Process demand: HP steam
hps_reserve	tph	Process demand: HP steam reserve
lpfg_demand	MW.th	Process demand: LP fuel gas
lps_demand	tph	Process demand: LP steam
mps_demand	tph	Process demand: MP steam
mps_reserve	tph	Process demand: MP steam reserve
pc_supply	tph	Process supply: process condensate return
prv1_a	–	Parameter: Enthalpy ratio: h_hpbfw /h_hps
prv1_b	–	Parameter: Enthalpy ratio: h_mps /h_hps
stg_ele_hl	MW	Parameter: STG: power generation: high limit
stg_ele_ll	MW	Parameter: STG: power generation: low limit
stg_lgc_hl	0/1	Calculated: STG: _LGC constraint: high limit
stg_lgc_ll	0/1	Calculated: STG: _LGC constraint: low limit
stg_stm_a	tph	Parameter: STG: inlet steam vs. power regression: intercept
stg_stm_b	tph/MW	Parameter: STG: inlet steam vs. power regression: slope
stg_stm_hl	tph	Parameter: STG: inlet steam: high limit
stg_stm_s	tph	Parameter: STG: standby mode: inlet steam flow
xi, yi	–	coordinates of an endpoint #i in the Lambda model

References

[1]   Chen, J., "What Is Henry Hub?", https://www.investopedia.com/terms/h/henry_hub.asp, (2022).

[2]   Martin, R. K. "Large Scale Linear and Integer Optimization, A Unified Approach", Kluwer Academic Publishers, Boston (1999).

[3]   Beale, E. M. L. and J. J. H. Forrest. "Global optimization using special ordered sets", Mathematical Programming, 10(1):52–69, (1976).

[4]   Anonymous, "LINGO the Modeling Language and Optimizer", Lindo Systems Inc, Chicago (2020).

[5]   Private communication with Frontline Systems Inc., (2023).

Part II: **Model deployment**

Part I of the book discussed the development aspects of a utility system model to create a global optimization model of a simple industrial process plant utility system. Part I focused on developing mixed integer linear programming models that guarantee global optimization.

Part I started from the basics of modeling and optimization and proceeded systematically to create individual equipment models, assembling individual equipment models to create subsystem models, and finally integrating the subsystem models to develop a complete model of a sample utility system. The model is representative of a simple process plant utility system.

Part II of the book focuses on the use and deployment aspects of the utility system optimization models. By deployment, we mean readying the model for users who know and understand their utility system operations but may not know or may not be interested in the nitty-gritties of the mathematical model. The focus in Part II shifts from developing to using the model with the ultimate aim of deriving benefits by implementing the results of global optimization to the utility system operations.

Although the optimization model developed thus far is usable, the focus in this part of the book is to add select features to make the model easier for use by a general user who is not a modeler and by automation. Doing so will require some changes to the model to enable automatic execution of the model with raw process data. As in Part I, we will proceed gradually and systematically addressing the following two uses:
– offline optimization, and
– online or real-time optimization

The next chapter will discuss offline optimization considerations, i.e., how to make the optimization system an easier-to-use tool for performing case studies to help improve the operations of the utility system. The following chapter will discuss online or real-time optimization considerations with the ultimate goal of unattended execution of the optimization cycle with live process data to reap the continuous benefits of global optimization.

https://doi.org/10.1515/9783111020679-010

9 Offline optimization

In Chapter 8, we developed an operations optimization model of the sample plant utility system. This optimization model determines the most cost-effective operations policy for specified process utility demands, equipment availability, equipment parameters, and external fuel and electrical power supply contracts.

This chapter discusses using this optimizer as a standalone tool for performing case studies. We will discuss the following use cases of the offline utility system optimizer:
- Ad hoc optimization
- Annual cost budgeting
- Energy contract evaluation
- Proposed investment evaluation

A key feature necessary for the abovementioned uses is the ability to store multiple sets of optimization data for the on-demand execution of selected cases. The following section will discuss a methodology for such case management.

9.1 Case management

Both Excel and LINGO support interfacing the optimization model to an external database management system for case management. However, in light of the fact that we are already using Excel for utility system optimization and in the spirit of simplicity, we will use Excel as a simple case management tool. In addition, since the LINGO modeling system provides an interface to Excel, we will also use the same Excel-based case management tool for optimizing selected cases using the LINGO solver.

To keep things organized, we will start with the Excel-based optimization model, UtilSys-v2, which has been developed so far, as a starting point for the offline utility system optimization model, **UtilSys-OFLO**. We will add a new worksheet, named CaseData, to the Excel workbook for case management capability. The CaseData worksheet has two sections: (1) the Input Data section, which serves as a repository of data for all cases of interest, and (2) the TOU rate data section, which is a repository of cost parameters relating to the various TOU electric power rates. In this arrangement, each row in a section has a unique key, a brief description of the case, and all necessary data for a particular case. Excel LOOKUP function selects the desired case data from the Input Data section using the case number on cell C2 of the "Model" worksheet as the key. Another Excel LOOKUP function fetches TOU electric power cost parameters using the key specified on cell C3 of the "Model" worksheet.

The selected input data items and TOU electric rate parameters are on row 1 of the CaseData worksheet. Each cell containing data has a cell name assigned to it, such as "hps_demand." Excel Name Manager shows all cell names and their locations.

https://doi.org/10.1515/9783111020679-011

Excel and the LINGO models will refer to the retrieved data items by these cell names. LINGO model, UtilSys-OFLO.lg4 will retrieve data from Excel using LINGO's @OLE() function. For details of the **@OLE()** function, refer to LINGO documentation [1, 2].

The CaseData worksheet has two tables, one dedicated to storing parameters relating to TOU electricity rates; is in columns AY through BD starting row 6 for five rows corresponding to the five different TOU rates. The other table is for all other data items in columns C through AV, starting row 6, with one row for each case.

Each case requires 66 input data items including 5 TOU electrical cost parameters as summarized in Tab. 9.1. The first 11 items pertain to process demands; items 12 through 43 specify the availability of the 16 pieces of equipment in the sample utility system. Data item number 44 is the cost of demineralized water. Items 45 through 61 are the parameters relating to the cost of purchased fuel, and the last five data items are the parameters relating to the TOU electric power contract. Note that there are two conventions to specify the equipment availability code: an alpha code A/R/U to denote Available/Required/Unavailable and a numeric code 1/2/3 – 1 for Available, 2 for Required, and 3 for Unavailable. The Excel model uses the alpha codes, while the LINGO model uses the numeric code as it cannot accept alpha codes. Air compressors AC1, 2, 3, and 4, being identical units, do not have individual ARU codes; instead, "ac1234_r" and "ac1234_a" specify their availability; the first data item specifies the number of air compressors that are required to operate in the optimal solution and the second item specifies the maximum number available to optimize.

Figure 9.1 shows part of the input data in the "CaseData" worksheet. The cursor in the figure is on cell C1, the retrieved value of the data item, *hps_demand*, the high-pressure, process steam demand. The cell name assigned to cell C1, hps_demand, shown in the text box on the upper left side, and the Excel LOOKUP function used to retrieve this value on the formula bar.

Figure 9.2 shows the TOU electric power rate parameters in the CaseData worksheet. The cursor is on cell AZ1, the retrieved value for the TOU rate parameter, *cbuy_-ener*, which is the electrical energy cost in $/MWh. Cell_name assigned to cell AZ1, cbuy_ener, is shown in the text box on the upper left side, and the formula bar shows the Excel LOOKUP function used to retrieve this value.

Figure 9.3 shows the model worksheet; cell C2 is the lookup value key for the input data and cell C3 is the lookup value for TOU electric power rate parameters. The cursor is placed on cell I4, showing the retrieved "hps_demand" value of 100 in light blue font. Several other retrieved input values are also in light blue font.

In LINGO, case data values are retrieved using LINGO's **@OLE()** interface function in the DATA section of the model [1]. The basic syntax is:

$$object\_list = @OLE('spreadsheet\_file\_name');$$

If the spreadsheet file name is omitted, LINGO uses the currently open Excel file.

Figure 9.4 shows the LINGO sample utility system model, **UtilSys-OFLO.lg4**. The first statement in the **DATA** section retrieves the 51 data items listed on the LHS of the

Tab. 9.1: Utility system model: UtilSys-OFLO: list of data items in CaseData worksheet.

#	Excel cell ID	Excel cell name	Sample value	UOM	Comment	#	Excel cell ID	Excel cell name	Sample value	UOM	Comment
1	C1	hps_demand	150	tph		34	Y1	ac1234_r	0	-	
2	D1	mps_demand	50	tph		35	Z1	ac1234_a	4	-	
3	E1	lps_demand	100	tph		36	AA1	ac5_aru	A	A/R/U	
4	F1	hps_reserve	50	tph		37	AA2	ac5_123	1	1/2/3	
5	G1	mps_reserve	50	tph	Process demands	38	AB1	cwp1_aru	A	A/R/U	Equipment avaialability
6	H1	cw_demand	15	ktph		39	AB2	cwp1_123	1	1/2/3	
7	I1	ca_demand	30	kM3/h		40	AC1	cwp2_aru	A	A/R/U	
8	J1	elec_demand	30	MW		41	AC2	cwp2_123	1	1/2/3	
9	K1	fo_demand	30	MW.th		42	AD1	cwp3_aru	A	A/R/U	
10	L1	lpfg_demand	30	MW.th		43	AD2	cwp3_123	1	1/2/3	
11	M1	hpfg_demand	30	MW.th		44	AE1	c_dmw	3	$/t	dmw cost
12	N1	bl1_aru	A	A/R/U		45	AF1	c_fo1	18	$/MWh.th	
13	N2	bl1_123	1	1/2/3		46	AG1	fo1_hl	20	MW.th	
14	O1	bl2_aru	A	A/R/U		47	AH1	c_fo2	19	$/MWh.th	
15	O2	bl2_123	1	1/2/3		48	AI1	fo2_hl	200	MW.th	
16	P1	bl3_aru	A	A/R/U		49	AJ1	c_lpf1	19	$/MWh.th	
17	P2	bl3_123	1	1/2/3		50	AK1	lpf1_hl	100	MW.th	
18	Q1	stg_aru	A	A/R/U		51	AL1	c_lpf21	19	$/MWh.th	
19	Q2	stg_123	1	1/2/3		52	AM1	lpf21_hl	100	MW.th	
20	R1	ctg_aru	A	A/R/U		53	AN 1	c_lpf22	18	$/MWh.th	Fuel cost parameters
21	R2	ctg_123	1	1/2/3		54	AO1	lpf22_hl	500	MW.th	
22	S1	bp1_aru	A	A/R/U	Equipment avaialability	55	AP1	c_hpf1	19	$/MWh.th	
23	S2	bp1_123	1	1/2/3		56	AQ1	c_hpf21	20	$/MWh.th	
24	T1	bp2_aru	A	A/R/U		57	AR1	hpf21_hl	20	MW.th	
25	T2	bp2_123	1	1/2/3		58	AS1	c_hpf22	19	$/MWh.th	
26	U1	bp3_aru	A	A/R/U		59	AT1	hpf22_hl	60	MW.th	
27	U2	bp3_123	1	1/2/3		60	AU1	c_hpf23	18	$/MWh.th	
28	V1	bp4_aru	A	A/R/U		61	AV1	hpf23_hl	120	MW.th	
29	V2	bp4_123	1	1/2/3		62	AZ1	cbuy_ener	50	$/MWh	
30	W1	gtg_aru	U	A/R/U		63	BA1	cbuy_dem	10	$/MW.peak	TOU elec
31	W2	gtg_123	3	1/2/3		64	BB1	cbuy_idem	100	$/MW.peak	power cost
32	X1	fc_aru	A	A/R/U		65	BC1	csel_elec	40	$/MWh	parameters
33	X2	fc_123	1	1/2/3		66	BD1	cur_pkdemand	20	MW.peak	

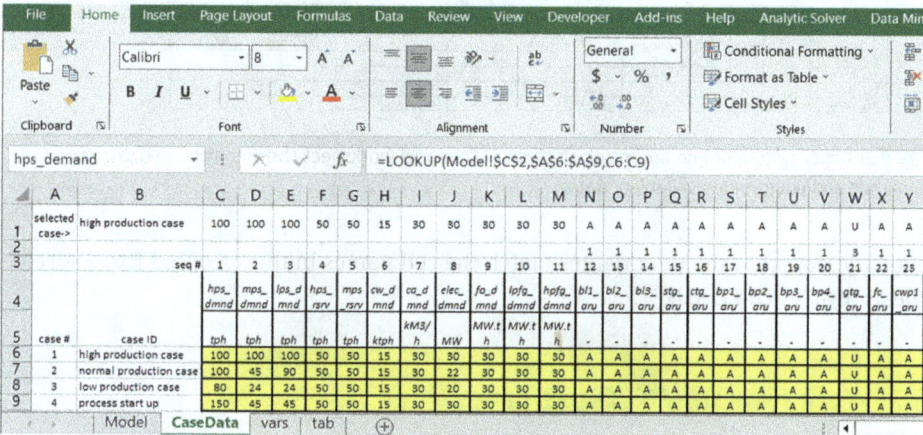

Fig. 9.1: Utility system offline optimizer: UtilSys-OFLO: CaseData worksheet showing process demands and equipment availability data.

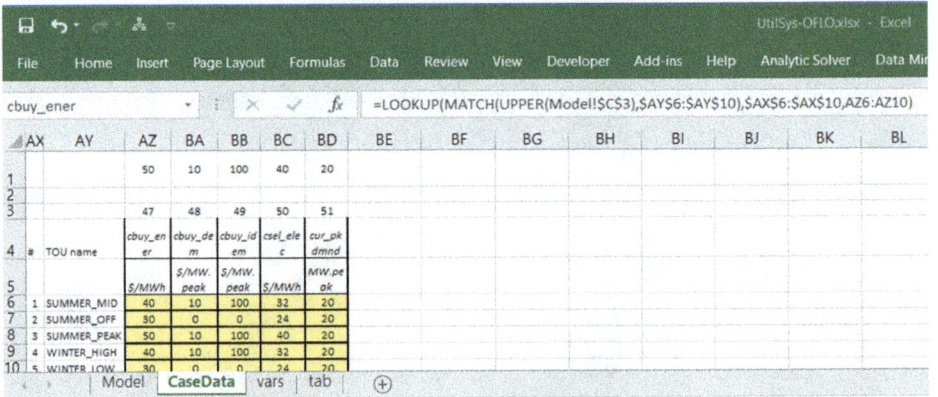

Fig. 9.2: Utility system offline optimizer: UtilSys-OFLO: CaseData worksheet showing TOU electrical Rate parameters.

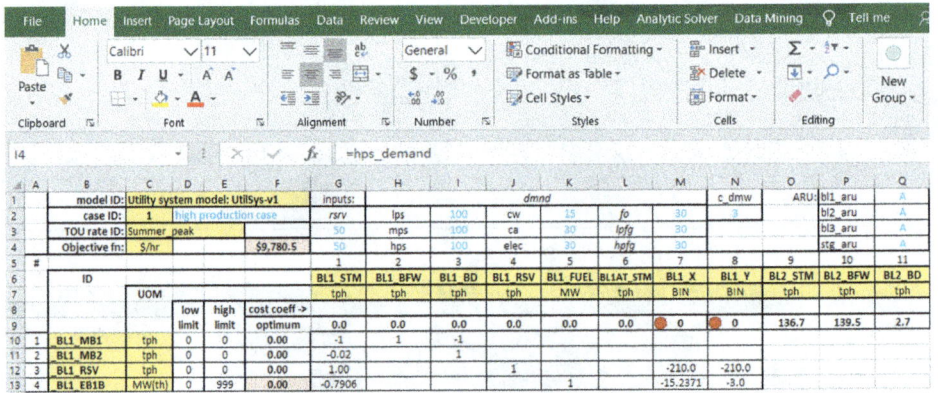

Fig. 9.3: Utility system offline optimizer: UtilSys-OFLO: model worksheet showing retrieved process demands and equipment availability data.

equality; note that the item names match the cell names in the specified Excel spreadsheet, "UtilSys-OFLO.xslx."

9.2 Work flow for OFLO use cases

Figure 9.5 shows a possible workflow for the use cases mentioned in the chapter introduction.

Ad hoc optimization: As the name suggests, it is an on-demand optimization of a selected case; the case could be a current operation, a past operation, or a proposed operation. The scenario of interest is defined in terms of process demands, equipment

```
MODEL:  TITLE Utility System Model for off-line optimization: UtilSys-OFLO;
! Ref: Nath, "Industrial Process Plants: Global Optimization of Utility Systems", Chapter 9, (2023);
!
DATA: hps_demand, mps_demand, lps_demand,   mps_reserve, hps_reserve, cw_demand, ca_demand, hpfg_demand,
      lpfg_demand, fo_demand,  elec_demand, bl1_123,     bl2_123,     bl3_123,   stg_123,   ctg_123,
      bp1_123,     bp2_123,    bp3_123,     bp4_123,     gtg_123,     fc_123,    cwp1_123,  cwp2_123,
      cwp3_123,    ac1234_r,   ac1234_a,    ac5_123,     c_dmw,       c_fo1,     fo1_hl,    c_fo2,
      fo2_hl,      c_lpf1,     lpf1_hl,     c_lpf21,     lpf21_hl,    c_lpf22,   lpf22_hl,  c_hpf1,
      c_hpf21,     hpf21_hl,   c_hpf22,     hpf22_hl,    c_hpf23,     hpf23_hl,  cbuy_ener, cbuy_dem,
      cur_pkdmnd,  cbuy_idem,  csel_elec = @OLE('C:\BookDG\ExcelModels\UtilSys-OFLO.xlsx');! get data;

    ! params;   @FILE('C:\BookDG\LingoModels\UtilSys-params.lng')             ! get params;
  ENDDATA

    ! calcs;    @FILE('C:\BookDG\LingoModels\UtilSys-calcs-v2.lng')           ! do calcs;

! Objective;    MIN = FO_CST + LPF1_CST + c_lpf21 *BUY_LPF21 + c_lpf22 *BUY_LPF22
                + lpf22_a *LPF22_X + c_hpf1 *BUY_HPF1 + c_hpf21 *BUY_HPF21 + c_hpf22 *BUY_HPF22
                + hpf22_a *HPF22_X + c_hpf23 *BUY_HPF23 + hpf23_a *HPF23_X + c_dmw *DEA_DMW
                + cbuy_ener *BUY_ELEC + DMND_CST + cbuy_idem *IDEMAND - csel_elec *SEL_ELEC;

                @FILE('C:\BookDG\LingoModels\stmSS-v2.lng')         @FILE('C:\BookDG\LingoModels\fuelSS-v1.lng')
                @FILE('C:\BookDG\LingoModels\cwSS-v2.lng')          @FILE('C:\BookDG\LingoModels\caSS-v1.lng')
                @FILE('C:\BookDG\LingoModels\Contracts-v1.lng')            ! read in subsystem constraints;

! Headers;
  [_HPS_HDR]    BL2_STM + BL3_STM = BL3AT_STM + PRV1_HPS + STG_STM + CTG_HPS + BP2T_STM + BP4T_STM
                   + CWP2T_STM + CWP3T_STM + hps_demand;
  [_MPS_HDR]    BL1_STM + PRV1_MPS + STG_STM + CTG_MPS + BP4T_STM + CWP2T_STM
                   = PRV2_STM + BL1AT_STM + mps_demand;
  [_LPS_HDR]    BL1AT_STM + BL3AT_STM + PRV2_STM + CTG_LPS + BP2T_STM + BDF_LPS + CWP3T_STM
                   = DEA_LPS + VENT + lps_demand;
  [_BFW_HDR]    BL1_BFW + BL2_BFW + BL3_BFW + PRV1_BFW = BP1_FLO + BP2_FLO + BP3_FLO + BP4_FLO;
  [_CW_HDR]     CWP1_FLO + CWP2_FLO + CWP3_FLO = CTG_CW + cw_demand;
  [_CA_HDR]     AC1234_CA + AC5_CA = VENT_CA + ca_demand;
  [_HPFG_HDR]   BUY_HPF1 + BUY_HPF2 + FC_FG = PRVFG + GTG_FG  + BL2_HPFG + BL3_HPFG + hpfg_demand;
  [_LPFG_HDR]   BUY_LPF1 + BUY_LPF2 + PRVFG
                   = FC_FG + FGFLARE + BL2_LPFG + BL3_LPFG + AC1234E_FG + lpfg_demand;
  [_FO_HDR]     BUY_FO = FOFLARE + BL1_FUEL + fo_demand;
  [_ELEC_HDR]   BUY_ELEC - SEL_ELEC + STG_ELE + CTG_ELE + GTG_ELE = FC_ELE + BL2AM_ELE + BL3AM_ELE
                + BP1M_ELE + BP2M_ELE + BP3M_ELE + BP4M_ELE + CWP1M_ELE + CWP3M_ELE + AC5M_ELE
                + elec_demand;

END MODEL UtilSys-OFLO
```

Fig. 9.4: Utility system offline optimizer model: UtilSys-OFLO: in LINGO.

availability, and purchased energy cost parameters. The optimizer gives the global optimum operations policy that could identify potential benefits and ways to achieve them. Analysis of additional cases could provide a better understanding of alternatives and insights into the optimum policy.

Annual cost budgeting: Utility systems typically conduct annual cost budgeting to estimate and budget the operating cost for the following year; usually, monthly estimates are produced to synchronize with the billing cycle. For a selected month, all unique operations cases are defined based on forecasted production, estimated utility demands, anticipated equipment availability, and purchased energy costs. The monthly operating cost is the weighted average of projected cases for the month. The annual budget is the aggregation of the projected monthly estimates. Although budgeting is more of an art and different groups approach budgeting differently, an Offline utility system optimizer is a valuable tool to quickly and consistently compute the optimal projected operating costs of selected cases.

Energy supply contract evaluation: External fuel suppliers often propose new or revised supply contracts with limited time validity, so the utility system must quickly evaluate and respond to the proposal. A utility system optimizer is very useful for such

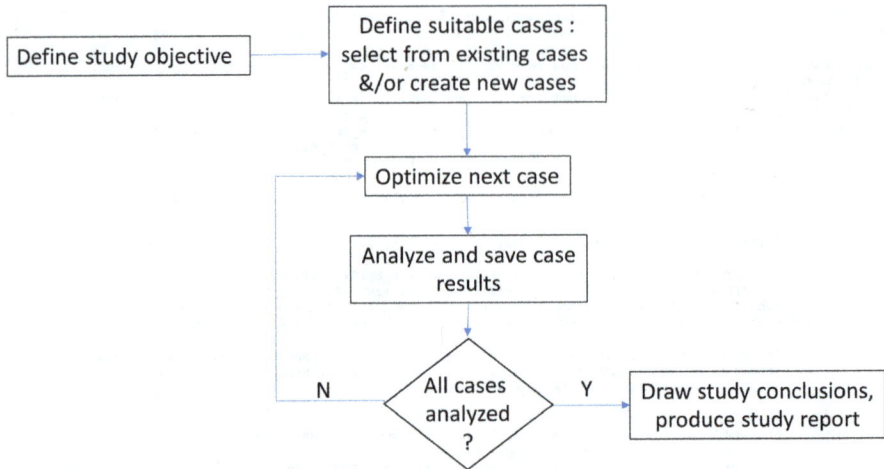

Fig. 9.5: Possible workflow for an offline utility system use case.

evaluations. The proposed supply contract could be analyzed by modeling the proposed supply contract in a modified version of the optimization model and then conducting case studies of typical operating scenarios with and without the proposed energy supply contract. A comparison of the two would give a consistent assessment of the impact of the proposed energy supply contract that would be useful in decision-making.

Proposed investment evaluation: A proposed utility system investment could be analyzed similarly to the assessment of a proposed energy supply contract using a modified version of the utility system model that includes a model of the proposed changes. Conducting a case study with typical operating scenarios with and without the proposed changes would give a realistic assessment of benefits due to the proposed investment. Such information would be helpful in business decision-making.

Summary

This chapter discusses the offline use of the sample utility system optimizer developed earlier. It presents several possible "use cases" and a simple Excel-based case management tool. The case management tool comprises a database of cases with all relevant data. The case management tool is Excel-based and is usable by both the Excel and the LINGO optimization models. A typical "use case" study comprises selecting multiple cases from the database and then running the optimizer for each to obtain the corresponding global optimal. Assembling the optimum information for all cases in the study forms a rational basis for decision-making. The chapter concludes with a possible workflow for conducting such case studies.

Nomenclature

Symbol	UOM	Description
AC1234_CA	kM3/h	Air compressor #1234: compressed air flow
AC1234M_FG	MW.th	Air compressor #1234: ICE drive: fuel consumption
AC5_CA	kM3/h	Air compressor #5: compressed air flow
AC5M_ELE	MW	Air compressor #5: motor drive: power consumption
BDF_LPS	tph	Blowdown flash: LP steam generation
BLj_BFW	tph	Boiler #j: boiler feed water inflow
BLj_FUEL	MW.th	Boiler #j: total fuel consumption
BLj_HPFG	MW.th	Boiler #j: HP fuel gas consumption
BLj_LPFG	MW.th	Boiler #j: LP fuel gas consumption
BLj_STM	tph	Boiler #j: steam generation
BLjAM_ELE	MW	Boiler #j: auxiliaries motor drive: electric power usage
BLjAM_X	0/1	Boiler #j: auxiliaries motor drive: 0 = OFF, 1 = ON
BLjAT_STM	tph	Boiler #j: Auxiliaries turbine: steam flow
BLjAT_X	0/1	Boiler #j: auxiliaries turbine drive: 0 = OFF, 1 = ON
BPj_FLO	tph	BFW Pump #j: boiler feed water inflow
BPjM_ELE	MW	BFW Pump #j: motor drive: power consumption
BPjT_STM	tph	BFW Pump #j: turbine: steam flow
BUY_ELEC	MW	Purchased energy: buy electric power
BUY_FO	MW.th	Purchased energy: buy fuel oil
BUY_HPFG	MW.th	Purchased energy: buy HP fuel gas
BUY_LPFG	MW.th	Purchased energy: buy LP fuel gas
CTG_CW	ktph	CTG: condenser: cooling water flow
CTG_ELE	MW	CTG: power generation
CTG_HPS	tph	CTG: inlet HP steam flow
CTG_LPS	tph	CTG: outlet LP steam flow
CTG_MPS	tph	CTG: outlet MP steam flow
CWPj_FLO	tph	CW Pump #j: boiler feed water inflow
CWPjM_ELE	MW	CW Pump #j: motor drive: power consumption
CWPjT_STM	tph	CW Pump #j: turbine: steam flow
DEA_DMW	tph	Deaerator: fresh demineralized water intake
DEA_LPS	tph	Deaerator: deaeration steam
FC_ELE	MW	Fuel compressor: power consumption
FC_FG	MW.th	Fuel compressor: fuel compressed
FGFLARE	MW.th	VENT: LPFG
FOFLARE	MW.th	VENT: fuel oil
GTG_ELE	MW	GTG: power generated
GTG_FG	MW.th	GTG: fuel consumed
PRV1_BFW	tph	PRV: de-superheating water flow
PRV1_HPS	tph	PRV: HP steam flow
PRV1_MPS	tph	PRV: MP steam flow
PRV2_STM	tph	PRV2: steam flow
PRVFG	MW.th	PRVFG: HPFG to LPFG
SEL_ELEC	MW	Purchased energy: sell electric power
STG_ELE	MW	STG: power generation
STG_STM	tph	STG: steam flow

(continued)

Symbol	UOM	Description
VENT	tph	VENT: steam flow
VENT_CA	kM3/h	VENT: compressed air
ac1234_a	0 .. 4	Input: Air compressors #1 ..4: number of available compressors
ac1234_r	0 .. 4	Input: Air compressors #1 ..4: number of required compressors
ac5_aru, ac5_123	A/R/U or 1/2/3	Input: Air compressor #5: avaialble,1/required,2/unavailable,3 spec
blj_aru, blj_123	A/R/U or 1/2/3	Input: BL #j: avaialble,1/required,2/unavailable,3 spec
c_buydem	$/MW.pk	Input cost: buy electric power: demand cost
c_buyener	$/MWh	Input cost: buy electric power: energy cost
c_buyidem	$/MW.pk	Input cost: buy electric power: incremental demand cost
c_dmw	$/ton	Input cost: buy demineralized water
c_foj	$/MWh.th	Input cost: buy fuel oil in segment j
c_hpffi	$/MWh.th	Input cost: buy HP FG: supplier #i
c_hpfij	$/MWh.th	Input cost: buy HP FG: supplier #i, in segment j
c_lpfi	$/MWh.th	Input cost: buy LP FG: supplier #i
c_lpfij	$/MWh.th	Input cost: buy LP FG: supplier #i, in segment j
ca_demand	kM3/h	Process demand: compressed air
ctg_aru, ctg_123	A/R/U or 1/2/3	Input: CTG: avaialble,1/required,2/unavailable,3 spec
cur_pkdemand	MW.pk	Input cost: buy electric power: current peak demand
cw_demand	ktph	Process demand: cooling water
cwpj_aru, cwpj_123	ARU or 123	Input: CW pump #j: avaialble,1/required,2/unavailable,3 spec
elec_demand	MW	Process demand: electrical power
fc_aru, fc_123	A/R/U or 1/2/3	Input: FC: avaialble,1/required,2/unavailable,3 spec
fo_demand	MW.th	Process demand: fuel oil
foj_hl	MW.th	Parameter: buy fuel oil: segment j: high limit
gtg_aru, gtg_123	A/R/U or 1/2/3	Input: CTG: available,1/required,2/unavailable,3 spec
hpfg_demand	MW.th	Process demand: HP fuel gas
hpfij_hl	MW.th	Input cost: buy HP FG: supplier #i: segment j: high limit
hpfi_hl	MW.th	Input cost: buy HP FG: supplier #i: high limit
hps_demand	tph	Process demand: HP steam
hps_reserve	tph	Process demand: HP steam reserve
lpfg_demand	MW.th	Process demand: LP fuel gas
lpfij_hl	MW.th	Input cost: buy LP FG: supplier #i: segment j: high limit
lpfj_hl	MW.th	Input cost: buy LP FG: supplier #i: high limit
lps_demand	tph	Process demand: LP steam
mps_demand	tph	Process demand: MP steam
mps_reserve	tph	Process demand: MP steam reserve
stg_aru, stg_123	A/R/U or 1/2/3	Input: STG: available,1/required,2/unavailable,3 spec

References

[1]   Anonymous, "LINGO The Modeling Language and Optimizer," Lindo Systems Inc, p. 446, (2020).

[2]   Details at Lindo Systems Inc.'s website: https://www.lindo.com

10 Online, real-time optimization, RTO

Utility systems operate in two very dynamic environments. On one side is the "process system." And the other is the external energy market. Utility demands from the process system change often and sometimes significantly as the "process system" adjusts and reacts to changing markets for feedstocks and products, and the uncertainties inherent in the process equipment health and operation. For utility production, the utility system itself needs to procure energy in the form of fuel and electricity from external suppliers. Although interaction with external energy suppliers is well-defined with binding contracts, prices can frequently change, hourly, for electric power, daily, for spot fuel purchases, or monthly as for most fuel contracts, except for the long-term contracts that are usually higher priced.

To accommodate the changing process demands and to mitigate the risk of unexpected utility system equipment malfunction, utility systems maintain a certain level of reserve capacity, i. e., keep some equipment in standby mode, and be ready to respond to changes on short notice.

The corporate entity and corporate buyers, not necessarily from the utility system operating departments, usually negotiate contracts with external suppliers. Consequently, no person in the utility operating department fully grasps the contractual agreements and their cost implications.

In such a dynamic environment with frequently changing process demands and purchased energy costs, determining the global optimal operations policy for the utility system is challenging. Reliance on prior operating experience for the least cost operations is of little help as situations not previously encountered often emerge. For these reasons, an automated global optimizer running continuously and reliably with live process data would be of immense value.

The offline optimizer, discussed in the previous chapter, cannot be effectively used to guide online operations as the offline optimizer requires manual data input and manual execution of the optimizer, which would be impractical for continuous online optimization. In a dynamic environment, such as the one faced by utility systems, frequent execution of the optimizer and automated implementation of the optimum policy is desirable. In such cases, Real-Time Optimization (RTO) or unattended, automatic execution of the optimization cycle with live process data is essential. Still, it could be problematic, especially when the optimum policy requires change(s) in equipment slate, as turning equipment on and off usually requires operator intervention and cannot be fully automated. How do we resolve this difficulty? We will take a dual-pronged approach to run two versions of online optimizations, one without any change in equipment slate, whose results are suitable for automated implementation, and the other with the flexibility to change the equipment slate, whose recommendations are issued as advisory to the operator for consideration and action. This way, partial benefits are accrued continuously with automated implementation and full benefits are achieved upon the operator taking

https://doi.org/10.1515/9783111020679-012

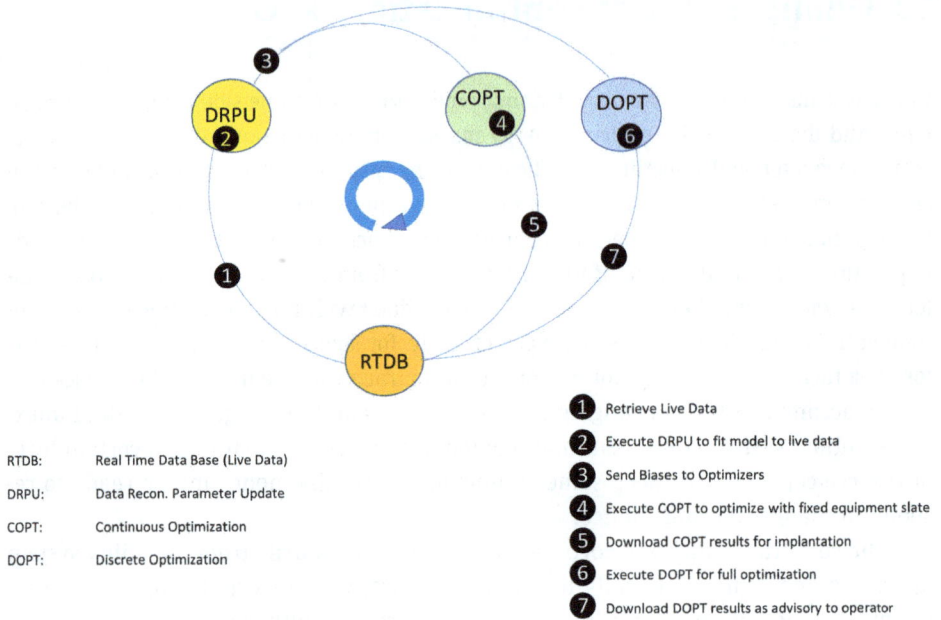

RTDB: Real Time Data Base (Live Data)
DRPU: Data Recon. Parameter Update
COPT: Continuous Optimization
DOPT: Discrete Optimization

1 Retrieve Live Data
2 Execute DRPU to fit model to live data
3 Send Biases to Optimizers
4 Execute COPT to optimize with fixed equipment slate
5 Download COPT results for implantation
6 Execute DOPT for full optimization
7 Download DOPT results as advisory to operator

Fig. 10.1: Proposed Real-Time Optimization cycle.

the recommended actions. Figure 10.1 shows such an RTO execution cycle. The optimization cycle comprises four elements and seven steps.

The four elements are:

1. **Real-Time Data Base (RTDB)** – a repository of live process measurements.
2. **The Data Reconciliation and Parameter Update model (DRPU)** – a modified version of the offline Utility System optimization model that uses the live measurements and determines certain parameters that fine-tune the model so that the tuned model matches the current observed operation of the utility system.
3. **The Continuous Optimization model (COPT)** – a modified version of the DRPU optimizer model, with the equipment slate restricted to the current state, which determines the optimum operations policy suitable for automated implementation.
4. **The Discrete Optimization model (DOPT)** – a modified version of the COPT optimizer model, with restrictions removed, determining the optimum operations policy suitable for advisory to the operator.

The seven steps performed or managed by the RTO executive are:

Step 1: Retrieve current process data (live data). Data could be pushed by RTDB or pulled by the RTO executive and passed on to the DRPU, COPT, and DOPT optimizers.

Step 2: Perform Data Reconciliation and Parameter Updating (DRPU) using the current measurements; this step aims to ensure that the model with updated parameters duplicates the current operation of the Utility System. This step also establishes a Base Case for comparison with COPT and DOPT results in steps 4 and 6.

Step 3: Transfer the DRPU parameters to the optimizers, COPT and DOPT. This transfer could be direct from DRPU to COPT and DOPT or alternately via the RTDB. Figure 10.1 shows a direct transfer of this information.

Step 4: Perform COPT Optimization with updated DRPU parameters but with the equipment slate fixed to the current slate, i.e., determine the minimum cost operations of the Utility System to meet the current process requirements while keeping the current equipment slate. The results of this optimization are suitable for automated implementation.

Step 5: Download the COPT optimization results to RTDB for automated implementation; if so desired, we will assume that such is the case.

Step 6: Perform DOPT Optimization with updated DRPU parameters but with the flexibility to change the equipment slate, i.e., determine the minimum cost operation of the utility system with possible changes to the equipment slate. Implementing the DOPT optimum policy that involves equipment slate changes, will require operator intervention, so it is an advisory to the operator.

Step 7: Present the DOPT optimization results as advisory to the operator for consideration for action. We recommend additional analysis to generate a multistaged implementation strategy if the optimal solution involves several equipment slate changes.

Before we discuss the details of the optimization execution cycle, we need to discuss several underlying concepts that will help us understand the methodology behind DRPU. We will do this in the next section.

Subsequent sections will discuss the steps mentioned above in detail and illustrate them with the sample plant utility system optimizer for test case #1 data.

10.1 DRPU concepts

Data Reconciliation and Parameter Updating is an essential step in RTO; this section will discuss several underlying concepts necessary to understand the DRPU methodology.

10.1.1 Existing automation

Utility systems transport utilities via headers. They are the first ones to sense changes in process demand. For example, as demand for a material utility increases, more material is drawn, and consequently, the header pressure drops; likewise, if the demand decreases, the header pressure rises. In most utility systems, the header pressure is maintained in a narrow band, and automation kicks in to protect the header pressure by increasing or decreasing the inlet or outlet flow from the header by a predefined strategy. For example, if HP steam demand rises, the HP header pressure drops, and if it drops below a preset limit, depending on the control scheme, the swing boiler responds by increasing HP steam production; this response is fast and reaches a new equilibrium very quickly. And, if the MP steam demand were to rise, depending on the control scheme, the MP boiler, if operating, may be the first to respond by increasing MP steam production; otherwise, the pressure reducing valve between the HP and MP headers will open up to increase the HP steam flow, which would trigger a drop in the HP header pressure – depending on the control strategy, the HP swing boiler may produce more HP steam to meet the increased MP steam demand.

For the increased fuel demand, the corresponding header pressure will drop and the swing external supply valve will open up to restore the fuel header pressure. For increased electrical demand, depending on the control scheme, more current will be drawn from the grid or, if possible, from one of the operating generators; here, usually, a board operator would be watching and would react if a new demand peak were possible, and initiate preplanned action.

In this way, automation of the Utility System ensures that changing process demand is satisfied as long as the requisite capacity exists, according to a preprogrammed strategy. However, since costs and process demands are constantly evolving, the preprogrammed actions may not be the most cost-effective; this is where online optimization could provide continuous guidance or action and keep the operations at or near the optimum.

At times, the optimal policy may be to start or stop a piece of equipment. Such an action requires operator intervention, and in such cases, the RTO system would issue an advisory to the operator of the recommended action(s) and let the operator consider the recommendations and make the necessary change(s).

10.1.2 Process measurements

Most modern plants are well-instrumented and measure numerous measurements frequently, so measured data is usually plentiful. Table 10.1 is a list of measurements at the sample utility system.

Note that there are a total of 67 online measurements. Of these, 11 are demand-related, 49 are utility system equipment operation-related, and seven are purchased energy-related. Of the 49 (utility system equipment operation-related) measurements, 26 are integers, and 23 are continuous variables. Also, note that there are 16 pieces of Utility System equipment with integers (some have multiple integers), 26 integers, and

Tab. 10.1: Sample utility system: measured variables.

#	Name in model	Measurement name	UOM	Sample value	#	Name in model	Measurement name	UOM	Sample value
Process demand:					34	CTG_MPS	ctg_mps	tph	90
1	hps_demand	hps_demand	tph	100	35	CTG_LPS	ctg_lps	tph	90
2	mps_demand	mps_demand	tph	100	36	CTG_ELE	ctg_ele	MW	9.3
3	lps_demand	lps_demand	tph	100	37	CTG_X	ctg_x	0/1	1
4	hps_rsrv	hps_rsrv	tph	50	38	CTG_Y	ctg_y	0/1	0
5	mps_rsrv	mps_rsrv	tph	50	39	BP1_X	bp1_x	0..2	0
6	ca_demand	ca_demand	kM3/h	30	40	BP2_X	bp2_x	0/1	1
7	cw_demand	cw_demand	ktph	15	41	BP2M_X	bp2m_x	0/1	0
8	elec_demand	elec_demand	MW	28	42	BP3_X	bp3_x	0..2	0
9	fo_demand	fo_demand	MW.th	30	43	BP4_X	bp4_x	0/1	1
10	lpfg_demand	lpfg_demand	MW.th	30	44	BP4M_X	bp4m_x	0/1	0
11	hpfg_demand	hpfg_demand	MW.th	30	45	VENT	vent	tph	0
Utility system operation					46	PRVFG	prvfg	MW.th	0
12	BL1_STM	bl1_stm	tph	0	47	FGFLARE	fgflare	MW.th	0
13	BL1_FUEL	bl1_fuel	mw.t	0	48	FOFLARE	foflare	MW.th	0
14	BL1_X	bl1_x	0/1	0	49	GTG_FG	gtg_fg	MW.th	0
15	BL1_Y	bl1_y	0/1	0	50	GTG_ELE	gtg_ele	mw	0
16	BL2_STM	bl2_stm	tph	135	51	GTG_X	gtg_x	0/1	0
17	BL2_LPFG	bl2_lpfg	mw.t	133	52	FC_FG	fc_fg	MW.th	0
18	BL2_HPFG	bl2_hpfg	mw.t	0	53	FC_X	fc_x	0/1	0
19	BL2_X	bl2_x	0/1	1	54	CWP1_X	cwp1_x	0/1	0
20	BL2_Y	bl2_y	0/1	0	55	CWP2_X	cwp2_x	0/1	1
21	BL3_STM	bl3_stm	tph	240	56	CWP3_X	cwp3_x	0/1	1
22	BL3_FUEL	bl3_fuel	mw.t	220	57	CWP3M_X	cwp3m_x	0/1	0
23	BL3AM_X	bl3am_x	0/1	0	58	AC1234_NX	ac1234_nx	0..4	0
24	BL3_X	bl3_x	0/1	1	59	AC5_X	ac5_x	0/1	1
25	BL3_Y	bl3_y	0/1	0	60	VENT_CA	vent_ca	KM3/H	0
26	BL3_Z	bl3_z	0/1	0	purchased energy:				
27	PRV1_HPS	prv1_hps	tph	0	61	BUY_FO	buy_fo	MW.th	30
28	PRV2_STM	prv2_stm	tph	58	62	BUY_LPF1	buy_lpf1	MW.th	100
29	STG_STM	stg_stm	tph	50	63	BUY_LPF2	buy_lpf2	MW.th	282.36
30	STG_ELE	stg_ele	MW	2.9	64	BUY_HPF1	buy_hpf1	MW.th	30
31	STG_X	stg_x	0/1	1	65	BUY_HPF2	buy_hpf2	MW.th	0
32	STG_Y	stg_y	0/1	0	66	BUY_ELEC	buy_elec	MW	18.5
33	CTG_HPS	ctg_hps	tph	190	67	SEL_ELEC	sel_elec	MW	0

16 continuous variables accounting for 42 measurements; the remaining seven are PRV and vent flows.

However, measurements are imperfect, so making material and energy balances using the process measurements, seldom close. This is primarily because measurement systems, like many other systems, require regular maintenance care, such as frequent calibrations, to maintain the accuracy of the measurement devices. Maintenance resources at a plant site typically are limited, and consequently, less critical measurements, such as those used for internal utility flow measurements, often receive less attention. Use of raw measurements in the RTO cycle will cause problems, as discussed below.

For example, consider the Fuel oil balance in the sample utility system. There is a single fuel oil supplier; its current measured amount is "buy_fo." In the sample utility system, there are three consumers of fuel oil: (a) boiler # 1, (b) process fuel oil users, and (c) flow to the flare; we have flow measurements for each as "bl1_fuel", "fo_demand" and "foflare," respectively. If the measurements were perfect, there would be a consistency, and fuel oil demand "fo_demand" would equal the sum of fuel oil consumption "bl1_fuel", "fo_demand" and "foflare"; that is,

$$\text{buy\_fo} = \text{bl1\_fuel} + \text{fo\_demand} + \text{foflare} \qquad (10.1)$$

Unfortunately, such is rarely the case due to errors in measurements, that is,

$$\text{buy\_fo} \neq \text{bl1\_fuel} + \text{fo\_demand} + \text{foflare} \qquad (10.2)$$

Therefore using the measured values in the Utility System model would be problematic; inconsistent measurements would cause the model to fail due to constraint violations. We will next discuss options to resolve this issue.

10.1.3 Data reconciliation

The problem of inconsistency in data measurements is well understood and has been studied by various researchers; it is often referred to as the problem of "Data Reconciliation" [1, 2]. A popular technique to reconcile data uses a weighted least squares minimization approach [3]. This technique computes adjustments to the measured values in proportion to their accuracies so that the adjusted values are consistent with model constraints. This "least squares" technique is a statistical technique based on the assumption that the measurement errors are randomly distributed around their respective true values. However, this assumption of random error distribution is valid only for measurements from well-maintained measurement devices, which is usually not the case for measurement devices commonly used in utility systems for internal utility flows, as such devices usually are orifice-type flow meters, for which systemic drift in measurements is the dominant error due to the erosion of the physical orifice. From this perspective, applying the popular least squares approach to utility system data reconciliation would be fundamentally flawed and thus not justified. This is not to say that the least squares technique has no applications; on the contrary, it is a reasonable technique when applied to measurements from well-maintained instruments where random errors are dominant, such as those encountered in custody transfer meters; such measurement systems are well maintained for obvious reasons. However, measurement drift is the dominant error for most internal flow measurements in the utility systems. Hence, the least squares technique is of limited use, if at all.

So, what can we do in this situation? We propose using a straightforward and practical technique; we will call it the "Method of Biases." This technique applies to measurements for which systemic drift is the dominant error. Although imperfect, it is simple and useful in reconciling the measured data and thus allow the use of optimization models with raw measurements.

For instance, for the fuel oil balance discussed above, a modified version of the material balance equation with a bias term to enforce consistency would be:

$$\text{buy\_fo} + \textbf{FO\_BIAS} = \text{bl1\_fuel} + \text{foflare} + \text{fo\_demand} \qquad (10.3)$$

where,

FO_BIAS is a variable the DRPU optimization calculates to enforce material balance closure. Equation (10.3) is one equation in one unknown, so there would be a unique value of **FO_BIAS**. A positive value of the bias indicates that consumption is more than supply, and an additional amount equal to **FO_BIAS** is required to achieve header balance. On the other hand, a negative value of the bias indicates excess supply, and additional consumption equal to the absolute value of **FO_BIAS** is required to achieve header balance.

By default, all MILP variables are nonnegative, which would be problematic in this situation as the solver would not allow negative values of **FO_BIAS**. How do we resolve this problem? One solution is to represent the bias, **FO_BIAS**, by a difference of two variables, one for positive and another for negative bias. Let these variables be **FO_XBP** and **FO_XBN**, respectively. Doing so, eq. (10.3) becomes

$$\text{buy\_fo} + \textbf{FO\_XBP} - \textbf{FO\_XBN} = \text{bl1\_fuel} + \text{foflare} + \text{fo\_demand} \qquad (10.4)$$

where

FO_XBP represents a bias that would be nonzero when consumption is greater than supply, and **FO_XBN** represents a bias that would be nonzero when supply is greater than consumption.

You may notice that eq. (10.4) has two unknowns, **FO_XBP** and **FO_XBN**, and one equation. So mathematically, there would be a family of solutions, i.e., an infinity of solutions, as long the difference between the two remains the same. However, for clarity, we prefer a solution with, at most, one nonzero value. We could ensure this by minimizing the sum of **FO_XBP** and **FO_XBN** in the Data Reconciliation objective function, i.e., add **FO_XBP** and **XO_XBN** in the minimization objective function; we will use this approach.

A fundamental assumption in this approach is that the dominant measurement error is a systemic bias due to mechanical erosion of the measurement orifice that slowly but continually degrades the measured value over time. This degradation is slow; over the short run, the assumption of constant bias is reasonable.

There are a couple of things to keep in mind when using this technique: (a) significant biases are indicative that some of the measurements have deteriorated significantly; so an investigation and an attempt to reduce significant biases is advisable, and (b) we should monitor the biases over time and sudden significant changes should be investigated as they could be indicative of possible device failure.

We will use the "Method of Biases" for the sample utility system data reconciliation.

10.1.4 Parameter updating

Depending on the available measurements, some equipment may have redundant measurements, i.e., more measured values than are strictly necessary. For example, for the steam turbogenerator model, STG, operating in normal mode, STG is producing power, and the measured value of STG indicator value, stg_x, is 1. In this situation, there is a single degree of freedom, i.e., all other variables can be calculated by specifying a value for either the inlet steam flow or the power generation. In other words, if steam flow is specified, power generation could be computed using the STG performance equation and vice versa. But what if steam flow and power generation were measured, and the two values were inconsistent with the performance equation? In other words, the following condition existed.

Measured steam flow≠calculated steam flow using the performance equation

$$(10.5)$$

$$stg\_stm \neq stg\_stm\_a * stg\_x + stg\_stm\_b * stg\_ele \qquad (10.6)$$

Where stg_stm, stg_ele, and stg_x are measured values for inlet steam flow, power generation, and operating status indicator, respectively, and stg_stm_a and stg_stm_b are performance equation parameters.

Specifying both the stream flow and power generation values in the model, in general, would be problematic, as the values generally would be inconsistent and would cause the model to fail due to the infeasibility of the performance equation constraint.

How do we deal with such situations of redundant measurements?

One way would be to use only one of the two measured values arbitrarily, but then the question is which one?

There is usually no rational basis for choosing one over the other. Upon reflection, you may notice a similar situation discussed in the previous section where Eq. (10.2) did not balance with measured values. Again, we could use the "Method of Biases" to resolve this inconsistency by modifying the performance constraint to;

$$stg\_stm + \textbf{STG\_XBP} - \textbf{STG\_XBN} = stg\_stm\_a * stg\_x + stg\_stm\_b * stg\_ele \qquad (10.7)$$

Again, we will add the bias variables, **STG_XBP** and **STG_XBN**, to the DRPU objective function so that, at most, one of the two would be nonzero. A positive value of **STG_XBP** would indicate that the measured value of steam flow, stg_stm, is too low, and adding **STG_XBP** amount of steam is necessary to enforce the constraint. A positive value of **STG_XBN** would indicate that the measured value of steam flow, stg_stm, is too high, and subtracting **STG_XBN** amount of steam is required.

We will use the "Method of Biases" to compute biases to reconcile data and update parameters in the DRPU model. Then, we will use the calculated biases in the optimization models; details are in the following sections.

10.2 Utility system real-time optimization execution cycle

As discussed in the chapter introduction and shown in Fig. 10.1, Real-Time Optimization comprises seven steps. The executive for the RTO cyclically executes each step. Online implementation of the RTO executive could use commercially available software to interact with the process database and schedule the various RTO steps. Such software is available from several sources; the author is familiar with one such system [4, 5].

We will discuss each of the seven steps in the following subsections using test case #1 data.

10.2.1 Step 1: Retrieve live data

Modern process plants have a Real-Time Data Base (RTDB) for storing current, historical, and related data. The RTO executive retrieves all required data from this RTDB. However, for the off-line testing of sample utility system, which is unconnected to an RTDB, we will use stored data sets in the worksheet called "CaseData."

The "CaseData" worksheet has the following three tables:

– Current measured data and purchased fuel cost parameters:
Each row of this table has a unique case number, such as 1. Figure 10.2 shows the first part of this table. Note that the cursor is on cell C2, which shows the retrieved value 100 for the measured value of "hps_demand," the high-pressure steam process demand. The cell name assigned to cell C2 is "hps_demand", as shown in the text box on the upper left side, and the LOOKUP function used to retrieve the value is on the formula bar.

– TOU electric rate contract parameters:
Each row of this table has a unique ID, such as "Summer_peak." Figure 10.3 shows the TOU electric rate contract parameters in cell range DF7 through DL11. Note that the cursor is placed on cell DH2, showing the retrieved value 50 for the TOU rate parameter, "cbuy_ener.," and the electrical energy cost in $/MWh. The cell name assigned to this cell is cbuy_ener, as shown in the text box on the upper left side, and the corresponding LOOKUP function used to retrieve this value is on the formula bar.

– DOPT ARU data:
Each row of this table has a unique ID, such as "DOPT." Figure 10.3 shows the DOPT ARU data, comprising the ARU status of various pieces of equipment in columns CL through DD, highlighted in a light green background.

Figure 10.4 shows the "DRPU" model worksheet and cell C2 is "case_num," the LOOKUP function key for the "current measured data and purchased fuel cost parameters" table. Cell C3 is "TOU_ID," the LOOKUP function key for the "TOU electrical rate parameters" table. The cursor is on cell I4, showing the retrieved value of "hps_demand" in blue font, with a light blue background; Fig. 10.4 also shows retrieved values of several other data items in blue font, with a light blue background.

Note that the ARU ID, LOOKUP function key for "DOPT ARU data," is specified on cell F3 of the DOPT worksheet.

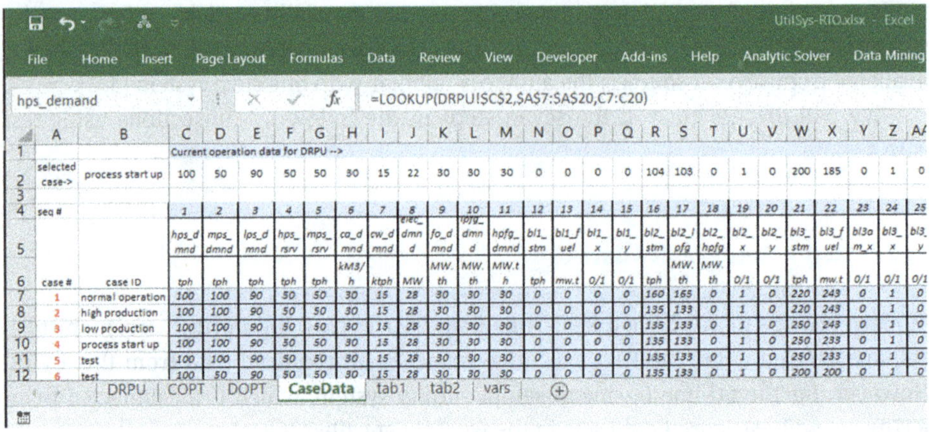

	A	B	C	D	E	F	G	H	I	J	K	L	M	N	O	P	Q	R	S	T	U	V	W	X	Y	Z	AA
1			Current operation data for DRPU -->																								
2	selected case->	process start up	100	50	90	50	50	30	15	22	30	30	30	0	0	0	0	104	103	0	1	0	200	185	0	1	0
3																											
4	seq #		1	2	3	4	5	6	7	8	9	10	11	12	13	14	15	16	17	18	19	20	21	22	23	24	25
5			hps_dmnd	mps_dmnd	lps_dmnd	hps_rsrv	mps_rsrv	ca_dmnd	cw_dmnd	elec_dmnd	fo_dmnd	lpfg_dmnd	hpfg_dmnd	bl1_stm	bl1_fuel	bl1_x	bl1_y	bl2_stm	bl2_lpfg	bl2_hpfg	bl2_x	bl2_y	bl3_stm	bl3_fuel	bl3om_x	bl3_x	bl3_y
6	case #	case ID	tph	tph	tph	tph	tph	kM3/h	ktph	MW	MW.th	MW.th	MW.th	tph	mw.t	0/1	0/1	tph	MW.th	MW.th	0/1	0/1	tph	mw.t	0/1	0/1	0/1
7	1	normal operation	100	100	90	50	50	30	15	28	30	30	30	0	0	0	0	160	165	0	1	0	220	243	0	1	0
8	2	high production	100	100	90	50	50	30	15	28	30	30	30	0	0	0	0	135	133	0	1	0	220	243	0	1	0
9	3	low production	100	100	90	50	50	30	15	28	30	30	30	0	0	0	0	135	133	0	1	0	250	243	0	1	0
10	4	process start up	100	100	90	50	50	30	15	28	30	30	30	0	0	0	0	135	133	0	1	0	250	233	0	1	0
11	5	test	100	100	90	50	50	30	15	28	30	30	30	0	0	0	0	135	133	0	1	0	250	233	0	1	0
12	6	test	100	50	90	50	50	30	15	28	30	30	30	0	0	0	0	135	133	0	1	0	200	200	0	1	0

Fig. 10.2: Sample utility system model: UtilSys-RTO: CaseData worksheet showing process measurements.

The LINGO model retrieves case data values using LINGO's **@OLE()** interface function [6]. The basic syntax of **@OLE()** function is:

$$object_list = @OLE(\text{'}spreadsheet_file_name\text{'});$$

where *object_list* is a comma-separated list of data item names to be retrieved.

If '*spreadsheet file name*' is omitted, LINGO uses the currently open Excel file.

All required data will come from the RTDB for an actual RTO implementation. In such implementations, data validation would be performed, i.e., at least some degree of data checking to confirm that each data item is reasonable.

In summary, the sample utility system has 67 data items related to the current operation, 18 fuel cost-related parameters, 17 items related to optimization, and five data items associated with the TOU electric power rate. In the Excel model, all data is retrieved using the Excel LOOKUP function; in LINGO, data is retrieved using LINGO's @OLE() interface function.

10.2.2 Step 2: Data reconciliation and parameter updating (DRPU)

The next step in the RTO cycle is to perform data reconciliation and parameter updating (DRPU) and to establish a base case operations that would serve as a reference for comparing the optimum operations determined by COPT and DOPT and developing implementation strategies for steering the current operations toward the optimum.

As discussed in the previous section, the model and the measurements are imperfect. Consequently, using redundant measured data to perform mass and energy balances will generally fail due to data inconsistency. Data reconciliation and parameter

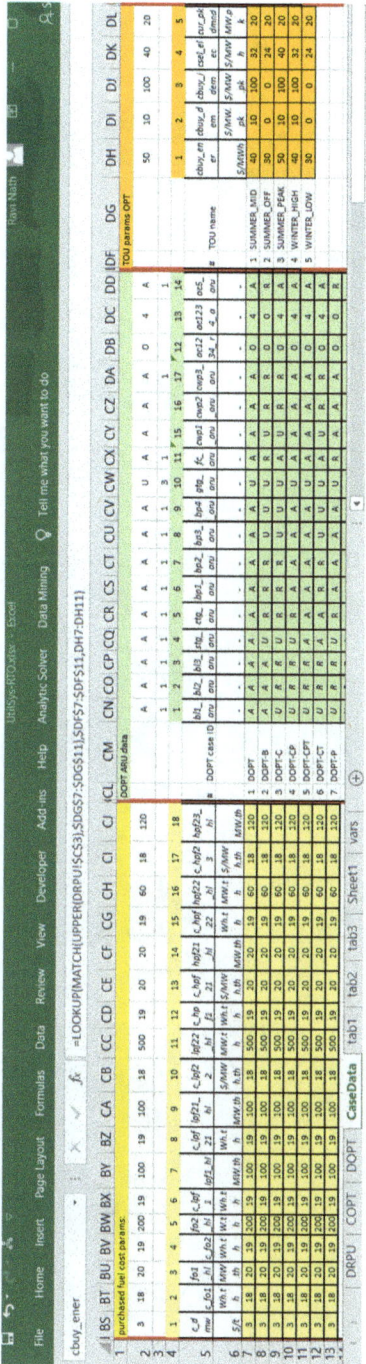

Fig. 10.3: Sample utility system model: UtilSys-RTO: CaseData worksheet, showing purchased energy cost parameters.

	B	C	D	E	F	G	H	I	J	K	L	M	N	O	P	Q
	model ID:	Utility system model: UtilSys-v1				inputs:				process demands			c_dmw	ARU:	bl1_aru	A
	case ID:	1	normal operation			rsrv	lps		cw		fo			3		bl2_aru	A
	TOU rate ID:	Summer	peak			50	mps		ca		lpfg					bl3_aru	A
#	Objective fn:	$/hr			$2,270.6		hps		elec		hpfg					stg_aru	A
	cost	$/hr			$8,421.5											bl2_aru	
	ID					BL1_STM	BL1_BFW	BL1_BD	BL1_RSV	BL1_FUEL	BL1AT_STM	BL1_X	BL1_Y	BL2_STM	BL2_BFW	BL2_BD
		UOM				tph	tph	tph	tph	MW.th	tph	B/N	B/N	tph	tph	tph
			low limit	high limit	cost coeff -> optimum											
9					0.0	0.0	0.0	0.0	0.0	0.0	0.0	0	0	104.0	106.1	2.1

Column index row: 1 (BL1_STM), 2 (BL1_BFW), 3 (BL1_BD), 4 (BL1_RSV), 5 (BL1_FUEL), 6 (BL1AT_STM), 7 (BL1_X), 8 (BL1_Y), 9 (BL2_STM), 10 (BL2_BFW), 11 (BL2_BD)

Input values: rsrv 50, lps 90, mps 45, hps 100 · process demands: cw 15, ca 30, elec 22 · fo/lpfg/hpfg 30/30/30 · c_dmw 3

#	ID	UOM	low limit	high limit	cost coeff -> optimum	BL1_STM	BL1_BFW	BL1_BD	BL1_RSV	BL1_FUEL	BL1AT_STM	BL1_X	BL1_Y	BL2_STM	BL2_BFW	BL2_BD
1	BL1_MB1	tph	0	0	0.00	-1	1									
2	BL1_MB2	tph	0	0	0.00	-0.02		1								
3	BL1_RSV	tph	0	0	0.00	1.00			1							
4	BL1_EB1B	MW.th	0	999	0.00	-0.7906				1	-210.0	-15.2371	-210.0			
5	BL1_EB1R	MW.th	0	999	0.00	-0.9535				1		12.4525	-3.0			
6	BL1_STM_LL	tph	0	999	0.00	1						-70.0				
7	BL1_STM_HL	tph	-999	0	0.00	1						-210.0	-3.0			
8	BL1AT_EB1	tph	0	0	0.00						1	-6.0				
9	BL1_...	tph	0	0	0.00						1	1	1			

Fig. 10.4: Sample utility system model: UtilSys-RTO: DRPU worksheet, showing the retrieved process data.

updating are necessary to resolve this discrepancy. We will achieve this functionality using the "Method of Biases."

The utility system model developed earlier, UtilSys-OFLO, will serve as a starting point. However, as discussed next, several modifications will be necessary to make it suitable for performing the DRPU functionality.

10.2.2.1 Data reconciliation of utility headers

The utility system model has three steam headers, three fuel headers, and an electrical header. Each header has redundant measurements and would require data reconciliation. We will achieve DRPU for each using the "Method of Biases," as detailed below. The Excel model in the "DRPU" worksheet of "UtilSys-RTO.xlsx" includes these model changes; for the LINGO, the utility system optimization model "UtilSys-RTO-DRPU.lg4" embodies these changes.

HPS header DRPU: There are measurements for HP steam generation in the two high-pressure boilers; high-pressure steam flows to PRV1, the turbogenerators, some pump turbine drivers, and the process steam demand. Flows to the boiler auxiliaries are not measured so, model predictions will be used instead. The utility system model will also predict boiler feed water (BFW) flows to the BFW pump. BFW pumps' status is known, and BFW flow is evenly split among the operating BFW pumps, so the model will also predict the HP steam usage by BFW pump turbine drives. Similarly, cooling water demand and the status of operating cooling water pumps are known; the model will predict the associated turbine driver steam flows. So, the HPS header balance equation could predict the high-pressure steam process demand. However, high-pressure process steam demand is also available from the process side measurements. Generally, the predicted HP steam demand from the header balance equation will not match the measured demand from the process side. So, data reconciliation will be required and will be performed using two bias variables, **HPS_XBP** and **HPS_XBN**, as shown below:

$$\mathrm{bl2\_stm} + \mathrm{bl3\_stm} + \mathbf{HPS\_XBP} - \mathbf{HPS\_XBN} = \mathbf{BL3AT\_STM} + \mathrm{prv1\_hps} + \mathrm{stg\_stm} + \mathrm{ctg\_hps}$$

$$+ \mathbf{BP2T\_STM} + \mathbf{BP4T\_STM} + \mathbf{CWP2T\_STM} + \mathbf{CWP3T\_STM} + \mathrm{hps\_demand}$$

$$(10.8)$$

Note that the variables in lowercase letters are measured values, and those in **bold** uppercase are model predictions; **HPS_XBP** and **HPS_XBN** are the bias variables.

MPS header DRPU: There are measurements for MP steam generation in Boiler #1, flows from the turbogenerators, PRVs, and the process MP steam demand, mps_demand; all other flows are model predictions. The modified MPS header constraint with biases is:

$$bl1\_stm + \textbf{PRV1\_MPS} + stg\_stm + ctg\_mps + \textbf{BP4T\_STM} + \textbf{CWP2T\_STM} + \textbf{MPS\_XBP}$$
$$- \textbf{MPS\_XBN} = prv2\_stm + \textbf{BL1AT\_STM} + mps\_demand$$

$$(10.9)$$

LPS header DRPU: There are measurements for PRV2 flow, CTG LP steam flow, steam vent, and lps_demand; all other flows are model predictions. The modified LPS header constraint with biases is:

$$\textbf{BL1AT\_STM} + \textbf{BL3AT\_STM} + prv2\_stm + ctg\_lps + \textbf{BP2T\_STM} + \textbf{BDF\_LPS} + \textbf{CWP3T\_STM}$$
$$+ \textbf{LPS\_XBP} - \textbf{LPS\_XBN} = \textbf{DEA\_LPS} + vent + lps\_demand$$

$$(10.10)$$

Fuel Oil header DRPU: There are measurements for the fuel oil purchase, fuel oil consumption in the MP boiler, fuel oil disposal to the flare, and measurement of fuel oil demand from the process side. The modified fuel oil constraint for data reconciliation is:

$$buy\_fo + \textbf{FO\_XBP} - \textbf{FO\_XBN} = foflare + bl1\_fuel + fo\_demand \qquad (10.11)$$

LP Fuel gas header DRPU: There are measurements for the LP fuel gas purchase, LP fuel flows to HP boilers, flow to fuel compressor, FC, flow to the LP fuel flare, and the process LP fuel gas demand. The modified LP fuel gas constraint for data reconciliation is:

$$buy\_lpf1 + buy\_lpf2 + prvfg + \textbf{LPFG\_XBP} - \textbf{LPFG\_XBN} = fc\_fg + fgflare + bl2\_lpfg$$
$$+ \textbf{BL3\_LPFG} + \textbf{AC1234E\_FG} + lpfg\_demand$$

$$(10.12)$$

HP Fuel gas header DRPU: There are measurements for the HP fuel gas purchase. HP fuel flows to HP boilers, HP fuel flare, the process HP fuel gas demand, and flow from fuel compressor, FC. The modified HP fuel gas constraint is:

$$buy\_hpf1 + buy\_hpf2 + fc\_fg + \textbf{HPFG\_XBP} - \textbf{HPFG\_XBN} = prvfg + gtg\_fg \quad (10.13)$$
$$+ \mathit{bl2\_hpfg} + \textbf{BL3\_HPFG} + hpfg\_demand$$

Electrical header DRPU: There are measurements for electric power purchase and selling back to the power grid, internal power generation from STG, CTG, GTG and the process power demand. The modified power balance constraint is:

buy_elec − sel_elec + stg_ele + ctg_ele + gtg_ele + **ELEC_XBP** − **ELEC_XBN** = **FC_ELE**

+ **BL2AM_ELE** + **BL3AM_ELE** + **BP1M_ELE** + **BP2M_ELE** + **BP3M_ELE** + **BP4M_ELE**

+ **CWP1M_ELE** + **CWP3M_ELE** + **AC5M_ELE** + elec_demand

$$(10.14)$$

10.2.2.2 Steam subsystem DRPU

The steam subsystem has redundant measurements for STG, CTG, Boilers 1, 2, and 3, and Boiler Feed Water Pumps 1, 2, 3, and 4. Following are the details of data reconciliation for each piece of equipment:

STG DRPU: The STG has redundant measurements since the inlet steam flow and the power generation are measured quantities. One measurement is sufficient and could be used to calculate the other using the STG performance equation. However, we have measurements for both, which, in general, could be inconsistent. We need bias variables to enforce consistency between the measurements and the performance equation. Note that there is another consideration: STG could operate either in the normal mode (producing power), indicated by stg_x = 1, or in the standby mode, indicated by stg_y = 1.

Either mode could be in operation. Hence, we will need four bias variables, two for each mode of operation;

stg_stm = (stg_stm_a + **STG_XBP** − **STG_XBN**) * stg_x + (stg_stm_s + **STG_YBP** − **STG_YBN**)

* stg_y + stg_stm_b * stg_ele

$$(10.15)$$

CTG DRPU: The CTG has measurements for the steam HP inlet, flows from the lower-pressure steam extractions, and power generation. Steam flows alone would be sufficient to calculate power generation using the CTG performance equation. However, we have independent power generation measurements, which, in general, could be inconsistent. Like the STG, we will need four bias variables to enforce consistency.

ctg_hps = (ctg_hps_a + **CTG_XBP** − **CTG_XBN**) * ctg_x + (ctg_hps_s + **CTG_YBP** − **CTG_YBN**)

* ctg_y + ctg_hps_b * ctg_ele + ctg_hps_c * ctg_mps + ctg_hps_d * ctg_lps

$$(10.16)$$

BL1 DRPU: Boiler #1 has four measurements: steam generation, fuel consumption, and normal and standby operations indicators. In the normal mode of operation, either steam generation or fuel consumption would be sufficient to calculate the other using the boiler performance equations. However, we have measurements for both, which, in general, could be inconsistent. We also have two operating modes and, therefore, need four bias variables to enforce consistency. Also, remember that the boiler model energy balance has the following two inequality constraints, _BL1_EB1B and _BL1_EB1R, to account for the two-segment performance curve; so, both constraints will need modifications, as shown below:

$$\text{bl1\_fuel} \geq (\text{bl1\_fg\_c} + \textbf{BL1\_XBP} - \textbf{BL1\_XBN})^* \text{bl1\_x} + (\text{bl1\_fg\_s} + \textbf{BL1\_YBP} - \textbf{BL1\_YBN})$$
$$^* \text{bl1\_y} + \text{bl1\_fg\_d}^* \text{bl1\_stm};$$

(10.17)

$$\text{bl1\_fuel} \geq (\text{bl1\_fg\_e} + \textbf{BL1\_XBP} - \textbf{BL1\_XBN})^* \text{bl1\_x} + (\text{bl1\_fg\_s} + \textbf{BL1\_YBP} - \textbf{BL1\_YBN})$$
$$^* \text{bl1\_y} + \text{bl1\_fg\_f}^* \text{bl1\_stm};$$

(10.18)

There is yet another consideration. Remember that the energy balance inequality constraints, _BL1_EB1B and _BL1_EB1R (reproduced below) worked because fuel is a cost, and the optimizer is minimizing the cost. As explained in Section 3.1.6, one of these constraints will dominate at the minimum cost solution (for a particular BL1_STM value). That constraint will become binding, and the corresponding inequality will become equality.

$$\textbf{BL1\_FUEL} \geq bl1\_fu\_c^* \textbf{BL1\_X} + bl1\_fu\_d^* \textbf{BL1\_STM} + bl1\_fu\_s^* \textbf{BL1\_Y} \qquad (\_\text{BL1\_EB1B})$$

$$\textbf{BL1\_FUEL} \geq bl1\_fu\_c^* \textbf{BL1\_X} + bl1\_fu\_d^* \textbf{BL1\_STM} + bl1\_fu\_s^* \textbf{BL1\_Y} \qquad (\_\text{BL1\_EB1R})$$

The complication is that in the DRPU model, all purchased energy is measured and thus fixed, i.e., there is no incentive for DRPU optimization to save fuel and thus for one of the constraints to become equality. This is problematic as the two-segment formulation correctly predicts fuel consumption only when one of the constraints becomes binding. So, what can we do? Here is an idea, but first, convert the inequalities (10.17) and (10.18) into equalities by calculating equation residuals, as shown below.

$$\textbf{BL1\_EB1BR} = \text{bl1\_fuel} - (\text{bl1\_fg\_c} + \textbf{BL1\_XBP} - \textbf{BL1\_XBN})^* \text{bl1\_x}$$
$$- (\text{bl1\_fg\_s} + \textbf{BL1\_YBP} - \textbf{BL1\_YBN})^* \text{bl1\_y} - \text{bl1\_fg\_d}^* \text{bl1\_stm}$$

(10.19)

$$\textbf{BL1\_EB1BR} = \text{bl1\_fuel} - (\text{bl1\_fg\_e} + \textbf{BL1\_XBP} - \textbf{BL1\_XBN})^* \text{bl1\_x}$$
$$- (\text{bl1\_fg\_s} + \textbf{BL1\_YBP} - \textbf{BL1\_YBN})^* \text{bl1\_y} - \text{bl1\_fg\_f}^* \text{bl1\_stm}$$

(10.20)

Where **BL1_EB1BR** and **BL1_EB1RR** are the constraint residuals or constraint values. We can now minimize the sum of the constraint residuals, **BL1_EB1BR** and **BL1_EB1RR**, in the DRPU objective function and force the binding constraint residual to the lowest possible value of 0, which is the desired outcome.

Note that the changes mentioned above to eqs. (10.19) and (10.20) are not required in the Excel model since the Excel model already calculates the residuals as shown in column F and allows the inclusion of the residuals in the objective function. However, such is not the case in LINGO, so these changes will only be necessary in the LINGO model.

BL2 DRPU: Similar to boiler #1, boiler #2 has redundant measurements and a two-segment performance curve. We will achieve DRPU by using four bias variables and adding the bias variables and the residuals of the energy balance constraints to the DRPU objective function. Note the Excel model already calculates the residuals; however, in the LINGO model, we will modify the energy balances to calculate the residuals as shown below:

$$\mathbf{BL2\_EB1BR} = bl2\_fuel - (bl2\_fg\_c + \mathbf{BL2\_XBP} - \mathbf{BL2\_XBN})^* bl2\_x$$
$$- (bl2\_fg\_s + \mathbf{BL2\_YBP} - \mathbf{BL2\_YBN})^* bl2\_y - bl2\_fg\_d^* bl2\_stm$$

(10.21)

$$\mathbf{BL2\_EB1RR} = bl2\_fuel - (bl2\_fg\_e + \mathbf{BL2\_XBP} - \mathbf{BL2\_XBN})^* bl2\_x$$
$$- (bl2\_fg\_s + \mathbf{BL2\_YBP} - \mathbf{BL2\_YBN})^* bl2\_y - bl2\_fg\_f^* bl2\_stm$$

(10.22)

BL3 DRPU: Similar to boiler #1 and boiler #2, boiler #3 also has redundant measurements, and a two-segment performance curve. We will achieve DRPU by using four bias variables and adding the bias variables and the residuals of the energy balance constraints to the DRPU objective function. Note the Excel model already calculates the residuals; however, in the LINGO model, we will modify the energy balances to calculate the residuals as shown below:

$$\mathbf{BL3\_EB1RR} = bl3\_fuel - (bl3\_fg\_c + \mathbf{BL3\_XBP} - \mathbf{BL3\_XBN})^* bl3\_x$$
$$- (bl3\_fg\_s + \mathbf{BL3\_YBP} - \mathbf{BL3\_YBN})^* bl3\_y - bl3\_fg\_d^* bl3\_stm$$

(10.23)

$$\mathbf{BL3\_EB1RR} = bl3\_fuel - (bl3\_fg\_e + \mathbf{BL3\_XBP} - \mathbf{BL3\_XBN})^* bl3\_x$$
$$- (bl3\_fg\_s + \mathbf{BL3\_YBP} - \mathbf{BL3\_YBN})^* bl3\_y - bl3\_fg\_f^* bl3\_stm$$

(10.24)

BFWP1 DRPU: Boiler feed water pump #1 has no redundant measurements; however, it has a two-segment performance curve to model internal recirculation at low flow rates correctly. We will achieve DRPU by adding the residuals of the energy balance constraints to the DRPU objective function. Note the Excel model already calculates the residuals; however, in the LINGO model, we will modify the energy balances to calculate the residuals as shown below:

$$\mathbf{BP1\_EB1BR} = \mathbf{BP1\_SP} - bp1\_sp\_a^* bp1\_x - bp1\_sp\_b^* \mathbf{BP1\_FLO}$$

(10.25)

$$\mathbf{BP1\_EB1RR} = \mathbf{BP1\_SP} - (bp1\_sp\_a + bp1\_sp\_b^* bp1\_des\_ll)^* bp1\_x$$

(10.26)

BFWP2 DRPU: Similar to the boiler feed water pump #1, the energy-balance constraints for the boiler feed water pump #2 will be modified as shown below:

$$\mathbf{BP2\_EB1BR} = \mathbf{BP2\_SP} - bp2\_sp\_a^* bp2\_x - bp2\_sp\_b^* \mathbf{BP2\_FLO}$$

(10.27)

$$\mathbf{BP2\_EB1RR} = \mathbf{BP2\_SP} - (bp2\_sp\_a + bp2\_x + bp2\_sp\_b^* bp2\_des\_ll)^* bp2\_x$$

(10.28)

BFWP3 DRPU: Similarly, for boiler feed water pump #3, we will modify the energy-balance constraints as shown below:

$$\mathbf{BP3\_EB1BR} = \mathbf{BP3\_SP} - \text{bp3\_sp\_a}^* \,\text{bp3\_x} - \text{bp3\_sp\_b}^* \,\mathbf{BP3\_FLO} \tag{10.29}$$

$$\mathbf{BP3\_EB1RR} = \mathbf{BP3\_SP} - (\text{bp3\_sp\_a} + \text{bp3\_sp\_b}^* \,\text{bp3\_des\_ll})^* \,\text{bp3\_x} \tag{10.30}$$

BFWP4 DRPU: Similar changes will be made for the boiler feed water pump #4:

$$\mathbf{BP4\_EB1BR} = \mathbf{BP4\_SP} - \text{bp4\_sp\_a}^* \,\text{bp4\_x} - \text{bp4\_sp\_b}^* \,\mathbf{BP4\_FLO} \tag{10.31}$$

$$\mathbf{BP4\_EB1RR} = \mathbf{BP4\_SP} - (\text{bp4\_sp\_a} + \text{bp4\_sp\_b}^* \,\text{bp4\_des\_ll})^* \,\text{bp4\_x} \tag{10.32}$$

For the Excel model, we will make the necessary changes to the "DRPU" worksheet of "UtilSys-RTO.xlsx." For the LINGO model, we will change the steam subsystem constraint file, "stmSS – RTO.lng."

10.2.2.3 Fuel subsystem DRPU

The fuel subsystem has only two major pieces of equipment: the proposed gas turbo-generator GTG and the existing fuel compressor FC. The fuel compressor has no redundant measurements, so DRPU is not required. However, we anticipate that the GTG will have redundant measurements when operational. The GTG energy balance has three inequalities as it has a three-segment performance curve. In the LINGO model, we must modify the constraints for each segment to calculate the residuals. The Excel model already calculates the residuals, so no changes are necessary. The DRPU objective will include these residuals.

GTG DRPU: Two bias variables, **GTG_XBP** and **GTG_XBN**, and performance residual calculations will be necessary for DRPU, as shown below;

$$\mathbf{GTG\_EB1GR} = \text{gtg\_fg} - (\text{gtg\_fg\_a} + \mathbf{GTG\_XBP} - \mathbf{GTG\_XBN})^* \,\text{gtg\_x} - \text{gtg\_fg\_b}^* \,\text{gtg\_ele} \tag{10.33}$$

$$\mathbf{GTG\_EB1BR} = \text{gtg\_fg} - (\text{gtg\_fg\_c} + \mathbf{GTG\_XBP} - \mathbf{GTG\_XBN})^* \,\text{gtg\_x} - \text{gtg\_fg\_d}^* \,\text{gtg\_ele} \tag{10.34}$$

$$\mathbf{GTG\_EB1RR} = \text{gtg\_fg} - (\text{gtg\_fg\_e} + \mathbf{GTG\_XBP} - \mathbf{GTG\_XBN})^* \,\text{gtg\_x} - \text{gtg\_fg\_f}^* \,\text{gtg\_ele} \tag{10.35}$$

For the Excel model, we will make the necessary changes to the "DRPU" worksheet of "UtilSys-RTO.xlsx." For the LINGO model, these changes will be made in the fuel subsystem constraint file, "FuelSS – RTO.lng."

10.2.2.4 Cooling water subsystem DRPU

The cooling water subsystem comprises three cooling water pumps. Cooling water pumps do not have any redundant measurements, so there is no need for DRPU as such. However, the pump energy balance equation is a two-segment curve to correctly model the internal recirculation at low flow rates. The energy balance comprises two inequality constraints. As for the boiler feed water pumps, we will need residuals for these two constraints, and explicitly penalize them in the DRPU objective function. The Excel model already calculates the residuals for all constraints, as shown in column F; so, no changes to constraints are necessary. However, we need to calculate residuals for the energy balance constraints for LINGO, as discussed below:

CWP1 modifications: For the LINGO model, the modified energy-balance constraints for cooling water pump #1 are:

$$\text{CWP1\_EB1RR} = \text{CWP1\_SP} - \text{cwp1\_sp\_a}^* \text{cwp1\_x} - \text{cwp1\_sp\_b}^* \text{CWP1\_FLO} \tag{10.36}$$

$$\text{CWP1\_EB1BR} = \text{CWP1\_SP} - \text{cwp1\_sp\_a}^* \text{cwp1\_x} - \text{cwp1\_sp\_b}^* \text{cwp1\_des\_ll}^*\text{cwp1\_x} \tag{10.37}$$

CWP2 modifications: Changes similar to the cooling water pump #1 are necessary for the cooling water pump #2 constraints in the LINGO:

$$\text{CWP2\_EB1RR} = \text{CWP2\_SP} - \text{cwp2\_sp\_a}^* \text{cwp2\_x} - \text{cwp2\_sp\_b}^* \text{CWP2\_FLO} \tag{10.38}$$

$$\text{CWP2\_EB1BR} = \text{CWP2\_SP} - \text{cwp2\_sp\_a}^* \text{cwp2\_x} - \text{cwp2\_sp\_b}^* \text{cwp2\_des\_ll}^*\text{cwp2\_x} \tag{10.39}$$

CWP3 modifications: Changes similar to cooling water pump #1 and cooling water pump #2 are required for the cooling water pump #3 in LINGO; the modified energy-balance constraints are:

$$\text{CWP3\_EB1RR} = \text{CWP3\_SP} - \text{cwp3\_sp\_a}^* \text{cwp3\_x} - \text{cwp3\_sp\_b}^* \text{CWP3\_FLO} \tag{10.40}$$

$$\text{CWP3\_EB1BR} = \text{CWP3\_SP} - \text{cwp3\_sp\_a}^* \text{cwp3\_x} - \text{cwp3\_sp\_b}^* \text{cwp3\_des\_ll}^*\text{cwp3\_x} \tag{10.41}$$

For the LINGO model, these changes are made to the cooling water subsystem constraint file, "cwSS-RTO.lng."

10.2.2.5 DRPU model

The Utility System model for DRPU is essentially the same as the model used for off-line optimization, UtilSys-OFLO, except for the following modifications.

- **Optimization Variables**: We need to add the 36 biases to the optimization variables in the DRPU model. Note that of the 67 measured values, 11 are inputs, and 56 are fixed, so they cannot be optimization variables. Since the original optimization problem has 155 variables, the DRPU model will have 155 – 56 + 36 = 135 optimization variables. Table 10.2 is a list of the DRPU optimization variables.

Tab. 10.2: Sample utility system model: UtilSys-RTO: DRPU: optimization variables.

#	Excel cell ID	Variable name	UOM	#	Excel cell ID	Variable name	UOM	#	Excel cell ID	Variable name	UOM
1	H9	BL1_BFW	tph	46	CG9	BF2	tph	91	ES9	LPF22_X	BIN
2	I9	BL1_BD	tph	47	CH9	BF3	tph	92	EU9	BUY_HPF21	MW.th
3	J9	BL1_RSV	tph	48	CI9	BF4	tph	93	EV9	BUY_HPF22	MW.th
4	L9	BL1AT_STM	tph	49	CJ9	BX1	BIN	94	EW9	BUY_HPF23	MW.th
5	P9	BL2_BFW	tph	50	CK9	BX2	BIN	95	EY9	HPF21_X	BIN
6	Q9	BL2_BD	tph	51	CL9	BX3	BIN	96	EZ9	HPF22_X	BIN
7	R9	BL2_RSV	tph	52	CM9	BX4	BIN	97	FA9	HPF23_X	BIN
8	S9	BL2_FUEL	MW.th	53	CN 9	BAVG	tph	98	FC9	DMND_CST	MW.pk
9	V9	BL2AM_ELE	MW	54	CO9	BDF_LPS	tph	99	FD9	IDEMAND	MW.pk
10	Z9	BL3_BFW	tph	55	CP9	BDF_CND	tph	100	FF9	HPS_XBP	tph
11	AA9	BL3_BD	tph	56	CR9	DEA_DMW	tph	101	FG9	HPS_XBN	tph
12	AB9	BL3_RSV	tph	57	CS9	DEA_LPS	tph	102	FH9	MPS_XBP	tph
13	AD9	BL3_LPFG	MW.th	58	DA9	FC_SP	MW.sp	103	FI9	MPS_XBN	tph
14	AE9	BL3_HPFG	MW.th	59	DC9	FC_ELE	MW	104	FJ9	LPS_XBP	tph
15	AF9	BL3AM_ELE	MW	60	DD9	CWP1_FLO	ktph	105	FK9	LPS_XBN	tph
16	AG9	BL3AT_STM	tph	61	DE9	CWP1_SP	MW.sp	106	FL9	FO_XBP	MW.th
17	AI9	BL3AT_X	0 / 1	62	DF9	CWP1M_ELE	MW	107	FM9	FO_XBN	MW.th
18	AJ9	BL3_W	0 / 1	63	DH9	CWP2_FLO	ktph	108	FN9	LPFG_XBP	MW.th
19	AN 9	MPS_RSV	tph	64	DI9	CWP2_SP	MW.sp	109	FO9	LPFG_XBN	MW.th
20	AO9	HPS_RSV	tph	65	DJ9	CWP2T_STM	tph	110	FP9	HPFG_XBP	MW.th
21	AQ9	PRV1_BFW	tph	66	DL9	CWP3_FLO	ktph	111	FQ9	HPFG_XBN	MW.th
22	AR9	PRV1_MPS	tph	67	DM9	CWP3_SP	MW.sp	112	FR9	ELEC_XBP	MW
23	BA9	CTG_CND	tph	68	DO9	CWP3M_SP	MW.sp	113	FS9	ELEC_XBN	MW
24	BB9	CTG_CW	ktph	69	DP9	CWP3M_ELE	MW	114	FT9	BL1_XBP	MW.th
25	BF9	BP1_FLO	tph	70	DR9	CWP3T_SP	MW.sp	115	FU9	BL1_XBN	MW.th
26	BG9	BP1_SP	MW.sp	71	DS9	CWP3T_STM	tph	116	FV9	BL1_YBP	MW.th
27	BH9	BP1M_ELE	MW	72	DT9	CWP3T_X	0/1	117	FW9	BL1_YBN	MW.th
28	BJ9	BP2_FLO	tph	73	DU9	CF1	ktph	118	FX9	BL2_XBP	MW.th
29	BK9	BP2_SP	MW	74	DV9	CF2	ktph	119	FY9	BL2_XBN	MW.th
30	BM9	BP2M_SP	MW.sp	75	DW9	CF3	ktph	120	FZ9	BL2_YBP	MW.th
31	BN 9	BP2M_ELE	MW	76	DX9	CX1	BIN	121	GA9	BL2_YBN	MW.th
32	BP9	BP2T_SP	MW.sp	77	DY9	CX2	BIN	122	GB9	BL3_XBP	MW.th
33	BQ9	BP2T_STM	tph	78	DZ9	CX3	BIN	123	GC9	BL3_XBN	MW.th
34	BR9	BP2T_X	0/1	79	EA9	CAVG	ktph	124	GD9	BL3_YBP	MW.th
35	BS9	BP3_FLO	tph	80	EB9	AC1234_CA	kM3/h	125	GE9	BL3_YBN	MW.th
36	BT9	BP3_SP	MW.sp	81	EC9	AC1234_SP	MW.sp	126	GF9	STG_XBP	tph
37	BU9	BP3M_ELE	MW	82	EE9	AC1234E_FG	MW.th	127	GG9	STG_XBN	tph
38	BW9	BP4_FLO	tph	83	EF9	AC5_CA	kM3/h	128	GH9	STG_YBP	tph
39	BX9	BP4_SP	MW.sp	84	EG9	AC5_SP	MW.sp	129	GI9	STG_YBN	tph
40	BZ9	BP4M_SP	MW.sp	85	EI9	AC5M_ELE	MW	130	GJ9	CTG_XBP	tph
41	CA9	BP4M_ELE	MW	86	EL9	FO_CST	$/hr	131	GK9	CTG_XBN	tph
42	CC9	BP4T_SP	MW.sp	87	EN 9	LPF1_CST	$/hr	132	GL9	CTG_YBP	tph
43	CD9	BP4T_STM	tph	88	EO9	BUY_LPF21	MW.th	133	GM9	CTG_YBN	tph
44	CE9	BP4T_X	0/1	89	EP9	BUY_LPF22	MW.th	134	GN9	GTG_XBP	MW.th
45	CF9	BF1	tph	90	ER9	LPF21_X	BIN	135	GO9	GTG_XBN	MW.th

Since the Excel model requires explicit specifications of the optimization variables, we need to adjust the variable specification by removing the 56 measured variables and adding the 36 bias variables. Figure 10.5 shows the optimization variables in the Premium Solver. Note that the variable list in Fig. 10.5 matches the list in Tab. 10.2.

However, the LINGO model does not require explicit specification of the optimization variables; so, no changes to the LINGO model are necessary. In LINGO, all unspecified variables automatically become optimization variables.

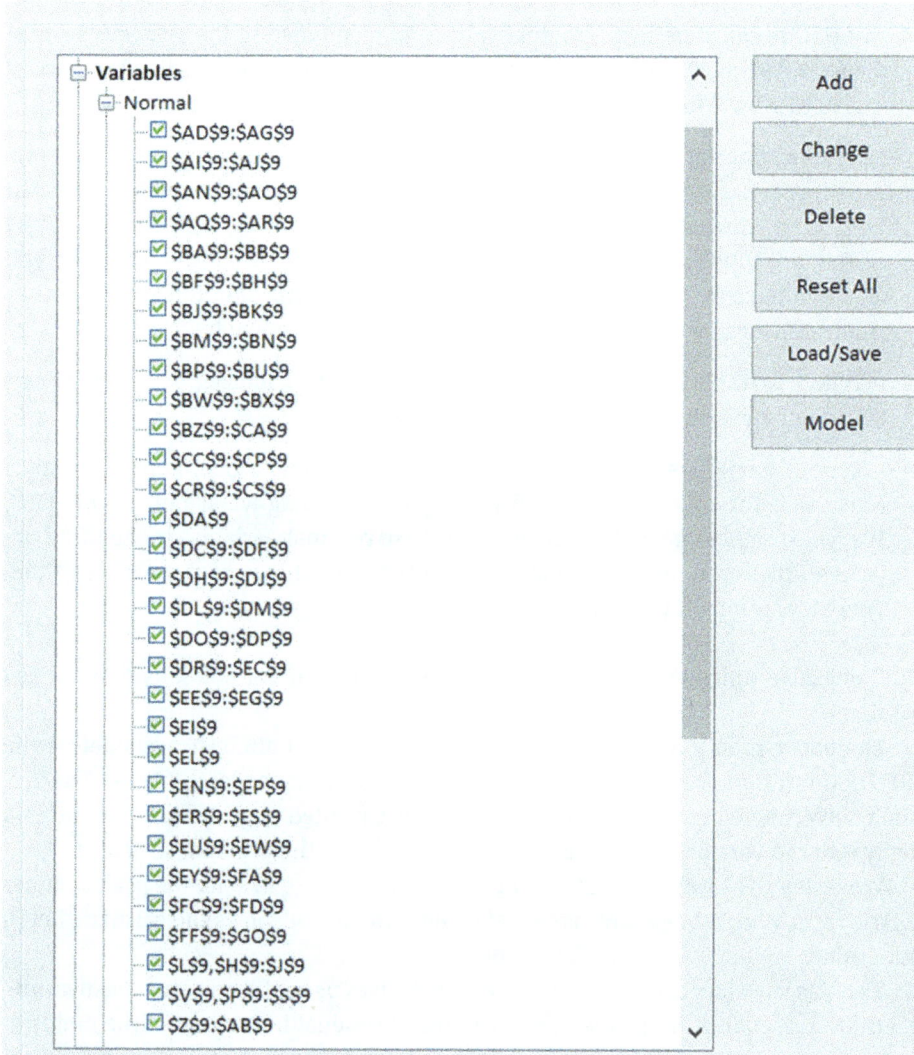

Fig. 10.5: Sample utility system model: UtilSys-RTO: DRPU variable specification in Excel Premium Solver.

- **Constraints**: Model constraints essentially remain unchanged in the DRPU model except for the addition of bias variables. In the Excel model, no additional changes are necessary as the model already computes the constraint residual values, and allows the inclusion of the residuals in the objective function.

   However, in the LINGO model, we need to modify the multisegment energy-balance constraints for boilers, boiler feed water pumps, cooling water pumps, and GTG to calculate their residual values. As mentioned in the previous section, the modified constraints for the steam, fuel, and cooling water subsystems are in 'stmSS-RTO.lng,' 'fuelSS-RTO.lng,' and 'cwSS-RTO.lng,' files, respectively.

- **Objective function**: All purchased energy flows are measured quantities; therefore, the corresponding variables are fixed values in the DRPU model. Note that fixing these variable values also fixes the operating cost; so, including operating costs in the DRPU objective function is unnecessary. For DRPU, the objective will be to minimize the sum of biases and constraint residuals of multisegment energy balance constraints. We will classify these variables as shown below:

   – **Type I variables** are the biases required to achieve balances, as discussed earlier in this subsection. There are 36 such variables.

   – **Type II variables** are the residual values of the multisegment energy-balance constraints for boilers, boiler Feed water pumps, cooling water pumps, and GTG. We will specify large cost coefficients for these residual variables in the DRPU objective function to ensure the binding constraints achieve zero residual values. There are 23 type II variables.

DRPU objective function comprises 36 type I and 23 type II variables, as summarized in Tab. 10.3.

   The above-mentioned objective function, variables, and constraints define the DRPU optimization problem. Figures. 10.6 and 10.7 show the model in LINGO. The Premium Solver model in Excel is too large to fit on a printed page. Figure 10.8 shows a highly reduced version from which one can only discern the problem structure.

   We tested the LINGO and Excel DRPU models with case #1 data for peak summer electric rates. Both models gave identical results. Table 10.4 shows the calculated optimum values of Type I and Type II variables.

   The bias variables are relatively small, which gives us confidence that the measurements are reasonable. Also, at least one constraint residual in each set of the multisegment energy balances has zero value, confirming adherence to the multisegment energy-balance curves. Table 10.4 and Fig. 10.9 give details of the DRPU optimization solution. The DRPU model also computes the total operating cost of the current operation, which is $8421.5/h.

Tab. 10.3: DRPU Optimization: variables in the objective function.

#	Excel cell ID	Variable name	UOM	Cost coefficient	#	Excel cell ID	LINGO Variable name	UOM	Cost coefficient
Type I variables:					Type II variables:				
1	FF9	HPS_XBP	tph	1	1	F13	BL1_EB1BR	MW.th	100
2	FG9	HPS_XBN	tph	1	2	F14	BL1_EB1RR	MW.th	100
3	FH9	MPS_XBP	tph	1	3	F23	BL2_EB1BR	MW.th	100
4	FI9	MPS_XBN	tph	1	4	F24	BL2_EB1RR	MW.th	100
5	FJ9	LPS_XBP	tph	1	5	F33	BL3_EB1BR	MW.th	100
6	FK9	LPS_XBN	tph	1	6	F34	BL3_EB1RR	MW.th	100
7	FL9	FO_XBP	MW.th	1	7	F67	BP1_EB1BR	MW.sp	100
8	FM9	FO_XBN	MW.th	1	8	F68	BP1_EB1RR	MW.sp	100
9	FN9	LPFG_XBP	MW.th	1	9	F73	BP2_EB1BR	MW.sp	100
10	FO9	LPFG_XBN	MW.th	1	10	F74	BP2_EB1RR	MW.sp	100
11	FP9	HPFG_XBP	MW.th	1	11	F83	BP3_EB1BR	MW.sp	100
12	FQ9	HPFG_XBN	MW.th	1	12	F84	BP3_EB1RR	MW.sp	100
13	FR9	ELEC_XBP	MW	1	13	F89	BP4_EB1BR	MW.sp	100
14	FS9	ELEC_XBN	MW	1	14	F90	BP4_EB1RR	MW.sp	100
15	FT9	BL1_XBP	MW.th	1	15	F118	GTG_EB1GR	MW.th	100
16	FU9	BL1_XBN	MW.th	1	16	F119	GTG_EB1BR	MW.th	100
17	FV9	BL1_YBP	MW.th	1	17	F120	GTG_EB1RR	MW.th	100
18	FW9	BL1_YBN	MW.th	1	18	F129	CWP1_EB1BR	MW.sp	100
19	FX9	BL2_XBP	MW.th	1	19	F130	CWP1_EB1RR	MW.sp	100
20	FY9	BL2_XBN	MW.th	1	20	F135	CWP2_EB1BR	MW.sp	100
21	FZ9	BL2_YBP	MW.th	1	21	F136	CWP2_EB1RR	MW.sp	100
22	GA9	BL2_YBN	MW.th	1	22	F141	CWP3_EB1BR	MW.sp	100
23	GB9	BL3_XBP	MW.th	1	23	F142	CWP3_EB1RR	MW.sp	100
24	GC9	BL3_XBN	MW.th	1					
25	GD9	BL3_YBP	MW.th	1					
26	GE9	BL3_YBN	MW.th	1					
27	GF9	STG_XBP	tph	1					
28	GG9	STG_XBN	tph	1					
29	GH9	STG_YBP	tph	1					
30	GI9	STG_YBN	tph	1					
31	GJ9	CTG_XBP	tph	1					
32	GK9	CTG_XBN	tph	1					
33	GL9	CTG_YBP	tph	1					
34	GM9	CTG_YBN	tph	1					
35	GN9	GTG_XBP	MW.th	1					
36	GO9	GTG_XBN	MW.th	1					

```
MODEL:  TITLE Utility System Model for real time DRPU: UtilSys-RTO-DRPU;
! Ref: Nath, "Industrial Process Plants: Global Optimization of Utility Systems", Chapter 10, (2023);
! ;
DATA: hps_demand, mps_demand, lps_demand, mps_reserve, hps_reserve, cw_demand, ca_demand, hpfg_demand,
      lpfg_demand, fo_demand,  elec_demand,bll_stm,    bll_fuel,    bll_x,      bll_y,      bl2_stm,
      bl2_lpfg,   bl2_hpfg,   bl2_x,      bl2_y,       bl3_stm,     bl3_fuel,   bl3am_x,    bl3_x,
      bl3_y,      bl3_z,      prv1_hps,   prv2_stm,    stg_stm,     stg_ele,    stg_x,      stg_y,
      ctg_hps,    ctg_mps,    ctg_lps,    ctg_ele,     ctg_x,       ctg_y,      bp1_x,      bp2_x,
      bp2m_x,     bp3_x,      bp4_x,      bp4m_x,      vent,        prvfg,      fgflare,    foflare,
      gtg_fg,     gtg_ele,    gtg_x,      fc_fg,       fc_x,        cwp1_x,     cwp2_x,     cwp3_x,
      cwp3m_x,    ac1234_nx,  ac5_x,      vent_ca,     buy_fo,      buy_lpf1,   buy_lpf2,   buy_hpf1,
      buy_hpf2,   buy_elec,   sel_elec,
      c_dmw,      c_fo1,      fo1_hl,     c_fo2,       fo2_hl,      c_lpf1,     lpf1_hl,    c_lpf21,
      lpf21_hl,   c_lpf22,    lpf22_hl,   c_hpf1,      c_hpf21,     hpf21_hl,   c_hpf22,    hpf22_hl,
      c_hpf23,    hpf23_hl,   cbuy_ener,  cbuy_dem,    cbuy_idem,   csel_elec,  cur_pkdmnd
                = @OLE('C:\BookDG\ExcelModels\UtilSys-RTO.xlsx');    ! retrieve current op data;

      ! params;  @FILE('C:\BookDG\LingoModels\UtilSys-params.lng')     ! retrieve equipment params;
    ENDDATA

! calcs;         @FILE('C:\BookDG\LingoModels\UtilSys-calcs-v2.lng')   ! perform initialization;

! Objective;    MIN      = LPFG_XBP  + LPFG_XBN  + HPFG_XBP  + HPFG_XBN  + ELEC_XBP  + ELEC_XBN
             + HPS_XBP   + HPS_XBN   + MPS_XBP   + MPS_XBN   + LPS_XBP   + LPS_XBN   + FO_XBP   + FO_XBN
             + BL1_XBP   + BL1_XBN   + BL1_YBP   + BL1_YBN   + BL2_XBP   + BL2_XBN   + BL2_YBP  + BL2_YBN
             + BL3_XBP   + BL3_XBN   + BL3_YBP   + BL3_YBN   + STG_XBP   + STG_XBN   + STG_YBP  + STG_YBN
             + CTG_XBP   + CTG_XBN   + CTG_YBP   + CTG_YBN   + GTG_XBP   + GTG_XBN
         +100*(BL1_EB1BR + BL1_EB1RR + BL2_EB1BR + BL2_EB1RR + BL3_EB1BR + BL3_EB1RR + GTG_EB1GR + GTG_EB1BR
             + GTG_EB1RR + BP1_EB1BR + BP1_EB1RR + BP2_EB1BR + BP2_EB1RR + BP3_EB1BR + BP3_EB1RR + BP4_EB1BR
             + BP4_EB1RR +CWP1_EB1BR +CWP1_EB1RR +CWP2_EB1BR +CWP2_EB1RR +CWP3_EB1BR +CWP3_EB1RR);

! retrieve subsystem models;
                @FILE('C:\BookDG\LingoModels\stmSS-RTO.lng')  @FILE('C:\BookDG\LingoModels\fuelSS-RTO.lng')
                @FILE('C:\BookDG\LingoModels\caSS-v1.lng')    @FILE('C:\BookDG\LingoModels\cwSS-RTO.lng')
                @FILE('C:\BookDG\LingoModels\Contracts-v1.lng')
```

Fig. 10.6: Sample utility system RTO model in LINGO: UtilSys-RTO-DRPU, part 1 of 2.

```
! compute current operating cost;
  [_TOC]        TOC = FO_CST + LPF1_CST + c_lpf21 * BUY_LPF21 + c_lpf22 * BUY_LPF22 + lpf22_a * LPF22_X
              + c_hpf1 * BUY_HPF1 + c_hpf21 * BUY_HPF21 + c_hpf22 * BUY_HPF22 + hpf22_a * HPF22_X
              + c_hpf23 *BUY_HPF23 + hpf23_a * HPF23_X + c_dmw * DEA_DMW + cbuy_ener * buy_elec
              + DMND_CST + cbuy_idem * IDEMAND - csel_elec * sel_elec;

! Header balances;
  [_HPS_HDR]    bl2_stm + bl3_stm + HPS_XBP - HPS_XBN
                    = BL3AT_STM + prv1_hps + stg_stm + ctg_hps + BP2T_STM + BP4T_STM  + CWP2T_STM + CWP3T_STM
                    + hps_demand;
  [_MPS_HDR]    bl1_stm + PRV1_MPS + stg_stm + ctg_mps + BP4T_STM + CWP2T_STM + MPS_XBP - MPS_XBN
                    = prv2_stm + BL1AT_STM + mps_demand;
  [_LPS_HDR]    BL1AT_STM + BL3AT_STM + prv2_stm  + ctg_lps + BP2T_STM + BDF_LPS + CWP3T_STM + LPS_XBP - LPS_XBN
                    = DEA_LPS + vent + lps_demand;
  [_BFW_HDR]    BL1_BFW + BL2_BFW + BL3_BFW + PRV1_BFW = BP1_FLO + BP2_FLO + BP3_FLO + BP4_FLO;
  [_CW_HDR]     CWP1_FLO + CWP2_FLO + CWP3_FLO = CTG_CW + cw_demand;
  [_CA_HDR]     AC1234_CA + AC5_CA = vent_ca + ca_demand;
  [_HPFG_HDR]   buy_hpf1 + buy_hpf2 + fc_fg + HPFG_XBP  - HPFG_XBN
                    = prvfg + gtg_fg + bl2_hpfg + BL3_HPFG + hpfg_demand;
  [_LPFG_HDR]   buy_lpf1 + buy_lpf2 + prvfg + LPFG_XBP  - LPFG_XBN
                    = fc_fg + fgflare + bl2_lpfg + BL3_LPFG + AC1234E_FG + lpfg_demand;
  [_FO_HDR]     buy_fo + FO_XBP - FO_XBN = foflare + bl1_fuel + fo_demand;
  [_ELEC_HDR]   buy_elec - sel_elec + stg_ele + ctg_ele + gtg_ele + ELEC_XBP - ELEC_XBN
                    = FC_ELE + BL2AM_ELE + BL3AM_ELE + BP1M_ELE + BP2M_ELE + BP3M_ELE + BP4M_ELE
                    + CWP1M_ELE + CWP3M_ELE + AC5M_ELE + elec_demand;

! save biases to file for use by Optimization;
DATA:
  @TEXT('C:\BookDG\LingoModels\UtilSys-Biases.lng') =
    HPS_XBP,    HPS_XBN,    MPS_XBP,    MPS_XBN,    LPS_XBP,    LPS_XBN,    FO_XBP,     FO_XBN,     LPFG_XBP,
    LPFG_XBN,   HPFG_XBP,   HPFG_XBN,   ELEC_XBP,   ELEC_XBN,   BL1_XBP,    BL1_XBN,    BL1_YBP,    BL1_YBN,
    BL2_XBP,    BL2_XBN,    BL2_YBP,    BL2_YBN,    BL3_XBP,    BL3_XBN,    BL3_YBP,    BL3_YBN,    STG_XBP,
    STG_XBN,    STG_YBP,    STG_YBN,    CTG_XBP,    CTG_XBN ,   CTG_YBP,    CTG_YBN,    GTG_XBP,    GTG_XBN;
ENDDATA

END MODEL UtilSys-RTO-DRPU
```

Fig. 10.7: Sample utility system RTO model in LINGO: UtilSys-RTO-DRPU, part 2 of 2.

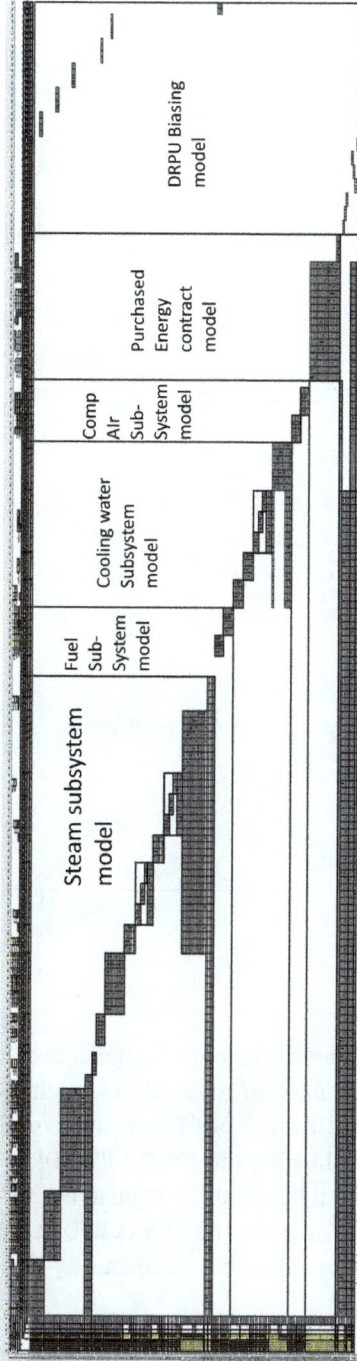

Fig. 10.8: Sample utility system RTO model in Excel: UtilSys-RTO: problem structure.

Tab. 10.4: Sample utility system model: UtilSys-RTO: DRPU, case #1 solution.

#	Excel cell ID	Variable name	UOM	DRPU value	#	Excel cell ID	LINGO Variable name	UOM	DRPU value
Type I variables:					Type II variables:				
1	FF9	HPS_XBP	tph	0.8	1	F13	BL1_EB1BR	MW.th	0.00
2	FG9	HPS_XBN	tph	0.0	2	F14	BL1_EB1RR	MW.th	0.00
3	FH9	MPS_XBP	tph	2.3	3	F23	BL2_EB1BR	MW.th	0.00
4	FI9	MPS_XBN	tph	0.0	4	F24	BL2_EB1RR	MW.th	13.48
5	FJ9	LPS_XBP	tph	0.0	5	F33	BL3_EB1BR	MW.th	0.00
6	FK9	LPS_XBN	tph	1.5	6	F34	BL3_EB1RR	MW.th	7.59
7	FL9	FO_XBP	MW.th	0.0	7	F67	BP1_EB1BR	MW.sp	0.00
8	FM9	FO_XBN	MW.th	0.0	8	F68	BP1_EB1RR	MW.sp	0.02
9	FN9	LPFG_XBP	MW.th	0.0	9	F73	BP2_EB1BR	MW.sp	0.00
10	FO9	LPFG_XBN	MW.th	2.0	10	F74	BP2_EB1RR	MW.sp	0.02
11	FP9	HPFG_XBP	MW.th	0.0	11	F83	BP3_EB1BR	MW.sp	0.00
12	FQ9	HPFG_XBN	MW.th	0.0	12	F84	BP3_EB1RR	MW.sp	0.00
13	FR9	ELEC_XBP	MW	0.0	13	F89	BP4_EB1BR	MW.sp	0.00
14	FS9	ELEC_XBN	MW	0.0	14	F90	BP4_EB1RR	MW.sp	0.00
15	FT9	BL1_XBP	MW.th	0.0	15	F118	GTG_EB1GR	MW.th	0.00
16	FU9	BL1_XBN	MW.th	0.0	16	F119	GTG_EB1BR	MW.th	0.00
17	FV9	BL1_YBP	MW.th	0.0	17	F120	GTG_EB1RR	MW.th	0.00
18	FW9	BL1_YBN	MW.th	0.0	18	F129	CWP1_EB1BR	MW.sp	0.00
19	FX9	BL2_XBP	MW.th	0.0	19	F130	CWP1_EB1RR	MW.sp	0.00
20	FY9	BL2_XBN	MW.th	2.0	20	F135	CWP2_EB1BR	MW.sp	0.00
21	FZ9	BL2_YBP	MW.th	0.0	21	F136	CWP2_EB1RR	MW.sp	0.74
22	GA9	BL2_YBN	MW.th	0.0	22	F141	CWP3_EB1BR	MW.sp	0.00
23	GB9	BL3_XBP	MW.th	0.0	23	F142	CWP3_EB1RR	MW.sp	0.74
24	GC9	BL3_XBN	MW.th	0.9					
25	GD9	BL3_YBP	MW.th	0.0					
26	GE9	BL3_YBN	MW.th	0.0					
27	GF9	STG_XBP	tph	0.0					
28	GG9	STG_XBN	tph	0.0					
29	GH9	STG_YBP	tph	0.0					
30	GI9	STG_YBN	tph	0.0					
31	GJ9	CTG_XBP	tph	1.1					
32	GK9	CTG_XBN	tph	0.0					
33	GL9	CTG_YBP	tph	0.0					
34	GM9	CTG_YBN	tph	0.0					
35	GN9	GTG_XBP	MW.th	0.0					
36	GO9	GTG_XBN	MW.th	0.0					

10.2.3 Step 3: Send biases to the optimizers

DRPU determines the bias values that will remain fixed in the optimizations. In the Excel DRPU worksheet, we will assign cell names to each of these biases, such as "hps_xbp." The COPT and DOPT optimizer models on COPT and DOPT worksheets will access these bias values using these cell names. In LINGO, the UtilSys-RTO-DRPU model determines the bias values and writes them to an external file named 'UtilSys-Biases.lng' using LINGO's @TEXT() interface function [7]. Later, the UtilSys-RTO-COPT and UtilSys-RTO-DOPT optimization models will access these bias values using LINGO's @FILE() interface function.

Utility system Model: UtilSys-RTO-DRPU. Case ID: 1, TOU rate ID: Summer_peak
Global optimal solution found.
Optimum objective function value (minimum operating cost, $/hr) $ 8,421

Purchased energy:

	contract	min	max	amount		COST, $./hr	Incr unit cst		imbalance
BUY DMW				120.5	tph	$ 361.5	$ 3.00		
BUY FO TIER 1&2	X55 PEN		200	30.000	MW.th	$ 550.0	$ 19.00		0.000
BUY LPFG - #1	TAKE/PAY		100	100.000	MW.th	$ 1,900.0	take or pay		
BUY LPFG #2-tier1		0	100	0.00	MW.th	$ -	$ 19.00		-2.000
BUY LPFG #2-tier2		1	500	240.00	MW.th	$ 4,420.0	$ 18.00		
BUY HPFG #1	SPOT			0.000	MW.th	$ -	$ 19.00		
BUY HPFG #2-tier1		0	20	0.000	MW.th	$ -	$ 20.00		0.000
BUY HPFG #2-tier2		0	60	0.000	MW.th	$ -	$ 19.00		
BUY HPFG #2-tier3		0	120	0.000	MW.th	$ -	$ 18.00		
BUY ELEC - energy charge				19.800	MW	$ 990.0	$ 50.00		
BUY ELEC - demand charge				20.000	MW-peak	$ 200.0	$ 10.00		
IDEMAND penalty				0.000	MW	$ -	$ 100.00		0.039
Sell ELEC				0.000	MW	$ -	$ (40.00)		
TOTAL OP COST						$ 8,421.5			

Process DEMAND				100	45	90		-67.5				15	30	30	30	30	22	
(unit inputs +ve, outputs -ve)				**HP**	**MP**	**LP**	**BD**	**PC/TC**	**DMW**	**HP-bfw**	**LP-bfw**	**CW**	**CA**	**FO**	**LPFG**	**HPFG**	**ELEC**	imbalance
Unit	A/R/U	min	max	tph	tph	tph	tph	tph	tph	tph		ktph	kM3/h	MW.th	MW.th	MW.th	MW	
BL1	A	70	210	0		0		0		0				0.0				0.0
-aux					0	0												
BL2	A	80	240	-104				-2		106				103.00	0.00			2.0
-aux																0.500		
BL3	A	100	300	-190				-4		194				175.00	0.00			0.9
-aux				5		-5										0.000		
STG	A	1.00	3.00	0	0											0.000	0.0
CTG	A	3.00	13.00	136	-70	-56		-10				0.4					-5.800	1.1
Bfw P1	A		250							-151	151						0.261	
Bfw P2	A		250	2		-2				-151	151						0.000	
Bfw P3	A		250							0	0						0.000	
Bfw P4	A		250	0	0					0	0						0.000	
PRV1				20	-22								2					
PRV2					70	-70												
BD Flash						-1	6											
Deaerator						54		78	120		-302							
steam VENT						0												
CW P1	A		8									0.0					0.000	
CW P2	A		8	21	-21							-7.7						
CW P3	A		8	10		-10						-7.7					0.000	
AC1-4	4	0	0										0.0		0.00			
AC5	A	20	55										-30.0				2.394	
Air Vent													0.0					
GTG	U		25													0.00	0.000	
FC	A	25	75											30.00	-30.00		0.484	
PRVFG															0.00	0.00		
Fuel FLARE													0.00		0.00			
Bias (-ve => supply)				-0.8	-2.3	1.5								0.00	2.00	0.00	-0.039	
sum column (net import)				0.0	0.0	0.0	0.0	0.0	120.5	0.0	0.0	0.0	0.0	30.00	340.00	0.00	19.80	

Fig. 10.9: Sample utility system model: UtilSys-RTO: DRPU, base case.

10.2.4 Step 4: Continuous optimization: COPT

The Utility System model for Continuous Optimization, COPT, is the same as the model used for DRPU optimization, except for the following modifications.

- **Optimization Variables**: The Utility System model has 16 pieces of equipment with 26 integer variables. In the COPT model, the equipment slate remains fixed to the current values, i.e., fixing the 26 integer variables and removing them from optimization. Also, the biases remain fixed at the values calculated by DRPU. The optimization variable count now becomes 155–26 = 129. Since the Excel model requires explicit specification of optimization variables, we modified the optimization variables accordingly. Table 10.5 shows the 129 COPT variables. Note that in the LINGO model, no changes are necessary since LINGO does not require explicit specification of the optimization variables. In a LINGO model, any variable without a specified value automatically becomes an optimization variable.
- **Objective function**: For the COPT model, we are back to the objective function of minimizing the operating cost, as in the offline optimizer, UtilSys-OFLO.
- **Constraints**: No changes are required. They remain unchanged.

Tab. 10.5: Sample utility system, UtilSys-RTO, COPT optimization variables.

#	Excel cell ID	Variable name	UOM	#	Excel cell ID	Variable name	UOM	#	Excel Cell ID	Variable name	UOM
1	G9	BL1_STM	tph	44	BK9	BP2_SP	MW.sp	87	DJ9	CWP2T_STM	tph
2	H9	BL1_BFW	tph	45	BM9	BP2M_SP	MW.sp	88	DL9	CWP3_FLO	ktph
3	I9	BL1_BD	tph	46	BN 9	BP2M_ELE	MW	89	DM9	CWP3_SP	MW.sp
4	J9	BL1_RSV	tph	47	BP9	BP2T_SP	MW.sp	90	DO9	CWP3M_SP	MW.sp
5	K9	BL1_FUEL	MW.th	48	BQ9	BP2T_STM	tph	91	DP9	CWP3M_ELE	MW
6	L9	BL1AT_STM	tph	49	BR9	BP2T_X	0/1	92	DR9	CWP3T_SP	MW.sp
7	O9	BL2_STM	tph	50	BS9	BP3_FLO	tph	93	DS9	CWP3T_STM	tph
8	P9	BL2_BFW	tph	51	BT9	BP3_SP	MW.sp	94	DT9	CWP3T_X	0/1
9	Q9	BL2_BD	tph	52	BU9	BP3M_ELE	MW	95	DU9	CF1	ktph
10	R9	BL2_RSV	tph	53	BW9	BP4_FLO	tph	96	DV9	CF2	ktph
11	S9	BL2_FUEL	MW.th	54	BX9	BP4_SP	MW.sp	97	DW9	CF3	ktph
12	T9	BL2_LPFG	MW.th	55	BZ9	BP4M_SP	MW.sp	98	DX9	CX1	BIN
13	U9	BL2_HPFG	MW.th	56	CA9	BP4M_ELE	MW	99	DY9	CX2	BIN
14	V9	BL2AM_ELE	MW	57	CC9	BP4T_SP	MW.sp	100	DZ9	CX3	BIN
15	Y9	BL3_STM	tph	58	CD9	BP4T_STM	tph	101	EA9	CAVG	ktph
16	Z9	BL3_BFW	tph	59	CE9	BP4T_X	0/1	102	EB9	AC1234_CA	kM3/h
17	AA9	BL3_BD	tph	60	CF9	BF1	tph	103	EC9	AC1234_SP	MW.sp
18	AB9	BL3_RSV	tph	61	CG9	BF2	tph	104	EE9	AC1234E_FG	MW.th
19	AC9	BL3_FUEL	MW.th	62	CH9	BF3	tph	105	EF9	AC5_CA	kM3/h
20	AD9	BL3_LPFG	MW.th	63	CI9	BF4	tph	106	EG9	AC5_SP	MW.sp
21	AE9	BL3_HPFG	MW.th	64	CJ9	BX1	BIN	107	EI9	AC5M_ELE	MW
22	AF9	BL3AM_ELE	MW	65	CK9	BX2	BIN	108	EJ9	VENT_CA	kM3/h
23	AG9	BL3AT_STM	tph	66	CL9	BX3	BIN	109	EK9	BUY_FO	MW.th
24	AI9	BL3AT_X	0 / 1	67	CM9	BX4	BIN	110	EL9	FO_CST	$/hr
25	AJ9	BL3_W	0 / 1	68	CN 9	BAVG	tph	111	EM9	BUY_LPF1	MW.th
26	AN 9	MPS_RSV	tph	69	CO9	BDF_LPS	tph	112	EN 9	LPF1_CST	$/hr
27	AO9	HPS_RSV	tph	70	CP9	BDF_CND	tph	113	EO9	BUY_LPF21	MW.th
28	AP9	PRV1_HPS	tph	71	CQ9	VENT	tph	114	EP9	BUY_LPF22	MW.th
29	AQ9	PRV1_BFW	tph	72	CR9	DEA_DMW	tph	115	EQ9	BUY_LPF2	MW.th
30	AR9	PRV1_MPS	tph	73	CS9	DEA_LPS	tph	116	ER9	LPF21_X	BIN
31	AS9	PRV2_STM	tph	74	CT9	PRVFG	MW.th	117	ES9	LPF22_X	BIN
32	AT9	STG_STM	tph	75	CU9	FGFLARE	MW.th	118	ET9	BUY_HPF1	MW.th
33	AU9	STG_ELE	MW	76	CV9	FOFLARE	MW.th	119	EU9	BUY_HPF21	MW.th
34	AX9	CTG_HPS	tph	77	CW9	GTG_FG	MW.th	120	EV9	BUY_HPF22	MW.th
35	AY9	CTG_MPS	tph	78	CY9	GTG_ELE	MW	121	EW9	BUY_HPF23	MW.th
36	AZ9	CTG_LPS	tph	79	CZ9	FC_FG	MW.th	122	EX9	BUY_HPF2	MW.th
37	BA9	CTG_CND	tph	80	DA9	FC_SP	MW.sp	123	EY9	HPF21_X	BIN
38	BB9	CTG_CW	ktph	81	DC9	FC_ELE	MW	124	EZ9	HPF22_X	BIN
39	BC9	CTG_ELE	MW	82	DD9	CWP1_FLO	ktph	125	FA9	HPF23_X	BIN
40	BF9	BP1_FLO	tph	83	DE9	CWP1_SP	MW.sp	126	FB9	BUY_ELEC	MW
41	BG9	BP1_SP	MW.sp	84	DF9	CWP1M_ELE	MW	127	FC9	DMND_CST	MW.pk
42	BH9	BP1M_ELE	MW	85	DH9	CWP2_FLO	ktph	128	FD9	IDEMAND	MW.pk
43	BJ9	BP2_FLO	tph	86	DI9	CWP2_SP	MW.sp	129	FE9	SEL_ELEC	MW

Figure 10.10 displays the LINGO COPT optimization model for the sample utility system, UtilSys-RTO-COPT. The corresponding Excel model is in a new worksheet named COPT. We tested this model with the case #1 data and Summer_peak TOU rates, the same as for the DRPU case. Both the Excel and the LINGO models gave identical solutions. Figure 10.11 shows details of the COPT solution. Note that the optimum operating cost with the equipment slate fixed to the current slate is $8325.3/h. It is an

```
MODEL:  TITLE Utility System Model for real time optimization: UtilSys-RTO-COPT;
! Ref: Nath, "Industrial Process Plants: Global Optimization of Utility Systems", Chapter 10, (2023);
DATA: hps_demand, mps_demand, lps_demand, mps_reserve,hps_reserve,cw_demand,
      ca_demand,  hpfg_demand,lpfg_demand,fo_demand,  elec_demand,                        ! 11 demands;
      b11_x,      b11_y,       b12_x,      b12_y,      b13am_x,     b13_x,      b13_y,      b13_z,      stg_x,
      stg_y,      ctg_x,       ctg_y,      bp1_x,      bp2_x,       bp2m_x,     bp3_x,      bp4_x,      bp4m_x,
      gtg_x,      fc_x,        cwp1_x,     cwp2_x,     cwp3_x,      cwp3m_x,    ac1234_nx,  ac5_x,! 26 int vars:
      c_dmw,      c_fo1,       fo1_hl,     c_fo2,      fo2_hl,      c_lpf1,     lpf1_hl,    c_lpf21,    lpf21_hl,
      c_lpf22,    lpf22_hl,    c_hpf1,     c_hpf21,    hpf21_hl,    c_hpf22,    hpf22_hl,   c_hpf23,    hpf23_hl,
      cbuy_ener,  cbuy_dem,    csel_elec,  cur_pkdmnd
      =@OLE('C:\BookDG\ExcelModels\UtilSys-RTO.xlsx');               ! cur op data from: UtilSys-RTO.xlsx;
      HPS_XBP,    HPS_XBN,     MPS_XBP,    MPS_XBN,    LPS_XBP,     LPS_XBN,    FO_XBP,     FO_XBN,     LPFG_XBP,
      LPFG_XBN,   HPFG_XBP,    HPFG_XBN,   ELEC_XBP,   ELEC_XBN,    BL1_XBN,    BL1_XBP,    BL1_YBP,    BL1_YBN,
      BL2_XBP,    BL2_XBN,     BL2_YBP,    BL2_YBN,    BL3_XBP,     BL3_XBN,    BL3_YBP,    BL3_YBN,    STG_XBP,
      STG_XBN,    STG_YBP,     STG_YBN,    CTG_XBP,    CTG_YBN,     CTG_YBP,    CTG_YBN,    GTG_XBP,    GTG_XBN
      = @FILE('C:\BookDG\LingoModels\UtilSys-Biases.lng');           ! 36 biases from UtilSys-Biases file:
      ! params:  @FILE('C:\BookDG\LingoModels\UtilSys-params.lng')
ENDDATA
      ! calcs:  @FILE('C:\BookDG\LingoModels\UtilSys-calcs-v2.lng');
! Objective:   MIN = FO_CST + LPF1_CST + c_lpf21 *BUY_LPF21 + c_lpf22 *BUY_LPF22
                + lpf22_a *LPF22_X + c_hpf1 *BUY_HPF1 + c_hpf21 *BUY_HPF21 + c_hpf22 *BUY_HPF22
                + hpf22_a *HPF22_X + c_hpf23 *BUY_HPF23 + hpf23_a *HPF23_X + c_dmw *DEA_DMW
                + cbuy_ener *BUY_ELEC + DMND_CST + cbuy_idem *IDEMAND - csel_elec *SEL_ELEC;
! include files: @FILE('C:\BookDG\LingoModels\stmSS-RTO.lng') @FILE('C:\BookDG\LingoModels\fuelSS-RTO.lng')
                @FILE('C:\BookDG\LingoModels\caSS-v1.lng')   @FILE('C:\BookDG\LingoModels\cwSS-RTO.lng')
                @FILE('C:\BookDG\LingoModels\Contracts-v1.lng')

! Headers:
  [_HPS_HDR]    BL2_STM + BL3_STM + HPS_XBP - HPS_XBN    = BL3AT_STM + PRV1_HPS + STG_STM + CTG_HPS
                + BP2T_STM + BP4T_STM + CWP2T_STM + CWP3T_STM + hps_demand;
  [_MPS_HDR]    BL1_STM + PRV1_MPS + STG_STM + CTG_MPS + BP4T_STM + CWP2T_STM + MPS_XBP - MPS_XBN
                = PRV2_STM + BL1AT_STM + mps_demand;
  [_LPS_HDR]    BL1AT_STM + BL3AT_STM + PRV2_STM + CTG_LPS + BP2T_STM + BDF_LPS + CWP3T_STM
                + LPS_XBP  - LPS_XBN = DEA_LPS + VENT + lps_demand;
  [_BFW_HDR]    BL1_BFW + BL2_BFW + BL3_BFW + PRV1_BFW = BP1_FLO + BP2_FLO + BP3_FLO + BP4_FLO;
  [_CW_HDR]           CWP1_FLO + CWP2_FLO + CWP3_FLO = CTG_CW + cw_demand;
  [_CA_HDR]           AC1234_CA + AC5_CA = VENT_CA + ca_demand;
  [_HPFG_HDR]   BUY_HPF1 + BUY_HPF2 + FC_FG + HPFG_XBP - HPFG_XBN
                = PRVFG + GTG_FG  + BL2_HPFG + BL3_HPFG + hpfg_demand;
  [_LPFG_HDR]   BUY_LPF1 + BUY_LPF2 + PRVFG + LPFG_XBP - LPFG_XBN
                = FC_FG + FGFLARE + BL2_LPFG + BL3_LPFG + AC1234E_FG + lpfg_demand;
  [_FO_HDR]           BUY_FO + FO_XBP - FO_XBN = FOFLARE + BL1_FUEL + fo_demand ;
  [_ELEC_HDR]   BUY_ELEC - SEL_ELEC + STG_ELE + CTG_ELE + GTG_ELE + ELEC_XBP - ELEC_XBN
                = FC_ELE + BL2AM_ELE + BL3AM_ELE + BP1M_ELE + BP2M_ELE + BP3M_ELE + BP4M_ELE
                + CWP1M_ELE + CWP3M_ELE + AC5M_ELE + elec_demand;
END MODEL UtilSys-RTO-COPT
```

Fig. 10.10: Sample utility system COPT model: UtilSys-RTO-COPT in LINGO.

improvement of $96.2/h, compared to the DRPU solution representing the base case operation.

10.2.5 Step 5: Download COPT results

Figure 10.11 shows the details of the COPT optimization solution.

A delta report comparing the COPT solution with the DRPU solution is also helpful for analysis as it highlights the optimization moves. Such a report is easily generated in Excel and shown in Fig. 10.12.

Analyzing the COPT solution report and the COPT-DRPU delta report reveals the optimization moves necessary to achieve the expected savings. These are,

1. **Changes to the boilers**:
Boiler #2: Decrease load and operate at the minimum load of 80 tph.
Boiler #3: Increase load accordingly.

Expect fuel consumption to be reduced by 0.96 MW.th.

Utility system Model: UtilSys-RTO-OPT-C. Case ID: 1, TOU rate ID: normal operation
Global optimal solution found.
Opmum objective function value (minimum operating cost, $/hr) $ 8,325.3

Purchased energy:

	contract	min	max	amount		COST, $/hr		incr unit cst	
BUY DMW				120.5	tph	$ 361.6		$ 3.00	
BUY FO TIER 1&:	XSS PEN	200		30.000	MW.th	$ 550.0		$ 19.00	
BUY LPFG - #1	TAKE/PAY	100		100.000	MW.th	$ 1,900.0		take or pay	
BUY LPFG #2-tier1	0	100			MW.th	$ -		$ 19.00	
BUY LPFG #2-tier2	1	500		240.96	MW.th	$ 4,437.2		$ 18.00	
BUY HPFG #1	SPOT			0.000	MW.th	$ (0.0)		$ 19.00	
BUY HPFG #2-tier1		20			MW.th	$ -		$ 20.00	
BUY HPFG #2-tier2		60			MW.th	$ -		$ 19.00	
BUY HPFG #2-tier3		120			MW.th	$ -		$ 18.00	
BUY ELEC - energy charge				17.531	MW	$ 876.6		$ 50.00	
BUY ELEC - demand charge				20.000	MW-peak	$ 200.0		$ 10.00	
IDEMAND penalty					MW	$ -		$ 100.00	
Sell ELEC					MW	$ -		$ (40.00)	
TOTAL OP COST						$ 8,325.3			

(right margin: -2.000, 0.039)

Process DEMAND

(unit inputs +ve, outputs -ve)				HP	MP	LP	BD	PC/TC	DMW	HP-bfw	LP-bfw	CW	CA	FO	LPFG	HPFG	ELEC	bias
DEMAND				100	45	90		-67.5				15	30	30	30	30	22	
Unit	A/R/U	min	max	tph	tph	tph	tph	tph	tph	tph		ktph	kM3/h	MW.th	MW.th	MW.th	MW	
BL1	A	70	210															
-aux																		
BL2	A	80	240	-80				-2		82					82.97			2.0
-aux																		0.500
BL3	A	100	300	-216				-4		220					195.99			0.9
-aux				5		-5												
STG	A	1.00	3.00															
CTG	A	3.00	13.00	158	-58	-90		-10				0.4					-8.069	0.261
Bfw P1	A		250							-151	151							
Bfw P2	A		250	2		-2				-151	151							
Bfw P3	A		250															
Bfw P4	A		250															
PRV1																		
PRV2					36	-36												
BD Flash						-1	6											
Deaerator						54			78	121	-302							
steam VENT						0												
CW P1	A		8															
CW P2	A		8	21	-21							-7.7						
CW P3	A		8	10		-10						-7.7						
AC1-4	4																	
AC5	A	20	55										-30.0				2.394	
Air Vent													0.0					
GTG	U		25															
FC	A	25	75											30.00		-30.00	0.484	
PRVFG																		
Fuel FLARE														0.00				
Bias (-ve => supply)				-0.8	-2.3	1.5										2.00	-0.039	
Net (-ve => export)				0.0	0.0	0.0		120.5					30.0		341.0	0.0	17.5	

Fig. 10.11: Sample utility system COPT model: UtilSys-RTO: COPT, case #1 solution report.

2. **Changes to turbogenerators**:
CTG: Increase the HP throttle flow by 22 tph, decrease MP extraction flow by 12 tph, and increase LP extraction by 34 tph. Expect electrical power generation to increase by 2.269 MW.

Depending on the controls at the utility system, the following setpoint changes may steer the current operations to the COPT optimum operation; the remaining changes should happen by existing automation.
1. **Boiler #2**: Reduce steam generation to 80 tph (the minimum).
2. **CTG**: Change HPS, MPS, and LPS setpoints to 158, 58, and 90 tph, respectively.

These setpoint changes could be sent to the RTDB for implementation or directly to boiler #2 and CTG controllers. Other expected changes should happen by automation.

Utility system Model: UtilSys-RTO: 'OPT-C vs DRPU' DELTA Report. Case ID: 1, TOU rate ID: normal operation
Global optimal solution found.
Opmum objective function value (minimum operating cost, $/hr) $ **(96.2)**

Purchased energy:

	contract min max	amount		COST, $./hr
BUY DMW		0.0	tph	0.1
BUY FO TIER 1&2			MW.th	
BUY LPFG - #1			MW.th	
BUY LPFG #2-tier1			MW.th	
BUY LPFG #2-tier2		0.96	MW.th	17.20
BUY HPFG #1		0.000	MW.th	0.000
BUY HPFG #2-tier1			MW.th	
BUY HPFG #2-tier2			MW.th	
BUY HPFG #2-tier3			MW.th	
BUY ELEC - energy charge		-2.269	MW	-113.438
BUY ELEC - demand charge			MW-peak	
IDEMAND penalty			MW	
Sell ELEC			MW	
TOTAL OP COST				$ (96.2)

Process DEMAND

(unit inputs +ve, outputs -ve)				HP	MP	LP	BD	PC/TC	DMW	HP-bfw	LP-bfw	CW	CA	FO	LPFG	HPFG	ELEC
Unit	A/R/U	min	max	tph	tph	tph	tph	tph	tph	tph		ktph	kM3/h	MW.th	MW.th	MW.th	MW
BL1	A	70	210														
-aux																	
BL2	A	80	240	24				0		-24					-20.03		
-aux																	
BL3	A	100	300	-26				-1		26					20.99		
-aux																	
STG	A	1.00															
CTG	A	3.00		22	12	-34		0				0.0					-2.269
Bfw P1	A		250							0	0						0.000
Bfw P2	A		250	0		0				0	0						
Bfw P3	A		250														
Bfw P4	A		250														
PRV1				-20	22					-2							
PRV2					-34	34											
BD Flash						0	0										
Deaerator						0			0		0						
steam VENT						0											
CW P1	A		8														
CW P2	A		8														
CW P3	A		8	0		0											
AC1-4	4																
AC5	A	20	55														
Air Vent													0.0				
GTG	U		25														
FC	A	25	75														
PRVFG																	
Fuel FLARE														0.00			
Bias (-ve => supply)																	

Fig. 10.12: Sample utility system model: UtilSys-RTO, COPT: delta report: comparison with DRPU solution.

10.2.6 Step 6: Discrete optimization: DOPT

The sample utility system model for DOPT is the same as the one used for COPT optimization, except that the 26 integer variables corresponding to the equipment states are no longer fixed and are available for optimization. Note that the biases calculated in the DRPU will remain fixed, as was the case with COPT. The optimization variable count now is back to 155, the same as that for the offline optimizer, UtilSys-OFLO. Since the Excel model requires explicit specification of variables, we will modify the variable list accordingly. No changes will be necessary in the LINGO model since LINGO does not require explicit specification of optimization variables; any variable with unspecified value automatically becomes an optimization variable.

Figure 10.13 is the DOPT optimization model in LINGO, UtilSys-RTO-DOPT. The Excel model is in a new worksheet named DOPT. We tested this model with the case #1 data and Summer_peak TOU rates, the same as for the DRPU case. Both the Excel and the LINGO solvers give identical results. Figure 10.14 summarizes the results of DOPT optimization. Note that in the DOPT case, the operating cost is $8049.3/h. It rep-

```
MODEL:  TITLE Utility System Model for Real Time Optimization: UtilSys-RTO-DOPT;
! Ref: Nath, "Industrial Process Plants: Global Optimization of Utility Systems", Chapter 10, (2023);
DATA: hps_demand, mps_demand, lps_demand, mps_reserve,hps_reserve,cw_demand, ca_demand, hpfg_demand,
      lpfg_demand,fo_demand,  elec_demand,                                                ! 11 demands;
      b11_123,    b12_123,    b13_123,    stg_123,    ctg_123,    bp1_123,    bp2_123,    bp3_123,   bp4_123,
      gtg_123,    fc_123,     cwp1_123,   cwp2_123,   cwp3_123,   ac1234_r,   ac1234_a,   ac5_123,!17 ARU 123s;
      c_dmw,      c_fo1,      fo1_hl,     c_fo2,      fo2_hl,     c_lpf1,     lpf1_hl,    c_lpf21,   lpf21_hl,
      c_lpf22,    lpf22_hl,   c_hpf1,     c_hpf21,    hpf21_hl,   c_hpf22,    hpf22_hl,   c_hpf23,   hpf23_hl,
      cbuy_ener,  cbuy_dem,   cbuy_idem,  csel_elec,  cur_pkdmnd                          ! 18+5=23 cost params;
      =@OLE('C:\BookDG\ExcelModels\UtilSys-RTO.xlsx');                                    ! 36 biases;
      HPS_XBP,    HPS_XBN,    MPS_XBP,    MPS_XBN,    LPS_XBP,    LPS_XBN,    FO_XBP,     FO_XBN,    LPFG_XBP,
      LPFG_XBN,   HPFG_XBP,   HPFG_XBN,   ELEC_XBP,   ELEC_XBN,   BL1_XBP,    BL1_XBN,    BL1_YBP,   BL1_YBN,
      BL2_XBP,    BL2_XBN,    BL2_YBP,    BL2_YBN,    BL3_XBP,    BL3_XBN,    BL3_YBP,    BL3_YBN,   STG_XBP,
      STG_XBN,    STG_YBP,    STG_YBN,    CTG_XBP,    CTG_XBN ,   CTG_YBP,    CTG_YBN,    GTG_XBP,   GTG_XBN
      = @FILE('C:\BookDG\LingoModels\UtilSys-Biases.lng');
      ! params;   @FILE('C:\BookDG\LingoModels\UtilSys-params.lng')
ENDDATA
      ! calcs;    @FILE('C:\BookDG\LingoModels\UtilSys-calcs-v2.lng')

! Objective;      MIN = FO_CST + LPF1_CST + c_lpf21 *BUY_LPF21 + c_lpf22 *BUY_LPF22
                  + lpf22_a *LPF22_X + c_hpf1 *BUY_HPF1 + c_hpf21 *BUY_HPF21 + c_hpf22 *BUY_HPF22
                  + hpf22_a *HPF22_X + c_hpf23 *BUY_HPF23 + hpf23_a *HPF23_X + c_dmw *DEA_DMW
                  + cbuy_ener *BUY_ELEC + DMND_CST + cbuy_idem *IDEMAND - csel_elec *SEL_ELEC;
! include files;  @FILE('C:\BookDG\LingoModels\stmSS-RTO.lng')
                  @FILE('C:\BookDG\LingoModels\fuelSS-RTO.lng')   @FILE('C:\BookDG\LingoModels\caSS-v1.lng')
                  @FILE('C:\BookDG\LingoModels\cwSS-RTO.lng')     @FILE('C:\BookDG\LingoModels\Contracts-v1.lng')
! Headers;
  [_HPS_HDR]      BL2_STM + BL3_STM + HPS_XBP - HPS_XBN    = BL3AT_STM + PRV1_HPS + STG_STM + CTG_HPS
                  + BP2T_STM + BP4T_STM + CWP2T_STM + CWP3T_STM + hps_demand;
  [_MPS_HDR]      BL1_STM + PRV1_MPS + STG_STM + CTG_MPS + BP2T_STM + CWP2T_STM + MPS_XBP - MPS_XBN
                  = PRV2_STM + BL1AT_STM + mps_demand;
  [_LPS_HDR]      BL1AT_STM + BL3AT_STM + PRV2_STM + CTG_LPS + BP2T_STM + BDF_LPS + CWP3T_STM
                  + LPS_XBP  - LPS_XBN = DEA_LPS + VENT + lps_demand;
  [_BFW_HDR]      BL1_BFW + BL2_BFW + BL3_BFW + PRV1_BFW = BP1_FLO + BP2_FLO + BP3_FLO + BP4_FLO;
  [_CW_HDR]          CWP1_FLO + CWP2_FLO + CWP3_FLO = CTG_CW + cw_demand;
  [_CA_HDR]          AC1234_CA + AC5_CA = VENT_CA + ca_demand;
  [_HPFG_HDR]     BUY_HPF1 + BUY_HPF2 + FC_FG + HPFG_XBP - HPFG_XBN
                  = PRVFG + GTG_FG + BL2_HPFG + BL3_HPFG + hpfg_demand;
  [_LPFG_HDR]     BUY_LPF1 + BUY_LPF2 + PRVFG + LPFG_XBP - LPFG_XBN
                  = FC_FG + FGFLARE + BL2_LPFG + BL3_LPFG + AC1234E_FG + lpfg_demand;
  [_FO_HDR]          BUY_FO + FO_XBP - FO_XBN = FOFLARE + BL1_FUEL + fo_demand ;
  [_ELEC_HDR]     BUY_ELEC - SEL_ELEC + STG_ELE + CTG_ELE + GTG_ELE + ELEC_XBP - ELEC_XBN
                  = FC_ELE + BL2AM_ELE + BL3AM_ELE + BP1M_ELE + BP2M_ELE + BP3M_ELE + BP4M_ELE
                  + CWP1M_ELE + CWP3M_ELE + AC5M_ELE + elec_demand;
END MODEL UtilSys-RTO-DOPT
```

Fig. 10.13: Sample utility system RTO model, UtilSys-RTO-DOPT, in LINGO.

resents an improvement of \$372.2/h compared to the current operation, the DRPU case, and a cost reduction of \$267/h, compared to the COPT optimum.

10.2.7 Step 7: Download DOPT results

Figure 10.14 shows that the DOPT results indicate an operating cost improvement of \$372.2/h, compared to the current operations, or an improvement of \$276/h, compared to the COPT operation.

Another helpful report would be a delta report comparing the DOPT operations with the current operations in the DRPU report; however, if automation is implemented in the COPT solution, then a delta report comparing the DOPT operations with the COPT operations would be more relevant. These comparison reports are easily generated in Excel and shown in Figs. 10.15 and 10.16, respectively. From now on, we will assume that the COPT solution implementation is proceeding.

Utility system Model: UtilSys-RTO-DOPT, Case ID: 1, TOU rate ID: normal operation
Global optimal solution found.
Optimum objective function value (minimum operating cost, $/hr) $ 8,049.3

Purchased energy:

	contract	min	max	amount	(unit)	COST, $/hr	incr unit cst	imbalance
BUY DMW				120.2	tph	$ 360.7	$ 3.00	
BUY FO TIER 1&XSS PEN			200	30.000	MW.th	$ 550.0	$ 19.00	-2.000
BUY LPFG - #1	TAKE/PAY		100	100.000	MW.th	$ 1,900.0	take or pay	
BUY LPFG #2-tier1			100		MW.th	$ -	$ 19.00	
BUY LPFG #2-tier2			500	220.35	MW.th	$ 4,066.3	$ 18.00	
BUY HPFG #1	SPOT				MW.th	$ -	$ 19.00	
BUY HPFG #2-tier1			20		MW.th	$ -	$ 20.00	
BUY HPFG #2-tier2			60		MW.th	$ -	$ 19.00	
BUY HPFG #2-tier3			120		MW.th	$ -	$ 18.00	
BUY ELEC - energy charge				19.446	MW	$ 972.3	$ 50.00	0.039
BUY ELEC - demand charge				20.000	MW-peak	$ 200.0	$ 10.00	
IDEMAND penalty					MW	$ -	$ 100.00	
Sell ELEC					MW	$ -	$ (40.00)	
TOTAL OP COST						**$ 8,049.3**		

Process DEMAND
(unit inputs +ve, outputs -ve)

Unit	A/R/U	min	max	HP	MP	LP	BD	PC/TC	DMW	HP-bfw	LP-bfw	CW	CA	FO	LPFG	HPFG	ELEC	fuel imbal
				100	45	90		-67.5				15	30	30	30	30	22	
		min	max	tph	tph	tph	tph	tph	tph	tph	tph	ktph	kM3/h	MW.th	MW.th	MW.th	MW	
BL1	A -aux	70	210															
BL2	A -aux	80	240												4.00			2.0
BL3	A -aux	100	300			-5	-6			282					250.92			0.9
STG	A	1.00	3.00	-276	-50												-3.000	
CTG	A	3.00	13.00	5														
Bfw P1	A		250	50														
Bfw P2	A		250	2		-2				-145	145							
Bfw P3	A		250	7	-7					-145	145							
Bfw P4	A		250	82						7								
PRV1					-90	-124												
					124													
PRV2					-21	-1												
						52												
BD Flash							6	68										
Deaerator									120		-289							
steam VENT																		
CW P1	A		8									-7.5						
CW P2	A		8	21	-21							-7.5						
CW P3	A		8	10		-10												
ACl-4	A	12	36															
AC5	A	20	55										-30.0					
Air Vent																		
GTG	U	25	75													30.00	0.484	
FC	A	25	25												3.43			
PRVFG															2.00			
Fuel FLARE																		
Bias (-ve => supply)				-0.8	-2.3	1.5											-0.04	
Net (-ve => export)				0.0	0.0	0.0			120.2				30.00	30.00	320.35	30.00	19.446	

Fig. 10.14: Sample utility system RTO model: UtilSys-RTO: DOPT, case #1 solution report.

Utility system Model: UtilSys-RTO: 'DOPT - DRPU' Δ Report. Case ID: 1, TOU rate ID: normal operation

Global optimal solution found.

Optimum objective function value (minimum operating cost, $/hr) $ (372.2)

Purchased energy:

	contract min	max	amount		COST, $/hr
BUY DMW			-0.3	tph	-0.8
BUY FO TIER 1&2				MW.th	
BUY LPFG - #1				MW.th	
BUY LPFG #2-tier1				MW.th	
BUY LPFG #2-tier2			-19.65	MW.th	-353.67
BUY HPFG #1				MW.th	
BUY HPFG #2-tier1				MW.th	
BUY HPFG #2-tier2				MW.th	
BUY HPFG #2-tier3				MW.th	
BUY ELEC - energy charge			-0.354	MW	-17.722
BUY ELEC - demand charge				MW-peak	
IDEMAND penalty				MW	
Sell ELEC				MW	
TOTAL OP COST				$	(372.2)

Process DEMAND

(unit inputs +ve, outputs -ve)

Unit	A/R/U	min	max	HP tph	MP tph	LP tph	BD tph	PC/TC tph	DMW tph	HP-bfw tph	LP-bfw tph	CW ktph	CA kiM3/h	FO MW.th	LPFG MW.th	HPFG MW.th	ELEC MW
BL1	A	70	210	104													
BL2	A	80	240				2			-106					-99.00		-0.500
BL3	A	100	300	-86			-2			88					75.92		
STG	A	1.00		50	-50	56		10									-3.000
CTG	A	3.00		-136	70	0											5.800
Bfw P1	A		250	0	0					151	-151	-0.4					-0.261
Bfw P2	A		250							6	-6						
Bfw P3	A		250	7	-7						145						
Bfw P4	A		250	62	-68	-54				-145							
PRV1					54	0	0	-10	0	6							
PRV2						-2					12						
BD Flash																	
Deaerator																	
steam VENT				0	0	0						0.2					
CW P1	A		8	0								0.2					
CW P2	A		8														
CW P3	A		8										-30.0		3.43		-2.394
AC1-4		12	36										30.0				
AC5	A	20	55													30.00	
Air Vent																	
GTG	U	25															
FC	A	25	75														
PRVFG																	
Fuel FLARE																	

Bias (-ve => supply)

Fig. 10.15: Sample utility system DOPT model, case #1: delta report: comparison with DRPU solution.

Utility system Model: UtilSys-RTO: DOPT vs COPT: Δ Report. Case ID: 1, TOU rate ID: Summer_peak
Global optimal solution found.
Optimum objective function value (minimum operating cost, $/hr) $ (276.1)

Purchased energy:

	amount		COST, $/hr
BUY DMW	120.2	tph MW.th	-0.9
BUY FO TIER 1&2	30.000	MW.th	
BUY LPFG - #1	100.000	MW.th	
BUY LPFG #2-tier1		MW.th	
BUY LPFG #2-tier2	220.35	MW.th	-370.87
BUY HPFG #1		MW.th	0.000
BUY HPFG #2-tier1		MW.th	
BUY HPFG #2-tier2		MW.th	
BUY HPFG #2-tier3		MW.th	
BUY ELEC - energy charge	19.446	MW	95.716
BUY ELEC - demand charge	20.000	MW-peak	
IDEMAND penalty		MW	
Sell ELEC		MW	
TOTAL OP COST			$ (276.1)

(unit inputs +ve, outputs -ve)

Unit	A/R/U	min	max	HP	MP	LP	BD	PC/TC	DMW	HP-bfw	LP-bfw	CW	CA	FO	LPFG	HPFG	ELEC
				tph	tph	tph	tph	tph	tph	tph	tph	ktph	kM3/h	MW.th	MW.th	MW.th	MW
BL1	A	70	210														
-aux																	
BL2	A	80	240	80			2			-82					-78.97		-0.500
-aux																	
BL3	A	100	300	-61			-1			62					54.93		
-aux																	
STG	A	1.00	250	50	-50	90		10									-3.000
CTG	A	3.00	250	-158	58					151	-151	-0.4					8.069
Bfw P1	A		250	0		0				6	-6						-0.261
Bfw P2	A		250	0						-145	145						
Bfw P3	A		250	7						7							
Bfw P4	A		250	82	-7	-88											
PRV1					-90	0											
PRV2					88	-2.1											
BD Flash						0	0										
Deaerator					0	0		-10	0		12						
steam VENT					0				0								
CW P1	A	8	8	0													
CW P2	A	8	8	0								0.2					
CW P3	A	8	8									0.2					
AC1-4	4	12	36										-30.0		3.43		-2.394
AC5	A	20	55										30.0				
Air Vent													0.0				
GTG	U	25	25													30.00	
FC	A	25	75												0.00		
PRVFG																	
Fuel FLARE																	
Bias (+ve => supply)																	

Fig. 10.16: Sample utility system DOPT model, case #1: delta report: comparison with COPT solution.

From these reports, we can identify the actions necessary to steer the operations toward the DOPT optimum. Please observe that the implementation of the DOPT optimization solution requires several discrete changes, as shown by the UP and DOWN icons in column 1 of the Delta report shown in Fig. 10.16. Such changes would require operator intervention; hence, an advisory would be issued to the operator to consider the recommended changes as they offer significant benefits. However, there are other considerations, as discussed later in this section.

The following changes would be required to achieve the DOPT global optimum:

1. **Changes to boilers**: No change in equipment slate is required.
   Boiler #2: Operate it on standby mode.
   Boiler #3: Operate it as the swing boiler.
2. **Changes to compressors**:
   Air compressors 1, 2, 3, and-4: Turn them ON.
   Air compressor #5: Turn it OFF.
3. **Changes to pumps**:
   BFW pump #4: Turn it ON and operate it with the turbine drive.
   BFW pump #1: Turn it OFF.
4. **Changes to turbogenerators**:
   STG: Turn it ON and operate for maximum power generation of 3 MW.
   CTG: Turn it OFF.

Asking the operator to make all the changes mentioned above in one shot will likely be overwhelming, and we should refrain from doing so.

A better strategy would be to present a list of more straightforward modifications to the operator with the ultimate goal of achieving the global optimum gradually. For instance, the list of options could be limited to the choices that require fewer equipment slate changes and let the operator exercise his judgment in picking one from the list. As RTO execution is cyclical, the RTO executive will ask the operator to make one move after another, and so on. These moves will take the utility system operations towards the global optimum by making slow but steady progress and involving the operator in the decision making process.

Generating multiple options would be easy since we already have a working DOPT model; each option would be just a restricted version of the DOPT case. The utility system model is relatively modest by LP/MILP standards, and execution is fast. Execution time-wise, the RTO executive could run several restricted DOPT cases without slowing the cycle. Based on the results of several restricted DOPT cases, the RTO executive would develop a more focused short list of actionable items for the operator.

We will use combinatorial enumeration [8] to generate various restricted DOPT cases. Note that the DOPT solution we saw earlier has four changes related to boilers, compressors, pumps, and turbogenerators. We will explore each area individually and name these four restricted cases as DOPT-B, DOPT-C, DOPT-P, and DOPT-T, respectively. We will also consider additional restricted cases that involve two areas at a time. There will be three such instances; we will name them DOPT-CP, DOPT-CT, and DOPT-PT. Finally, there will be a case that involves slate changes to all three areas; we will call it DOPT-CPT.

We will define each restricted DOPT optimization case by specifying the ARU status of "Available" to the equipment in the areas where changes in equipment slate are allowed; equipment in all other operating areas will remain fixed to the current slate. For example, for the DOPT-C case, the ARU status of "Available" will be specified for the compressors FC, AC1234, and AC5, and the remaining equipment will remain fixed to its current state.

Table 10.6 summarizes these eight restricted DOPT cases along with the COPT and DOPT cases; note that "Available" equipment is highlighted by an "A" in brown font on a pale-yellow background. Since the GTG is not operational, it will remain unavailable in all cases. We will assume that automation is implementing the COPT solution, so restricted DOPT cases will be compared to the COPT optimum.

Tab. 10.6: Restricted DOPT cases for developing the DOPT implementation strategy.

ARU spec	COPT	DOPT-B	DOPT-C	DOPT-P	DOPT-T	DOPT-CP	DOPT-CT	DOPT-PT	DOPT-CPT	DOPT
bl1_aru	U	A	U	U	U	U	U	U	U	A
bl2_aru	R	A	R	R	R	R	R	R	R	A
bl3_aru	R	A	R	R	R	R	R	R	R	A
stg_aru	U	U	U	U	A	U	A	A	A	A
ctg_aru	R	R	R	R	A	R	A	A	A	A
bp1_aru	R	R	R	A	R	A	R	A	A	A
bp2_aru	R	R	R	A	R	A	R	A	A	A
bp3_aru	U	U	U	A	U	A	U	A	A	A
bp4_aru	U	U	U	A	U	A	U	A	A	A
gtg_aru	U	U	U	U	U	U	U	U	U	U
fc_aru	R	R	A	R	R	A	A	R	A	A
cwp1_aru	U	U	U	A	U	A	U	A	A	A
cwp2_aru	R	R	R	A	R	A	R	A	A	A
cwp3_aru	R	R	R	A	R	A	R	A	A	A
ac1234_r	0	0	0	0	0	0	0	0	0	0
ac1234_a	0	0	4	0	0	4	4	0	4	4
ac5_aru	R	R	A	R	R	A	A	R	A	A

These restricted cases are easy to set up in Excel and LINGO using the existing DOPT models; only the "ARU" specifications change from case to case. Note that the ARU specifications for the cases shown in Tab. 10.6 are also stored in the "CaseData" worksheet of the UtilSys-RTO workbook and retrieved using the "ARU id" keyword specified on cell F3 of the DOPT worksheet. The LINGO model will retrieve all data from the UtilSys-RTO Excel file.

Table 10.7 shows the results of the eight restricted DOPT cases.

A possible strategy would be implementing DOPT-B boiler optimization, common to all DOPT cases, and achieving 44% of the maximum savings. Then, implement DOPT-C, compressor optimization or DOPT-T, turbogenerator optimization, and to reach 63–65% of the maximum savings. Next, implement DOPT-CT, combined com-

Tab. 10.7: Summary of the results of the restricted DOPT cases.

Restricted case ID	Opertaing cost, $/hr	Savings, $/hr	Savings, % of max	Comment
COPT	$ 8,325.3			Base case
DOPT-B	$ 8,203.9	$ 121.4	44%	BL2 on standby (BL3 becomes swing boiler)
DOPT-C	$ 8,146.0	$ 179.3	65%	BL2 on standby, swap AC1234 for AC5
DOPT-P	$ 8,203.9	$ 121.4	44%	BL2 on standby, swap BP4 for BP1
DOPT-T	$ 8,152.4	$ 172.9	63%	BL2 on standby, turn STG ON
DOPT-CP	$ 8,146.2	$ 179.1	65%	BL2 on standby, swap AC1234 for AC5, swap BP4 for BP1
DOPT-CT	$ 8,052.7	$ 272.6	99%	BL2 on standby, swap AC1234 for AC5, swap STG for CTG
DOPT-PT	$ 8,152.6	$ 172.7	63%	BL2 on standby, swap BP4 for BP1, turn STG ON
DOPT-CPT	$ 8,049.3	$ 276.0	100%	BL2 on standby, swap AC1234 for AC5, swap STG for CTG, swap BP4 for BP1
DOPT	$ 8,049.3	$ 276.0	100%	Same as DOPT-CPT

pressor, and turbogenerator optimizations to achieve 99% of the maximum savings. One could stop at this point or optionally go for the maximum savings by implementing the complete DOPT optimization. Figure 10.17 shows this implementation strategy.

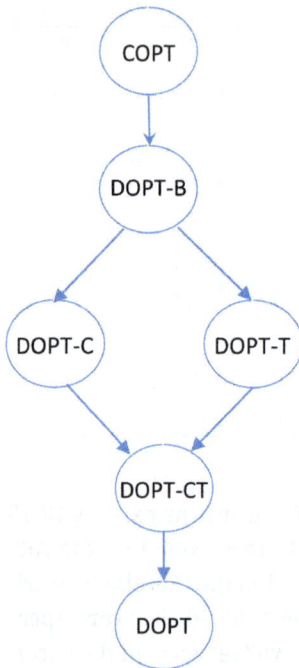

Fig. 10.17: COPT to DOPT implantation strategy.

We have developed the strategy in Fig. 10.17 by observing the solutions of the various restricted DOPT cases in Tab. 10.7. In an actual implementation, a logic-based program must determine this strategy. A summary of the restricted DOPT cases, as in Tab. 10.7, and recommendations based on the summary reports, such as in Fig. 10.17, must be automatically generated, and the RTO executive must present implementation options to the operator periodically. Although developing the logic-based program is not trivial, it is a necessary and doable activity. However, it is site-specific, so we still need a general methodology. Hopefully, the example presented here will help you develop an implementation strategy at your site. Looking to the future, I am hopeful that the emerging Artificial Intelligence (AI) technologies will help simplify this task.

Summary

This chapter discussed deploying the Utility System Optimization model for Real-Time Optimization (RTO). RTO is cyclically executed at a high frequency. It comprises seven steps. The RTO executive uses live data that is usually imperfect, so Data Reconciliation and Parameter Updating (DRPU) Optimization is necessary to resolve this problem. A simple yet practical method called the "Method of Biases" is introduced for DRPU; this method computes biases that modify the model to fit the measured data perfectly and thus readies the optimization model. Two optimizations, COPT and DOPT, are performed. The results of COPT are suitable for implementation via automation. Results of DOPT usually require several changes in the equipment slate, which the operator may be reluctant to make in one go. We outline a gradual implementation approach, leading to the DOPT global optimization. The complete RTO cycle is illustrated with a test case for the sample utility system model.

Nomenclature

Symbol	UOM	Description
AC1234_CA	kM3/h	Air compressor #1234: compressed air flow
AC1234_NX; ac1234_nx	0 .. 4	Air compressor #1234: number of operating pumps; measured value
AC1234_SP	MW.sp	Air compressor #1234: shaft power required
AC1234M_FG	MW.th	Air compressor #1234: ICE drive: fuel consumption
AC5_CA	kM3/h	Air compressor #5: compressed air flow
AC5_SP	MW.sp	Air compressor #5: shaft power required
AC5_X; ac5_x	0/1	Air compressor #5: operational status: 0 = OFF, 1 = ON; measured value
AC5M_ELE	MW	Air compressor #5: motor drive: power consumption
BAVG	tph	BFW pump flow equalization: average BFW flow
BDF_CND	tph	Blowdown flash: LP condensate discharge
BDF_HPC	tph	Blowdown flash: inlet HP blowdown
BDF_LPS	tph	Blowdown flash: LP steam generation
BFj	tph	BFW pump flow equalization: flow corresponding to j operating pumps
BL3_W	0/1	Boiler #3: LPFG indicator: 1 = using LPFG, 0 = using HPFG
BL3_Z	0/1	Boiler #3: HPFG indicator: 1 = using HPFG, 0 = using LPFG
BLj_BD	tph	Boiler #j: blowdown flow

(continued)

Symbol	UOM	Description
BLj_BFW	tph	Boiler #j: boiler feed water inflow
BLj_EB1BR	MW.th	Boiler #j performance eq.: segment #1: residual
BLj_EB1RR	MW.th	Boiler #j performance eq.: segment #2: residual
BLj_FUEL; blj_fuel	MW.th	Boiler #j: total fuel consumption; measured value
BLj_HPFG, blj_hpfg	MW.th	Boiler #j: HP fuel gas consumption; measured value
BLj_LPFG; blj_lpfg	MW.th	Boiler #j: LP fuel gas consumption; measured value
BLj_RSV	tph	Boiler #j: steam reserve
BLj_STM; blj_stm	tph	Boiler #j: steam generation; measured value
BLj_X; blj_x	0/1	Boiler #j: operational status: 0 = OFF, 1 = ON; measured value
BLj_XBN	MW.th	Boiler #j energy balance: normal operation: calculated negative bias
BLj_XBP	MW.th	Boiler #j energy balance: normal operation: calculated positive bias
BLj_Y; blj_y	0/1	Boiler #j: standby status: 1 = in standby mode, 0 otherwise; measured value
BLj_YBN	MW.th	Boiler #j energy balance: standby operation: calculated negative bias
BLj_YBP	MW.th	Boiler #j energy balance: standby operation: calculated positive bias
BLjAM_ELE	MW	Boiler #j: auxiliaries motor drive: electric power usage
BLjAM_X; bljam_x	0/1	Boiler #j: auxiliaries motor drive: 0 = OFF, 1 = ON; measured value
BLjAT_STM	tph	Boiler #j: Auxiliaries turbine: steam flow
BLjAT_X	0/1	Boiler #j: auxiliaries turbine drive: 0 = OFF, 1 = ON
BPj_EB1BR	MW.th	BFW pump #j performance eq.: segment #1: residual
BPj_EB1RR	MW.th	BFW pump #j performance eq.: segment #2: residual
BPj_FLO	tph	BFW Pump #j: boiler feed water inflow
BPj_SP	MW.sp	BFW Pump #j: shaft power required
BPj_X; bpj_x	0/1	BFW Pump #j: operational status: 0 = OFF, 1 = ON; measured value
BPjM_ELE	MW	BFW Pump #j: motor drive: power consumption
BPjM_SP	MW.sp	BFW Pump #j: motor drive: shaft power produced
BPjM_X; bpjm_x	0/1	BFW Pump #j: motor drive: operational status: 0 = OFF, 1 = ON; measured value
BPjT_SP	MW.sp	BFW Pump #j: turbine drive: shaft power produced
BPjT_STM	tph	BFW Pump #j: turbine: steam flow
BPjT_X	0/1	BFW Pump #j: turbine drive: operational status: 0 = OFF, 1 = ON
BUY_ELEC; buy_elec	MW	Purchased energy: buy electric power; measured value
BUY_FO; buy_fo	MW.th	Purchased energy: buy fuel oil; measured value
BUY_HPF2j	MW.th	HP fuel gas purchase from supplier #2: in tier #j
BUY_HPFj; buy_hpfj	MW.th	Purchased energy: buy HP fuel gas from supplier #j; measured value
BUY_LPF2j	MW.th	LP fuel gas purchase from supplier #2: in tier #j
BUY_LPFj; buy_lpfj	MW.th	Purchased energy: buy LP fuel gas from supplier #j; measured value
BXj	0/1	BFW pump flow equalization: Indicates that j pumps are operating.
CAVG	tph	CW pump flow equalization: average CW flow
CFj	tph	CW pump flow equalization: flow corresponding to jj operating pumps
CTG_CND	tph	CTG: condensate from condenser
CTG_CW	ktph	CTG: condenser: cooling water flow
CTG_ELE; ctg_ele	MW	CTG: power generation; measured value
CTG_HPS; ctg_hps	tph	CTG: inlet HP steam flow; measured value
CTG_LPS; ctg_lps	tph	CTG: outlet LP steam flow; measured value
CTG_MPS; ctg_mps	tph	CTG: outlet MP steam flow; measured value

(continued)

Symbol	UOM	Description
CTG_X; ctg_x	0/1	CTG: operational status: 0 = OFF, 1 = ON; measured value
CTG_XBN	tph	CTG performance eq.: normal operation: calculated negative bias
CTG_XBP	tph	CTG performance eq.: normal operation: calculated positive bias
CTG_Y; ctg_y	0/1	CTG: standby status: 1 = in standby mode, 0 otherwise; measured value
CTG_YBN	tph	CTG performance eq.: standby operation: calculated negative bias
CTG_YBP	tph	CTG performance eq.: standby operation: calculated positive bias
CWPj_EB1BR	MW.th	CW pump #j performance eq.: segment #1: residual
CWPj_EB1RR	MW.th	CW pump #j performance eq.: segment #2: residual
CWPj_FLO	tph	CW Pump #j: boiler feed water inflow
CWPj_SP	MW.sp	CW Pump #j: shaft power required
CWPj_X; cwpj_x	0/1	CW Pump #j: operational status: 0 = OFF, 1 = ON; measured value
CWPjM_ELE	MW	CW Pump #j: motor drive: power consumption
CWPjM_SP	MW.sp	CW Pump #j: motor drive: shaft power produced
CWPjM_X; cwpjm_x	0/1	CW Pump #j: motor drive: operational status: 0 = OFF, 1 = ON; measured value
CWPjT_SP	MW.sp	CW Pump #j: turbine drive: shaft power produced
CWPjT_STM	tph	CW Pump #j: turbine: steam flow
CWPjT_X	0/1	CW Pump #j: turbine drive: operational status: 0 = OFF, 1 = ON
CXj	0/1	CW pump flow equalization: Indicates that j pumps are operating.
DEA_BFW	tph	Deaerator: boiler feed water production
DEA_DMW	tph	Deaerator: fresh demineralized water intake
DEA_LPS	tph	Deaerator: deaeration steam
DMND_CST	$/hr	Electrical power demand charge
DMND_CST	$/h	electric power purchase: demand charges
ELEC_XBN	MW.th	ELEC Header energy balance: calculated negative bias
ELEC_XBP	MW.th	ELEC Header energy balance: calculated positive bias
FC_ELE	MW	Fuel compressor: power consumption
FC_FG; fc_fg	MW.th	Fuel compressor: fuel compressed; measured value
FC_SP	MW.sp	Fuel compressor: turbine drive: shaft power produced
FC_X; fc_x	0/1	Fuel compressor: operational status: 0 = OFF,; measured value
FGFLARE; fgflare	MW.th	VENT: LPFG; measured value
FO_BIAS	MW.th	FO Header mass balance: calculated bias
FO_CST	$/h	Fuel oil purchase: cost
FO_XBN	MW.th	FO Header mass balance: calculated positive bias
FO_XBP	MW.th	FO Header mass balance: calculated negative bias
FOFLARE; foflare	MW.th	VENT: fuel oil; measured value
FUEL_CST	$/h	VENT: fuel oil
FUELj_X	0/1	indicator for fuel purchase in tier #j region
GTG_EB1BR	MW.th	GTG performance eq.: segment #2: residual
GTG_EB1GR	MW.th	GTG performance eq.: segment #1: residual
GTG_EB1RR	MW.th	GTG performance eq.: segment #3: residual
GTG_ELE; gtg_ele	MW	GTG: power generated; measured value
GTG_FG; gtg_fg	MW.th	GTG: fuel consumed; measured value
GTG_X; gtg_x	0/1	GTG: operational status: 0 = OFF, 1 = ON; measured value
HPF2j_X	0/1	HP fuel gas purchase from supplier #2: in tier #j: indicator
HPFG_XBN	MW.th	HPFG Header energy balance: calculated negative bias

(continued)

Symbol	UOM	Description
HPFG_XBP	MW.th	HPFG Header energy balance: calculated positive bias
HPS_RSV	tph	HP Boilers: total steam reserve
HPS_XBN	tph	HP steam header mass balance: calculated negative bias
HPS_XBP	tph	HP steam header mass balance: calculated positive bias
IDEMAND	MW.pk	incremental demand peak
LPF1_CST	$/h	LP fuel gas purchase from supplier #1: cost
LPF2j_X	0/1	LP fuel gas purchase from supplier #2: in tier #j: indicator
LPFG_XBN	MW.th	LPFG Header energy balance: calculated negative bias
LPFG_XBP	MW.th	LPFG Header energy balance: calculated positive bias
LPS_XBN	tph	LP steam header mass balance: calculated negative bias
LPS_XBP	tph	LP steam header mass balance: calculated positive bias
MPS_RSV	tph	MP Boiler: steam reserve
MPS_XBN	tph	MP steam header mass balance: calculated negative bias
MPS_XBP	tph	MP steam header mass balance: calculated positive bias
PRV1_BFW	tph	PRV: de-superheating water flow
PRV1_HPS; prv1_hps	tph	PRV: HP steam flow; measured value
PRV1_MPS	tph	PRV: MP steam flow; measured value
PRV2_STM; prv2_stm	tph	PRV2: steam flow
PRVFG; prvfg	MW.th	PRVFG: HPFG to LPFG; measured value
SEL_ELEC; sel_elec	MW	Purchased energy: sell electric power; measured value
STG_ELE; stg_elec	MW	STG: power generation
STG_STM; stg_stm	tph	STG: steam flow
STG_X; stg_x	0/1	STG: operational status: 0 = OFF, 1 = ON; measured value
STG_XBN	tph	STG performance eq.: normal operation: calculated negative bias
STG_XBP	tph	STG performance eq.: normal operation: calculated positive bias
STG_Y; stg_y	0/1	STG: standby status: 1 = in standby mode, 0 otherwise; measured value
STG_YBN	tph	STG energy balance: standby operation: calculated negative bias
STG_YBP	tph	STG performance eq.: standby operation: calculated positive bias
VENT; vent	tph	VENT: steam flow; measured value
VENT_CA; vent_ca	kM3/h	VENT: compressed air; measured value
a, b, c, ci	–	generic constants
ac1234_a	0 .. 4	Input: Air compressors #1 ..4: number of available compressors
ac1234_r	0 .. 4	Input: Air compressors #1 ..4: number of required compressors
ac5_aru	ARU or 123	Input: Air compressor #5: "avaialble'/"required"/"unavailable" spec
c_dmw	$/ton	Input cost: buy demineralized water
c_fo	$/MWh.th	Input cost: buy fuel oil
c_fuelj	$/MWh.th	fuel cost
c_hpfg	$/MWh.th	Input cost: buy LPFG
c_lpfg	$/MWh.th	Input cost: buy LPFG
c_selelec	$/MWh	Input cost: sell electric power
ca_demand	kM3/h	Process Demand: compressed air
cur_pkdemand	MW.pk	Input cost: buy electric power: current peak demand
cw_demand	ktph	Process Demand: cooling water

(continued)

Symbol	UOM	Description
cwpj_aru	ARU or 123	Input: CW pump #j: "avaialble'/"required"/"unavailable" spec
elec_demand	MW	Process Demand: electrical power
fo_demand	MW.th	Process Demand: fuel oil
fuel_demand	MW.th	fuel demand
hpfg_demand	MW.th	Process Demand: HP fuel gas
hps_demand	tph	Process Demand: HP steam
hps_reserve	tph	Process Demand: HP steam reserve
lpfg_demand	MW.th	Process Demand: LP fuel gas
lps_demand	tph	Process Demand: LP steam
mps_demand	tph	Process Demand: MP steam
mps_reserve	tph	Process Demand: MP steam reserve
stg_stm_a	tph	Parameter: STG: inlet steam vs. power regression: intercept
stg_stm_b	tph/MW	Parameter: STG: inlet steam vs, power regression: slope

References

[1] Narasimhan, S. and C. Jordache. "Data Reconciliation and Gross Error Detection, an Intelligent Use of Process Data", Elsevier, (1999).

[2] Romagnoli, J. A. and M. C. Sanchez. "Data Processing and Reconciliation for Chemical Process Operations", Academic Press, (2000).

[3] Stanley, G. M. and R. S. H. Mah. Observability and redundancy in process data estimation, Chemical Engineering Science, 36:259–272, (1981).

[4] Horn, B., et. al. "Platform for advanced control applications", IFAC Proceedings Volumes, 38 (1):161–166, (2005).

[5] Anonymous, *"URT Users Guide"*, Honeywell Forge APC Rel 150.1, (2020).

[6] Anonymous, *"LINGO The Modeling Language and Optimizer"*, Lindo Systems Inc, p. 504, (2020).

[7] Anonymous, *"LINGO The Modeling Language and Optimizer"*, Lindo Systems Inc, p. 489, (2020).

[8] "Combinatorics", In Wikipedia. https://en.wikipedia.org/wiki/Combinatorics, (2023).

Appendix A: List of models

We have discussed and developed many LP/MILP models in this book using the Microsoft Excel Solver from Frontline System Inc. and the LINGO Solver from LINDO Systems Inc. We have used the Basic Solver that comes standard with Excel for smaller models with up to 50 constraint rows and an upgraded version called the Premium Solver for larger models. We have used Microsoft Office Professional Plus 2016, Premium Solver version V2023 Q2 (23.2.1.0), and LINGO Solver version Win64, Release 20.0.21 (23 August 2023), on a Windows 10 Pro 64 bit computer.

Following is a list of all the models discussed in the book. All model files are available from the publisher's website for this book.

https://doi.org/10.1515/9783111020679-013

#	Model files, .xlsx, .lg4	Component ID	Obj fn	UOM	Vars	Ints	Rows	Lingo include file .lng	Book section
1	Enginola	-	2100.0	$/day	2	0	3	-	1.6
2	BL1-v1	Boiler #1	1691.5	$/hr	5	1	6	-	3.1.4
3	BL1-v2	Boiler #1	1691.5	$/hr	7	2	9	-	3.1.5
4	BL1-v3	Boiler #1	1697.4	$/hr	7	2	10	-	3.1.6
5	BL1-v4	Boiler #1	-	$/hr				BL1-v4	3.1.6
6	BL2-v1	Boiler #2	1931.2	$/hr	9	2	11	BL2-v1	3.2
7	BL3-v1	Boiler #3	1949.8	$/hr	11	3	14	BL3-v1	3.3
8	BLSS-v1	Boilers	3201.6	$/hr	29	7	35	BL1-v4 BL2-v1 BL3-v1	3.4
9	M01-v1	motor	1.05	MW	3	1	4	-	4.1.1
10	T01-v1	turbine	2.39	tph	3	1	4	-	4.1.2
11	E01-v1	ICE	0.88	MW.th	3	1	4	-	4.1.3
12	P01-v1	pump	0.226	MW.sp	3	1	4	-	4.1.4
13	P01MD-v1	pump w motor drive	0.238	MW	4	1	6	P01MD-v1	4.1.5
14	P01TD-v1	pump w turbine drive	2.19	tph	4	1	6	P01TD-v1	4.1.6
15	P01DD-v1	pump w dual drives	14.3	$/hr	9	2	10	P01DD-v1	4.1.7
16	BLSS-v2	Boilers with auxiliaries	3318.6	$/hr	35	8	40	BL1-v4 BL2-v1 BL3-v1	4.1.8
17	C01-v1	compressor	0.364	MW.sp	3	1	4	-	4.1.9
18	C01MD-v1	compressor w motor drive	0.383	MW	4	1	5	C01MD-v1	4.1.9
19	C01TD-v1	compressor w turbine drive	3.360	tph	4	1	5	C01TD-v1	4.1.9
20	C01DD-v1	compressor w dual drives	33.6	$/hr	9	2	10	C01DD-v1	4.1.9
21	C01ED-v1	compressor w ICE drive	0.889	MW.th	4	1	5	C01ED-v1	4.1.9
22	STG-v1	STG	0.0	$/hr	4	2	5	STG-v1	5.1
23	CTG-v1	CTG	0.0	$/hr	9	2	12	CTG-v1	5.2
24	GTG-v1	GTG	0.0	$/hr	3	1	4	GTG-v1	5.3
25	PRV-v1	PRV with desuperheating	9.2	tph	3	0	3	-	6.1.2
26	BDF-v1	blowdown flash	8.2	tph	4	0	4	-	6.2
27	DEA-v1	deaerator unit	16.5	tph	4	0	4	-	6.3
28	stmSS-v1	Steam subsystem	6321.3	$/hr	82	18	93	BL1-v4 BL2-v1 BL3-v1 BP1MD-v1 BP2DD-v1 BP3MD-v1 BP4DD-v1	6.4

#	Model files, .xlsx, .lg4	Component ID	Obj fn	UOM	Vars	Ints	Rows	Lingo include file .lng	Book section
29	FuelSS-v1	Fuel subsystem	1709.1	$/hr	13	2	14	FuelSS-v1	7.2
30	CwSS-v1	Cooling water subsystem	135.8	$/hr	17	4	23	CwSS-v1	7.4
31	CaSS-v1	Compressed air sunsystem	65.3	$/hr	9	2	11	CaSS-v1	7.6
32	UtilSys-v1	Utility system	9969.3	$/hr	123	26	146	UtilSys-params UtilSys-calcs StmsSS-v1 FuelSS-v1 CwSS-v1 CaSS-v1	7.7
33	Fuel_Cst_2T-v1	two tier contract	1100.0	$/hr	5	2	6	-	8.1.4
34	Fuel_Cst_3T-v1	three-tier contract	1440.0	$/hr	7	3	8	-	8.1.5
35	Fuel_Cst_Lambda-v1	three-tier contract, λ method	-	-	-	-	-	-	8.1.5
36	UtilSys-v2	Utility system	9780.5	$/hr	155	38	192	UtilSys-params UtilSys-calcs StmsSS-v2 FuelSS-v1 CwSS-v2 CaSS-v1	8.4
37	UtilSys-OFLO	Utility system	9780.5	$/hr	155	38	192	UtilSys-params UtilSys-calcs StmsSS-v2 FuelSS-v1 CwSS-v2 CaSS-v1	9.1
38	UtilSys-RTO-DRPU	Utility system	2270.6	-	135	12	192	UtilSys-params UtilSys-calcs StmsSS-RTO FuelSS-RTO CwSS-RTO CaSS-v1 Contracts-v1	10.2.2.5
39	UtilSys-RTO-COPT	Utility system	8325.3	$/hr	129	12	192	UtilSys-params UtilSys-biases UtilSys-calcs StmsSS-RTO FuelSS-RTO CwSS-RTO CaSS-v1 Contracts-v1	10.2.4
40	UtilSys-RTO-DOPT	Utility system	8049.3	$/hr	155	38	192	UtilSys-params UtilSys-biases UtilSys-calcs StmsSS-RTO FuelSS-RTO CwSS-RTO CaSS-v1 Contracts-v1	10.2.6

Index

https://doi.org/10.1515/9783111020679-014

www.ingramcontent.com/pod-product-compliance
Lightning Source LLC
Chambersburg PA
CBHW061401210326
41598CB00035B/6059